The Marine Plant Biomass
of the Pacific Northwest Coast

The Marine Plant Biomass
of the Pacific Northwest Coast

Edited by
Robert W. Krauss

College of Science
Oregon State University

OREGON STATE UNIVERSITY PRESS

Library of Congress Cataloging in Publication Data

Main entry under title:

Marine plant biomass of the Pacific Northwest coast.

    1. Marine algae culture--Northwest Coast of North
America--Congresses. 2. Oceanography--Northwest Coast
of North America--Congresses. 3. Ocean engineering--
Congresses. I. Krauss, Robert Wallfar, 1921-
SH390.5.U6M37   630   77-25049
ISBN 0-87071-447-3

Preface

Mankind is facing the spectre of a planet stripped of the resources necessary to sustain the human populations that currently exist—not to mention those unborn multitudes which seem certain to be added in the coming decades. Time is rapidly running out for scientists and engineers to find new resources to supply demands for the myriad of components necessary to sustain the global economy and to ensure a reasonable standard of living. The next 50 years will witness a more thorough examination of the natural resource inventory of the planet than has ever been attempted before. The search for renewable resources and the development of technologies to enhance them will certainly be major preoccupations in the future. Scientists, engineers, and industrialists will need to join hands more frequently to examine potentials for enhancing supplies of raw materials wherever they can be found. This volume represents a first step in examining an untapped renewable resource—the marine plant biomass of the Pacific Northwest Coast.

This book grew out of a symposium convened on March 2, 1977 at Salishan Lodge, Gleneden Beach, Oregon. Participating in this conference were 49 scholars with many years of experience in disciplines which were pertinent to a balanced study of this resource. The authors who contributed to this volume were drawn from a broad spectrum of fields including oceanography, marine botany, engineering, geology, business, law, government, and education. The symposium was sponsored by Oregon State University with financial support from the Pacific Northwest Regional Commission which is headed by the governors of the States of Washington, Oregon, and Idaho.

The plans for the symposium were developed after Governor Robert W. Straub of Oregon responded to inquiries of Edward Hall and Maurice Commanday of Chromalloy Ocean Resources as to the possibility of the Oregon Coast serving as a site for developing plant biomass resources. Mr. Dale Mallicoat in Governor Straub's office called Oregon State University and conversations were begun in the fall of 1976 to explore this potential resource. Meetings were held in the College of Science between representatives of Chromalloy and selected faculty members from the College of Science, and the Schools of Oceanography and Engineering. At the conclusion of these discussions, it was decided to assemble a conference for the purpose of bringing together expertise from many parts of the world, as well as Oregon, to review the considerations that must be taken into account when exploiting such a resource. I accepted the task of assembling a group of compatible scientists and scholars who could bring their special perspectives to bear on this topic. The selection of authors was done after consultations with my numerous associates in the field of phycology— but especially with Dr. Joyce Lewin of the Friday Harbor Laboratories who has been most generous in her suggestions and support during the course of preparing the program. The Pacific Northwest Regional Commission funded travel costs as well as some of the secretarial expenses. The College of Science, the office of John Byrne, the Dean of Research, and the Oregon State University Press have contributed both financial and intellectual resources toward the effort.

The book, in a way, is an experiment in interdisciplinary communication. It is comprised of four major themes. The first six chapters deal with a review of the physical, chemical, and biological characteristics of the Pacific Northwest Coast. The data in these chapters are presented by specialists in the field of ocean geology, chemistry, and biology, and in several instances comprise the first attempts at a general review of the state of the science in the region. Chapters 7 through 13 deal with studies on the culture of marine macrophytes currently underway in the Pacific Northwest as well as in southern California, New England, Florida, the Philippines, Hawaii, and Europe. Chapters 14 through 17 are directed to potential items of value which can be obtained from algal species from the perspectives of analytical chemistry, pharmacy, and the very large international seaweed industry. Included in the final four chapters are introductions to oceanic engineering problems as well as a review of the systems approach to exploiting a resource which is part of a complex and delicate ecological matrix—a matrix in many ways similar to the legal hurdles which may be faced by an ocean-based industry as treated in the final chapter.

No apology need be made for the heterogeneity of the volume. It was intended that this work should present an interdisciplinary approach to a major problem-area and expose those elements of both basic and applied research which needed attention. Examination of the chapters will reveal a very pressing need for additional data and research. Many of the participants who did not contribute formal papers at this symposium wrote extensive suggestions for a major research program to bring more finely honed knowledge to bear on the problems that have surfaced. I considered concluding the volume with a research plan built on those contributions. For many reasons it is probably inappropriate at this time. Grant proposals are in the minds of many who participated in the conference and it would be doing a disservice to expose their strategies in this publication. However, it would be neglectful not to mention the high level of enthusiasm and anticipation felt by all participants for the work that still needs to be done. The quickened spirit of the bench researchers, and the improved understanding in government and industry of the promise of this new resource, made it clear to all that the years ahead would surely see additional efforts to answer the many questions that have been raised.

It is abundantly clear that there are both advantages and disadvantages in attempting to exploit the marine plant resources in the Northwest. The algal global harvest amounts to billions of dollars annually. It is inevitable that the Northwest portion of the Pacific rim will play a role in this commerce in the years ahead. However, the time is short. Economics of many nations, including that of the United States, are weakening. Every effort must be expended by scientists to identify new resources before a severe crisis in energy and raw materials makes the cost of such research unsupportable.

In the many details which attend publication activities, the editor has had the devoted assistance of the permanent staff of the Dean's Office of the College of Science. Special credit must be assigned to Penny Hardesty without whose diligence and steady support, the whole effort would have been impossible. She and Jean Haynes, my Administrative Assistant, were instrumental in moving both the symposium and the publication to completion. They were supported by Patricia Potts and Beverly McFarland who contributed talent to the editorial task. Although many hands assisted in the typing, most of that difficult task was executed by Judi Bowman and Patty Williams. Gratitude is certainly due them for their contributions to the final product and its speedy completion.

The editor took as few liberties as possible with the chapters. However, in most cases, some dialogue did take place in the process of pouring them into an appropriate mold for publication. In no cases was the thinking of the author distorted and each is responsible for the genius exhibited in his presentation. If the text serves as a starting point for a more thorough study of the renewable marine-plant resources of the Northwest Pacific rim, it will have served its purpose. It will have justified the efforts of each of the authors who focused attention, for a time, on a special regional problem.

September 1977   *Robert W. Krauss*

Contents

Contributing Authors

Elliot L. Atlas

Department of Chemistry
Texas A & M University
College Station, Texas 77843

Richard P. Benner

1000 Friends of Oregon
Suite 407
519 S. W. 3rd Street
Portland, Oregon 97204

George H. Constantine

Department of Pharmacology & Toxicology
Oregon State University
Corvallis, Oregon 97331

James A. DeBoer

University of Texas Marine Science Institute
Port Aransas Marine Laboratory
Port Aransas, Texas 78373

Maxwell S. Doty

Department of Botany
University of Hawaii
Honolulu, Hawaii 96822

Louis I. Gordon

School of Oceanography
Oregon State University
Corvallis, Oregon 97331

Robert A. Grace

Department of Civil Engineering
University Hawaii
Honolulu, Hawaii 96822

Peter L. Haaker

California Department of Fish & Game
Marine Resources Division
350 Golden Shore
Long Beach, California 90802

Edward N. Hall

Chromalloy Ocean Resources
P. O. Box 1067
2100 West 139th Street
Gardena, California 90249

Doyle A. Hanan

California Department of Fish & Game
Marine Resources Division
350 Golden Shore
Long Beach, California 90802

Adriana Huyer

School of Oceanography
Oregon State University
Corvallis, Oregon 97331

Robert W. Krauss — College of Science, Oregon State University, Corvallis, Oregon 97331

LaVerne D. Klum — School of Oceanography, Oregon State University, Corvallis, Oregon 97331

Tore Levring — Marine Botanical Institute, Carl Skottsbergs Gata 22, S-413 19 Gothenburg, Sweden

Joyce Lewin — Department of Oceanography, University of Washington, Seattle, Washington 98195

C. David McIntire — Department of Botany & Plant Pathology, Oregon State University, Corvallis, Oregon 97331

James R. Moss — Agro-Mar, Inc., 6405 Chartres Dr., Rancho Palos Verdes, California 90274

Thomas J. Mumford, Jr. — Division of Marine Land Management, Department of Natural Resources, Olympia, Washington 98504

John H. Nath — Department of Mechanical Engineering, Oregon State University, Corvallis, Oregon 97331

Michael Neushul — Marine Science Institute, Department of Biological Sciences, University of California, Santa Barbara, California 93017

Wheeler J. North — W. M. Keck Engineering Laboratories, California Institute of Technology, Pasadena, California 91109

Glenn W. Patterson — Department of Botany, University of Maryland, College Park, Maryland 20742

Harry K. Phinney — Department of Botany & Plant Pathology, Oregon State University, Corvallis, Oregon 97331

John H. Ryther — Woods Hole Oceanographic Institution, Woods Hole, Massachusetts 02543

Robert L. Smith — School of Oceanography, Oregon State University, Corvallis, Oregon 97331

Richard D. Tomlinson

METRO
Water Quality Laboratories
410 W. Harrison Street
Seattle, Washington 98119

J. Robert Waaland

Department of Botany
University of Washington
Seattle, Washington 98195

Kenneth C. Wilson

California Department of Fish & Game
Marine Resources Division
350 Golden Shore
Long Beach, California 90802

Participants & Guests

Armond J. Bryce

Marine Biomass Programs
Re-entry & Environmental Systems Division
General Electric Company
P. O. Box 8518
Philadelphia, Pennsylvania 19101

John V. Byrne

Research Office
Oregon State University
Corvallis, Oregon 97331

Maurice R. Commanday

Chromalloy Ocean Resources
P. O. Box 1067
2100 West 139th Street
Gardena, California 90249

Jean Dano

Pierefitte - Auby
46 Rue Jacques Dulud
92202 Neuilly-sur-Seine
France

Lynda Goff

Division of Natural Science
Applied Science Building
University of California
Santa Cruz, California 95064

Judith E. Hansen

Sea Grant Program
University of California
Santa Cruz, California 95064

George H. Keller

School of Oceanography
Oregon State University
Corvallis, Oregon 97331

Robert Kunz

LA Water Treatment Company
17400 East Chestnut
City of Industry, California 97149

Herschel C. Loveless

Chromalloy - Washington Bureau
918 - 16th Street, N.W.
Washington, D.C. 20036

Dale Mallicoat

Pacific Northwest Regional Commission
Salem, Oregon 97310

Thomas C. Moore

Department of Botany & Plant Pathology
Oregon State University
Corvallis, Oregon 97331

Richard Y. Morita

Department of Microbiology
Oregon State University
Corvallis, Oregon 97331

George F. Papenfuss

Department of Botany
University of California
Berkeley, California 94720

Ralph S. Quatrano

Department of Botany & Plant Pathology
Oregon State University
Corvallis, Oregon 97331

Robert F. Scagel

Department of Botany
University of British Columbia
Vancouver 8, British Columbia
Canada

Paul Scherer

Marietta
5626 Bell Station Road
Glenn Dale, Maryland 20769

Bela Szelenyi

Stauffer Chemical
5800 Perkins Road
Oxnard, California 93030

William Q. Wick

Sea Grant Program
Oregon State University
Corvallis, Oregon 97331

Dedicated to

George Frederik Papenfuss

for

A lifetime in the study of marine plants

A vessel gathering the marine alga, *Macrocystis*, off the coast of California. Worldwide, progress is being made in research and engineering toward sustained-yield harvests of marine plants.

Photo courtesy of the Stauffer Chemical Company.

-1-

Marine Plants—A Renewable Resource for Mankind

ROBERT W. KRAUSS

A useful starting point for a book dealing with the potential value of marine plants to the welfare of society is to refer to some words of Garrett Hardin spoken in Corvallis in May of 1975 in his keynote address to a symposium on "Radio Ecology and Energy Resources" (4). Dr. Hardin agonized, as is his custom, over the future of an enlarging humanity, which appears to be increasing its dependence on atomic power plants to satisfy ever-growing demands for energy. He saw a paradox in the dilemma of seeking more and more energy at greater and greater risk. One paragraph encapsulates the thrust of his perceptions:

> Why do we push atomic energy? Because we say there is a
> shortage of energy. But shortage is not an absolute term.
> It is a relative one. We have an energy shortage relative
> to something else -- mainly relative to the number of people.
> To speak of a shortage of energy is to utter a half truth.
> We can just as truly speak of a longage of people. Why do
> we prefer one half truth to the other? It is especially
> dangerous to speak of a shortage of energy because in a
> world in which population never stops growing, an energy
> shortage can never be satisfied. Not if the world is
> finite in some real sense.

This book is written in that same context. A naturally balanced ecology probably could not support well the numbers of people already on this planet. Surely they could scarcely be supported with the high standard of living, with the same level of education, or with the political freedoms now taken for granted in the United States and most of the western world. A glance at the newspaper on any day is sufficient to establish that we are not living in a balanced ecology. Just how unbalanced is a matter of debate, but the current rate of exploitation of non-renewable as well as renewable raw materi-

als has created an expectation of the good life for a greater number
of human beings than the planet ultimately can support. It is not
our mission to be directly concerned with the growing world population, but no one can talk of utilizing a segment of our renewable
resources to support human beings without being aware of the real
problem--a "longage" of people. Regrettably, the solution to that
problem is in the hands of politicians rather than scientists--
although it may not always be so. This volume is designed to address
the logic of attempting to enrich the store of resources by use of
the plants of the ocean, especially in the Pacific Northwest, and to
examine an opportunity to ease some of the pressures for survival on
this nation and others in the years ahead.

In any study of the economic feasibility of exploiting a natural
resource, questions of price--the value of the product versus the
cost of production--become dominant. As populations rise, resources
become depleted and prices rise because of competition for those
resources. This phenomenon is quite independent of any contrived
inflationary policy of governments. The prices of products in the
biomass, varying from potentially priceless drugs to products valued
only for their energy, are rising now and will continue to rise.
This is a new driving force in the affairs of men that scientists
must begin to appreciate--the increasing cost of everything.

It is the fundamental increase in real prices that has raised
the questions to be addressed in this volume. Are there compounds
of value in the renewable biomass of the ocean that can be retrieved
with present or new technology? Can the plant biomass be manipulated
without endangering the fish and shellfish which are already an
economic resource of great value? Are there ways in which the ocean
can be made to produce more harvestable biomass or will there be
limits based upon the nutrients, illuminance, and temperature prevailing in coastal waters? Can other activities of mankind fit into
a feedback loop that will in some way enhance the productivity of the
coast? Is it feasible to attempt to cultivate and harvest plants in
a wild and powerful oceanic environment without high risk and potential
loss of investment?

Because of the very nature of the questions being asked and the
primitive state of our knowledge, this volume is more concerned with
identifying unsolved problems than it is with supplying answers to
questions that might be raised. The contributors to this volume
write from diverse backgrounds. Some bring the strictly scientific
preoccupation--a concern with basic questions and the solution to
fundamental problems in biology. Others have commercial concerns,
and for them science is not so much a mechanism for widening their
philosophical horizons as it is for providing a way to unlock soon
new economic resources from the sea.

At this stage, we are more interested in identifying problems.
The advantage to humans of harnessing algal productivity has attracted
for years the attention of the scientists of many nations. Often
advances have been slow simply because we did not understand clearly
what the problems were. This volume attempts to assemble the perspectives of many disciplines so that we shall appreciate the work

that must be done, the engineering that must be attempted, and the societal problems that must be faced if we are to know the targets and be better able to take aim.

The photosynthetic efficiencies and rapid growth rates of algae have been known for over 100 years. From the early observations in 1871 on the growth of unicellular algae in cultures supplying minimal inorganic nutrients to the time when John Burlew edited Algal Culture from Laboratory to Pilot Plant (3), the potential of utilizing algae for the production of foods, drugs, or other compounds has been recognized (5). Dr. Vannevar Bush was president of the Carnegie Institution of Washington when the first major efforts to understand the problems of culturing unicellular algae such as *Chlorella* for useful purposes began some 25 years ago. In the 1950s, concern was for additional sources of protein and the cultivation of algae was envisioned as an ideal mechanism for supplying such food. I should like to quote from Vannevar Bush's forward to Algal Culture from Laboratory to Pilot Plant, coming shortly after his triumphs during World War II in fathering the development of atomic power. His words had a prophetic ring:

> The need of the world for additional sources of high pro-
> tein food is so great--especially in overpopulated areas--
> that serious effort in tracking down every promising lead
> is certainly warranted. Such great advances in technology
> have already come from the coupling of engineering with
> biology that it seems inevitable that the production of
> food--at least in certain areas--will eventually be
> carried out by process industries. The large-scale
> culture of algae may well become the first of them. In
> regions of the world where population is especially dense
> and fertile land is limited, it is entirely possible that
> the process industry methods for producing food may
> furnish a respite from the threat of famine and so con-
> tribute toward more salutory conditions for civilized
> living. If algal culture can serve such a purpose, it
> is well worth the development for that reason alone.

Those who have followed the efforts to create a system for inexpensive production of unicellular algae know that the realization of this prophecy has eluded us. The costs of the technology have been too great to match economically the value of the yields. However, much has been learned about the culture of algae. Now that our attention is to be directed primarily to the large forms of algae, the thallus forms often called seaweeds or kelps growing in natural rather than primarily man-made environments, it is well to consider the lessons that have been learned about the culture of the unicells.

In writing the introduction to the book on algal culture, John Burlew (3) chose to extrapolate from the photosynthetic equation, and its known efficiencies, to the amount of land that would be required for algal factories to obtain the needed protein for all of mankind. His calculations showed that, employing unicellular algae alone, a million acres--about the size of the state of Rhode Island--would be enough to provide the world population with its

total protein needs. This was an optimistic prediction. The reader
need not be troubled with a careful analysis of why we are not now
utilizing such a small area to feed so many people. It is sufficient
to say that there are major problems blocking such developments. We
can expect similar problems to be confronted in utilizing the larger
marine plants off the coast of the Pacific Northwest or elsewhere.
However, some commercial culture of unicellular algae is in fact
underway in several parts of the world. The culture of unicellular
algae, coupled with waste disposal (7), is a major research topic
at the University of California at Berkeley, and many proponents
believe they can evolve a methodology which is economically feasible.

The commercial utilization of large marine algae has always been
ahead of that of unicellular algae. Today's marine plant industry
is a billion dollar international operation with products, primarily
polysaccharides, common on the industrial scene, and other compounds,
such as the mannitols, involved in the pharmaceutical industry. Large
and efficient companies on both the east and west coasts of the United
States, and in Europe and Asia, have engaged in types of ocean farming during the last century. Many species of red, brown, and green
algae are in commerce today. The coast of Europe in the northern
latitudes has long served as a source of marine plants which are
utilized extensively or intensively locally or are shipped abroad.
The history of the involvement of algae as part of a staple food
diet of both Europeans and Asiatics is documented in literature (6,
8). Contributing to this volume are representatives of successful
commercial enterprises who are not preoccupied with beginnings, for
those beginnings are already on record. Here the future must be
considered to determine what science must do, what commerce must do,
and what governments must do if utilization of the algal resource is
to be expanded in the Pacific Northwest and elsewhere.

The northwest coast of the United States is the area to which
primary attention will be given in this volume. No significant kelp
or seaweed industry now exists in the Northwest, although numerous
species of potential value are present. The coast in the Pacific
Northwest is climatically similar to the coast in northwest Europe
although there are differences in coastal morphology and substrate.
The Pacific Northwest Regional Commission and the governors of the
states of Oregon and Washington are interested in potential new
resources for these states facing common natural resource needs.
However, much must be learned from the status of algal harvests in
California, the eastern United States and Europe and Asia. The
uniqueness of the Northwest can be viewed best with perspectives
from an industry functioning so effectively elsewhere.

Much can be said about the potential use of marine plants to
solve the energy crisis (2). Few topics are of greater pressing
concern than the need for fuel. Today the utilization of algae for
energy may appear remote. The problems connected with producing a
sufficient biomass within coastal waters capable of producing benthic
algae loom large and the costs of large-scale culture are at present
staggering. However, final judgment would be premature. This
volume will deal in part with the possibility of utilizing algae as

a source of fuel, but the more immediate probability of harvesting
comparatively small algal crops as a source of organic compounds
of high value to society will be the primary occupation. This is
an age where steering wheels of automobiles are made of soy beans,
where paint is made of wheat or milk, and where hulls of boats are
made of petroleum or sand. We may well be overlooking a group of
compounds in algae that can play unexpected roles in commerce and
relieve pressures on other renewable resources.

A previous symposium held almost exactly a year ago in Washington treated the problem of capturing the sun through bioconversion
(1). That program makes interesting reading. In it, projections
were made of vast areas of ocean on the west coast that could
potentially supply food and fuel to the nation. However, the extrapolations made in that symposium need to be addressed in greater
detail by scientists especially knowledgeable about the algae and
the environments in which they grow. In the last analysis, the
solutions to many of the most obvious obstacles will be made by
phycologists who are responsible for understanding the growth and
reproduction of the varieties and the species employed, and for
designing the methodology used to maximize the biomass. Experience
with land-based agriculture is consonant with this view. Botanists
and agronomists establish the parameters for optimum growth and
yield. Engineers and industry combine to match technology to the
crop. Once we learn how to produce the biomass, the chemical engineering technology for its processing to useful forms will follow
almost automatically. *The major problem is how to generate the biomass and manage a sustained yield from it.* The reader should be
especially interested in assessing the potential of the ocean environment to supply necessary nutrients, temperature, illuminance, and
substrates to sustain the type of standing crop that seems to be
necessary if some of the long-range projections we have heard are
to be achieved. The ecosystem in which this biomass exists and to
which it is, in a sense, susceptible must also be given a detailed
scrutiny.

Those with their eyes on algae as a source of fuel project that
marine plants could contribute as much as 20 quads of the nation's
total current need of approximately 100 quads (1 quad = 1 x 10^{12}
B.T.U.). The production of 20 quads would require nearly 100 million
metric tons of algae. Such production would require a highly efficient scrubbing of the nutritional components which exist naturally
or which, to some degree, might be replenished by man.

It may be useful to make extrapolations concerning one of the
resources off the Oregon coast and the probability of supporting a
significant algal population. Nitrogen is commonly a limiting factor
in algal growth, and the adequacy of its supply will determine how
large a crop can be grown. The estimate of nitrogen flow during a
period of upwelling through one square meter of ocean surface as
calculated by Atlas in Chapter 4 of this volume makes possible a
projection of the potential algal biomass. Data indicate that the
on-shore nitrogen drift during a typical upwelling period is approximately 31.5 kg of nitrogen per meter parallel to the coast per day.

Assuming approximately 740,800 meters of Oregon coastline between parallels 42° and 46° north latitude, 23,335,200 kg of nitrogen move into Oregon's coastal photic zone per day. Employing a conservative analysis in typical kelp of 2% of the dry weight as nitrogen, 1 kg of nitrogen will produce 50 kg of algae assuming it is absorbed without loss and plays its normal metabolic role. Thus, there could be 1,166,760 metric tons of algae produced per day along the Oregon coast, during a period of upwelling if nitrogen were the only limiting factor. However, such strong upwellings occur during only 25% of the days in an upwelling season which normally lasts six months in the summer. If there are only 45 days of adequate upwelling during the summer season to sustain growth, the total algal crop could amount to 52,500,000 metric tons. Therefore, of the total U.S. energy supply requiring 500 million metric tons of algae, over 10% could theoretically be produced off the coast of Oregon during the upwelling season, if upwelling alone were the source of nitrogen!

   Clearly such estimates represent performance under ideal conditions and are based on the known supply of only one of the factors which might be limiting. However, they do suggest the potential that does exist if all other conditions could be optimal. Reality must be retained in any projection. Certainly modest cultivation of some algae for commercial purposes will need to precede their employment as a significant source of fuel substrates to replace petroleum or gas. This volume presents much to keep extrapolators within bounds. It should serve as a guide to the science that exists or needs to be created. It also raises warning signals to those who wish to exploit the plant biomass. It should provide many with an understanding of what the commercial concerns may be and identifies many questions for the solution of which governmental funding may be required. It is designed not as a textbook nor as a coherent assembly of all the knowledge we have. Rather it should serve as a starting point for a new approach to the management and exploitation of a very great and increasingly precious renewable resource of the Pacific Northwest coast.

REFERENCES

1.   Anon. 1976. Proceedings of a Conference on Capturing the Sun through Bioconversion. Washington Center for Metropolitan Studies, 1717 Massachusetts Ave., N.W., Washington, D.C. 20036. 861 pp.
2.   Budhraja, V.S., B.N. Anderson, G.E. Hoffman, F.J. Nickels, R.H. Schneider, and D.H. Walsh. 1976. Ocean Food and Energy Farm Project Subtask No. 6: Systems Analysis. Vols. I to VII, ERDA/USN/1027-76/1. 349 pp. + appendices. National Technical Information Service, Dept. of Commerce, Springfield, Virginia 22161.
3.   Burlew, J.S. (ed.) 1953. Algal Culture from Laboratory to Pilot Plant. Carnegie Institution of Washington Pub. 600, Washington, D.C. 357 pp.

4. Cushing, C.E., Jr. (ed.) 1976. Radioecology and Energy
       Resources. Proceedings of the Fourth National Symposium
       on Radioecology. Dowden, Hutchinson, and Ross, Inc.,
       Stroudsberg, Pennsylvania 18360. ix + 401 pp.
5. Krauss, R.W. 1962. Mass culture of algae for food and other
       organic compounds. Am. J. Bot. 49:425-435.
6. Levring, T., H.A. Hoppe, and O.J. Schmid. 1969. Marine Algae.
       A Survey of Research and Utilization. Crom, DeGruyter and
       Co., Hamburg, Germany. 421 pp.
7. Oswald, W.J., and C.G. Golueke. 1968. Large scale production
       of algae. *In*: Mateles, R.I., and S.R. Tannenbaum (eds.)
       Single-Cell Protein. The M.I.T. Press, Massachusetts
       Institute of Technology, Cambridge, Massachusetts, pp. 271-
       305.
8. Stephenson, W.A. 1968. Seaweed in Agriculture and Horticulture.
       Faher and Faher, London, England. 231 pp.

-2-
Coastal Morphology and Geology
of the Ocean Bottom—The Oregon Region

LaVERNE D. KULM

The Oregon coastline and the adjacent continental shelf and slope have been studied in detail by Oregon State oceanographers and other workers since 1960. The major geological features are shown in Fig. 1. A few estuaries have been investigated in enough detail so that the basic estuarine circulation system and the sedimentary deposits are known in a general fashion. A number of studies have described the severe effects of erosion along the Oregon coastline (6, 8, 37) and the influence of coastal geology on these erosional processes. The effects of coastal erosion are most apparent to those persons living at or near the coastline and have an important bearing upon the socio-economic climate of the region. Several sedimentological studies were also conducted on the Oregon continental shelf (4, 11, 20, 28, 32, 42, 43, 49) during the 1960s and early 1970s. The main objective of these related studies was to determine the sedimentary facies on the continental shelf and relate the facies to (1) the estuarine circulation system and sediment input from the coastal drainages, (2) the modern and ancient hydraulic regime of the shelf, and (3) the benthic activities that modify these deposits. A concurrent investigation, emphasizing the geologic history of the regime, was conducted by Kulm and Fowler (27) on the shallow structure and stratigraphy of the continental shelf.

Unless otherwise indicated by references, the bulk of this chapter was taken largely from the papers of Kulm *et al.* (28) on Oregon shelf sediments and Kulm and Fowler (27) on geologic history of the Oregon shelf. Some passages and paragraphs have been extracted verbatim from the published work of the author and his co-workers and integrated here to give an overview of the geology of the region.

MORPHOLOGY

The Oregon coastline is relatively straight except where it is indented by estuaries and interrupted by erosion-resistant headlands.

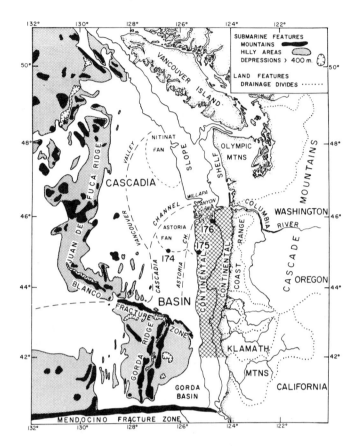

Figure 1. *Map of the Oregon continental margin, coastal region and surrounding features.*

Volcanic rocks generally form these headlands whereas crystalline and sedimentary rocks occur along the coastline in other areas (8). Well developed emergent marine terraces are prominent along the central and southern Oregon Coast. Near Cape Blanco (Fig. 2) the youngest terraces are elevated about 60 m above sea level, dropping in elevation to the north and south. The terraces consist of uncon-solidated beach and dune sands associated with former Pleistocene stillstands.

The morphology of the Oregon continental terrace (shelf and slope) has been described by Byrne (5, 6, 7), Maloney (32), Carlson (10), and Spigai (49). The combined width of the terrace ranges from 75 km off Cape Blanco to 135 km off the Columbia River (Fig. 2). The shelf varies in width from about 17 to 74 km, has an inclination of 0°08' to 0°43', and a depth at its outer edge of 145 to 183 meters. These dimensions are narrower, steeper, and deeper than the average values given for the continental shelves of the world (45). Promi-nent submarine banks (i.e., Nehalem, Stonewall, Heceta, and Coquille) of exposed bedrock occur near the outer edge of the shelf and have

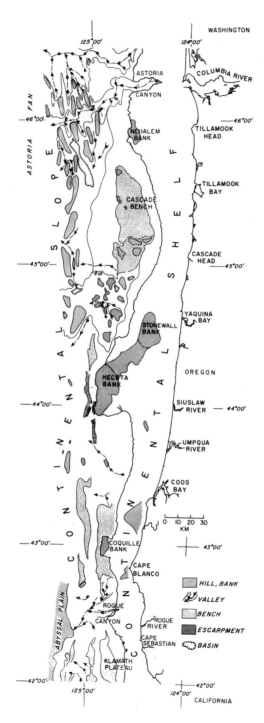

Figure 2. *Morphology of the Oregon continental shelf and slope and location of coastal rivers and headlands (from Kulm et al., 28).*

as much as 75 m of relief. Bedrock also crops out on the inner shelf, especially between Coos Bay and the Rogue River.

The upper continental slope (180-190 m) is characterized by benches and low relief hills (Fig. 2). The largest bench (Cascade Bench) occurs between Cascade Head and Tillamook Bay and is 400-600 m deep. Another prominent bench lies between 500 and 700 m and is located between Cape Sebastian and the California border (49). It is a continuation of the Klamath Plateau found on the upper continental slope off northern California (46). Off central Oregon (Heceta Bank) the upper slope forms a steep escarpment. Astoria and Rogue River canyons and numerous small submarine valleys cross the outer shelf and upper slope. They are the most important avenues of sediment dispersal by turbidity currents originating on the continental terrace (10, 35, 49).

The irregular width of the shelf, submarine banks, abundant benches, and linear ridges on the slope, all suggest strong structural control of the morphology and general configuration of the terrace.

COASTAL EROSION

Studies by Byrne (6, 8) and North (37) show how effective erosion is along portions of the Oregon coast. Numerous sea cliffs, sea stacks, and landslides are testimonials to the erosive power of sea and swell conditions associated with storms generated in the northeast Pacific Ocean. When these physical forces are coupled with certain geologic conditions along the coastal region, erosion of this region is severe and rapid.

Byrne (8) has devised an erosional classification for the northern Oregon coast which shows how various geologic parameters influence coastal erosion (Fig. 3). The relative resistance to erosion is related chiefly to lithology, structure, and stratigraphy. He believe that lithology, either by itself or in conjunction with structure or stratigraphy, has the most influence on erosion.

Landslides are most likely to occur where the relatively soft sedimentary rocks are exposed at the coastline. Periods of excessive rainfall, which occur frequently in the winter months, cause these types of lithologies to slip off their frequently seaward-dipping bedd planes onto the beaches where the material is redistributed by wave action. Coastal terraces consisting of unconsolidated sand also retreat landward rapidly under wave attack. In contrast, the hard igneous rocks commonly found at protruding headlands are much more resistant to wave erosion.

Broad, sandy buffer zones (unconsolidated sediments) along the coastline (e.g., coastal dunes and broad beaches) inhibit erosion as long as the littoral sands are allowed to move seasonally north (winter) and south (summer) without interruption by man-made barriers such as jetties. Areas deprived of these buffer materials often display excessive erosion in the winter months.

Because the Oregon coast is affected by severe winter storms, coastal land erosion by waves is an important mechanism for redistributing sediment along the coastline. Significant heights of storm waves range from 5.8 to 9.6 m (34). Estimated rates of erosion on

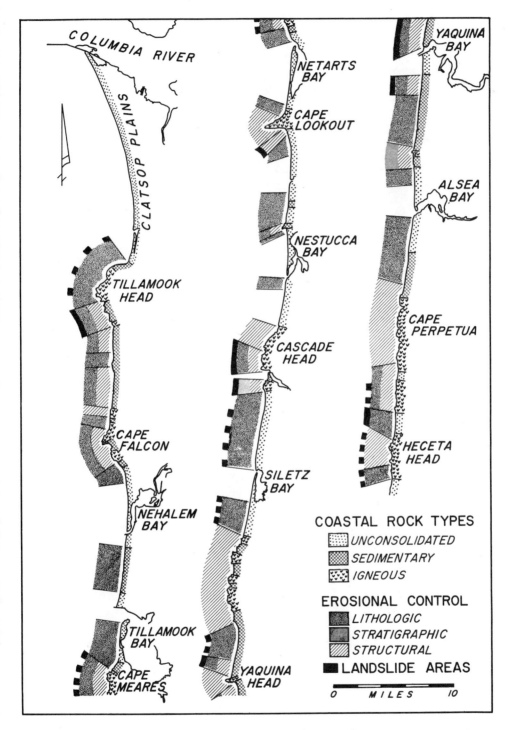

Figure 3.   *Erosional classification of the northern Oregon coast.
All symbols apply to the coastline (from Byrne, 8).*

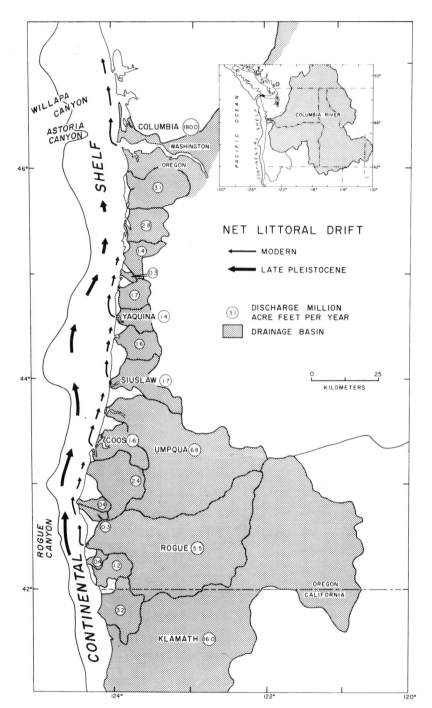

Figure 4. *Coastal drainage basins and adjacent continental shelf off Oregon. Littoral drift summarized according to Scheidegger et al. (44).*

the northern Oregon coast range from 0.6 m per year for sedimentary rocks overlain with consolidated marine terrace deposits to 16 m per year for unconsolidated sands and gravels along the coastline (6). Runge (43) calculated 595,000 cubic meters of coastal sediment potentially could be added to the continental shelf each year.

SEDIMENT SOURCES

Coastal Drainages

The bulk of the sediment supplied to the Oregon shelf is derived from the Columbia River and from the southern coastal rivers such as Rogue and Umpqua (Fig. 4). The Columbia River, which has the third largest flow of all rivers in the United States (22), has a drainage basin 12 times the area of all neighboring basins along the Oregon coast (Fig. 4). Peak discharge of the Columbia River, occurring during May and June, is caused by melting snow in the Rockies and Cascades. Peak discharge occurs in the smaller coastal rivers from October through March as a result of local precipitation. Approximately 11×10^6 cubic meters of suspended sediment is transported annually to the Pacific Ocean by the Columbia River (50). An average annual bedload of 1,360,916 cubic meters was measured at a station at Vancouver, Washington (30). Although the sediment load of the smaller coastal streams is not monitored, these streams probably supply an amount proportional to their drainage area. All streams have their headwaters in mountainous regions that receive large amounts of rainfall during the winter months.

Heavy mineral studies of the sand in these coastal drainages and on the continental shelf show that the Columbia River drainage basins and the Klamath Mountains of southern Oregon and northern California are the principal sources of coarse sediment for the shelf (43, 44).

Estuarine Circulation Versus Sediment Supply

Drowned river estuaries, which occur at the mouths of the coastal rivers, act as filters for sediment being transported between the fluvial environment and the open marine environment (26, 40). The two-layered and partly mixed circulation systems characteristic of Oregon estuaries during winter and spring (3) produce a net upstream flow along the bottom, which transports marine sand from the longshore drift into these embayments (26, 30). In the smaller estuaries, during peak fresh water discharge in the winter and early spring, a partly mixed or two-layered circulation system inhibits the transport of fluvial sands out of the estuary, although a portion of the suspended fine-grained sediment continues to diffuse seaward. Aerial photos of the surface plumes of muddy water and optical studies of the suspended material in the continental shelf waters (15, 21, 38) indicate that a significant part of the fine sediment is transported out of the estuary and across the shelf. During peak flow in May or June, the Columbia River pushes the salt wedge seaward toward the mouth of the estuary (30), allowing the suspended material to pass through the

surf zone and diffuse directly to the shelf where the surface plume
moves south (38). During the winter months, the plume moves north
(15).

The filtering process, which traps the coarse sediments (fine
sand and coarser [26]) in the estuaries, is most efficient when the
fresh water discharge is at a maximum (two-layered system) and there
is a corresponding high sediment input from the stream. During periods of low river runoff and corresponding low sediment supply, the
estuaries become the well-mixed system, with a net flow seaward at
all levels (3). Because the peak discharge of the Columbia River
occurs several months after the peak discharge of the smaller coastal
streams, there is a seasonal change in the primary source of shelf
sediment, the coastal stream dominating in the winter months and the
Columbia dominating in late spring and early summer.

CONTINENTAL SHELF SEDIMENTS

Sediment Composition

The Oregon shelf sands are grouped into three main genetic
classes: detrital, biogenic, and authigenic (4, 11, 32, 49, 51).
Detrital sands consist chiefly of quartz, feldspar, and rock fragments. They generally are found on the inner continental shelf and
portions of the outer shelf. Quartz:feldspar ratios average 1:1
(4) and are similar to the ratios found in adjacent beach sands (25).
These deposits are classified as arkosic sand by Williams, Turner,
and Gilbert (52). All detrital sands have a fresh surface texture
and lack the typical iron oxide coating that is found on relict sand
deposits (16).

Biogenic components, such as sponge spicules, diatoms, radiolarians, and planktonic and benthic foraminifera, constitute as much
as 32% of the sand fraction of the fine-grained muds (32), but a
minor portion of the total deposit. Sponge spicules and diatoms are
most abundant in the muds of the Heceta swale, whereas benthic foraminifera are abundant in the vicinity of the submarine banks. Biogenic constituents form a minor part of the sand deposits on the
shelf.

Glauconite is the principal authigenic constituent of the shelf
sands (43). The highest concentrations range up to 98% of the coarse
fraction and occur on the outer edge of the shelf and topographic
highs.

Sediment Facies

Definition. Sedimentary facies on the Oregon continental shelf
have been determined from the textural analysis of more than 900
surface and 100 subsurface samples. These data are contained in
several studies (4, 11, 32, 42, 43, 49) and are summarized by Kulm
et al. (28).

Distribution. The distribution of sedimentary facies on the
Oregon shelf is shown in Fig. 5. The sand facies occurs mainly on

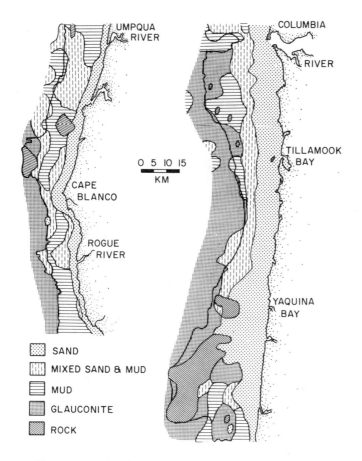

Figure 5. *Sedimentary facies of the Oregon continental shelf.*
Glauconite pattern denotes those sediments with greater than 15%
glauconite. The continuous solid line that intersects the facies
patterns is the 183-meter contour at the edge of the shelf (modified
from Kulm et al., 28).

the inner shelf or in a few small patches on the outer edge of the
shelf. This facies extends seaward from the shore to depths of 90
to 100 meters from the Columbia River to the Siuslaw River and to a
depth of only 50 meters from the Siuslaw River to the California
border.

It is apparent that the mud facies is irregular and patchy and
concentrated in areas near rivers that have a fairly high discharge
(Fig. 5). Although this facies occurs largely at mid-shelf depths,
it may extend to the outer edge of narrower portions of the continental shelf, as in southern Oregon. The mud layer is usually less
than 10 cm thick off the Rogue River (Fig. 6), 1-3 cm thick or absent
off central Oregon (Fig. 5), and more than 40 cm thick south of the
Columbia River. The rate of sedimentation in the large mud patch
south of the Columbia River (Fig. 5) is only about 6 cm/1000 yrs

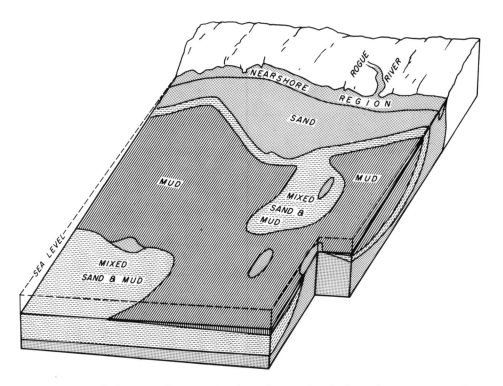

Figure 6. *Model of sediment facies determined from box cores taken of the southern Oregon shelf. Vertical scale is approximately 40 cm (from Kulm et al., 28).*

based upon a Mazama ash horizon (36) encountered in a gravity core.

The mixed sand and mud facies, which is created by burrowing benthic organisms mixing the basal sand into the overlying mud, may occur anywhere between the sand facies and the outer edge of the continental shelf (Fig. 5). It usually occurs between the sand and the mid-shelf mud facies except off the Umpqua River and the California border. Just south of the Umpqua River, the mixed sand-and-mud facies extends seaward from the sand facies on the inner shelf and transects the mid-shelf mud facies. South of the Rogue River, the mixed sand-and-mud facies actually may exist as a narrow belt on the inner shelf, but it may not have been detected because of the small number of samples collected in this area.

There is no systematic depth dependency in the facies distribution over the Oregon continental shelf. However, where the shelf is widest, (northern and central Oregon) the sand facies extends seaward to water depths of 90-100 m. Where the shelf is narrowest and somewhat steeper (southern Oregon), the sand facies retreats to a water depth of 50 m.

Wave Influence on Sediment Facies

Hundreds of bottom photographs taken on the Oregon shelf show

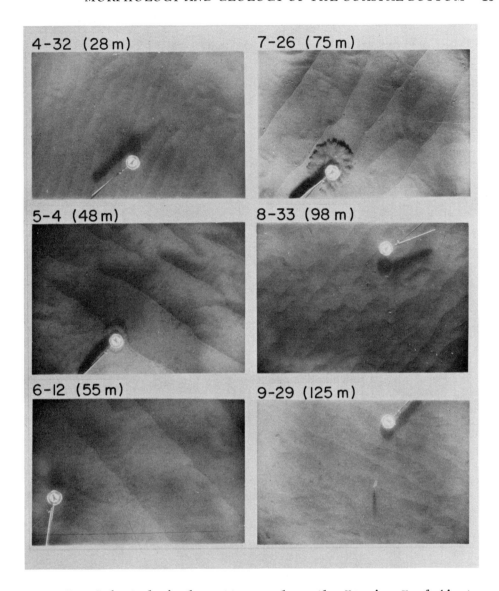

Figure 7. *Selected ripple patterns along the Yaquina Head (just north of Yaquina Bay, see Fig. 2) camera profile. Water depths indicated in meters (i.e., 28m) (from Komar et al., 24).*

symmetrical ripples in the sedimentary deposits (Fig. 7) which are produced by oscillatory currents associated with the orbital motion of surface waves (24). The Oregon coast is noted for the severity of its wave conditions as shown in Fig. 8. Such extreme long-period wave conditions are capable of stirring the bottom sediments across the entire width of the continental shelf (24). This wave stirring causes the sands and muds to be resuspended and helps to create the bottom turbid layer described below. Long-period winter waves are

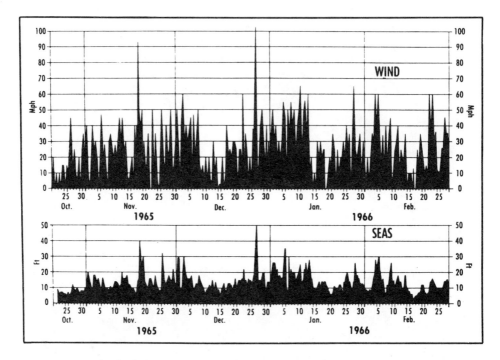

Figure 8. *Wind and wave conditions recorded from an oil rig off the coast of Oregon (from Rogers, 41).*

capable of stirring the bottom to water depths of 200 m whereas the shorter period summer waves stir the bottom to a depth of about 90 m.

Periodic high speed unidirectional currents ranging up to 70 cm/sec provide the energy for sand erosion and transport on the inner shelf and mid-shelf region (47) and probably are related, in part, to the extreme wave conditions during the winter months. Even though there is sufficient energy for resuspending and transporting sand-sized material, the exact nature of the sand dispersal system is not known.

FINE SEDIMENT DISPERSAL SYSTEM

Definition of Turbid Layers

The distribution pattern of the mud facies on the central and northern Oregon continental shelf (Fig. 5) is related, in part, to the dispersal of fine-grained, suspended material across the continental shelf from the coastal region (21, 38). Profiles of turbidity from the surface to the bottom were recorded over the northern and central shelf using a beam transmissometer (Fig. 9); bottom current measurements were made in selected areas of the shelf in conjunction with the turbidity measurements (21). These measurements were made in February, April, and May when sediment supply from coastal drainage is high to moderate. Although the turbid regions were not

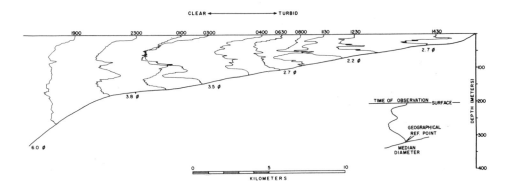

Figure 9. *Turbidity measurements made along a profile taken at 45° 11' N. Transect was made February 9, 1971. Comparisons of turbidity between profiles can be made only on a relative basis (from Harlett and Kulm, 21).*

sampled, the suspended particles probably consist of clay and silt-sized particles of biogenic and terrigenous debris similar to that described in the section on Composition of Shelf Sediments.

Transmissometer measurements show that suspended materials are concentrated at three surfaces (21): (1) the seasonal thermocline, (2) the permanent pycnocline, and (3) the bottom (Fig. 10). Three rather distinct layers, the surface layer, mid-water layer, and bottom layer, are defined above these surfaces. They are described in the following sections.

Surface Layer

In shallow water, the surface turbid layer at the seasonal thermocline is usually the most turbid of the three layers. During times when surface waves are large, turbidity increases nearshore with no apparent layering, becoming so intense that it completely obliterates the compass in bottom photographs taken 2 m above the sea floor. This indicates that a large amount of suspended material is moving seaward from the surf zone during storms. Sediment finer than very fine sand (< 62μ) is absent on Oregon beaches and must either diffuse seaward or re-enter the estuaries. On the northern shelf, sediment in the Columbia River plume dominates the surface layer in the late spring and early summer. During these months, the plume extends offshore and to the south; Pak (38) traced the plume as far south as Yaquina Bay on the central coast during May. Since the maximum discharge of the Columbia River occurs in late spring and early summer, its influence on sedimentation is magnified at this time. The surface layer tends to retain its identity across the shelf, although its intensity decreases as distance from the source increases.

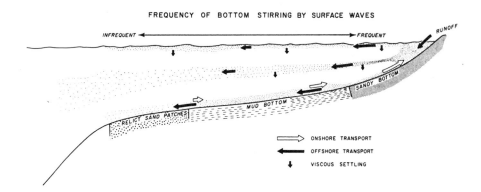

Figure 10. *Conceptual model of suspended sediment transport across the northern Oregon shelf. Turbid layers are indicated by stipple pattern in water column (from Harlett and Kulm, 21).*

Mid-Water Layer

The mid-water layer of turbid water occurs at the permanent thermocline, in which the temperature decreases about 1.5°C (21). Because of the density increase associated with the temperature decrease in the permanent thermocline, the permanent pycnocline and the permanent thermocline coincide. Drake (14) has noted similar concentrations of suspended material at levels with much smaller temperature changes. Pak *et al.* (39) found a scattering maximum at the level of the permanent pycnocline during upwelling, which they interpreted as biogeneous material sinking with upwelled water.

Unlike the surface layer, the mid-water layer becomes vertically thicker and less intense with increasing distance from the shore. Near the shelf edge the layer becomes diffuse (Fig. 9). Near the surf zone, the mid-water layer receives material diffused from the mid-water levels and the water column above. This material moves laterally across the shelf on the gently sloping pycnocline, continually receiving a contribution from the surface waters. Furthermore, during upwelling a significant amount of biogeneous material is added where the denser upwelled water sinks in the middle part of the shelf (39). Observations made at a time-series station west of the Columbia River show that the mid-water layer migrates vertically, apparently in response to a similar migration of the permanent thermocline. The intensity of the layer remains nearly constant although its vertical extent changes slightly.

Bottom Layer

Where the bottom sediments consist of mud, the bottom layer of turbid water is almost as intense as the surface layer. Here the turbidity increases sharply within 10 to 20 meters of the bottom and generally decreases with distance from the shore. In regions where

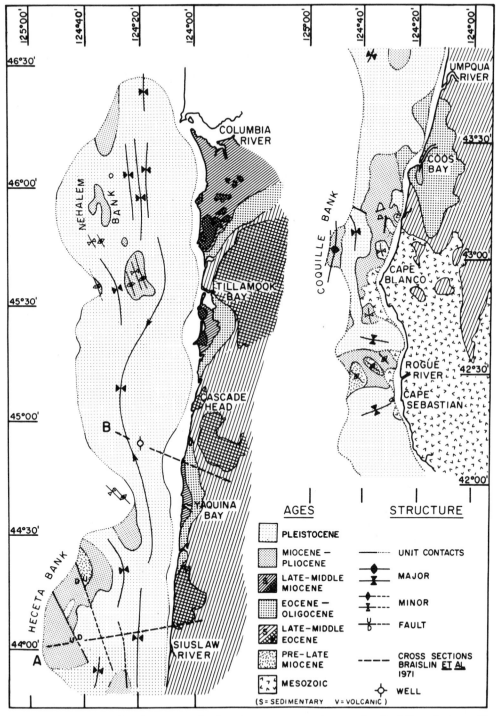

Figure 11. *Geologic map of the Oregon continental shelf and immediate onshore region (from Kulm and Fowler, 27).*

there is only a small amount of fine material in the surface sediment or on topographic highs, the bottom layer is much reduced and
in places non-existent. Turbidity measurements from time-series
observations over the mud facies west of the Columbia River (Fig. 5)
show a well-developed bottom layer which increases and decreases
vertically in response to concomitant increasing and decreasing bottom current strength at a nearby station.

     The bottom turbid layer apparently receives suspended material
from the surf zone and from the water column above, as well as from
the bottom itself. Surface waves passing over the continental shelf
generate sufficient orbital velocities to resuspended bottom sediments, including fine sand, to water depths of at least 125 meters
(24). Bottom photographs show that turbidity in the bottom layer is
greater during winter months, and sand ripples are more often present,
indicating that orbital wave motion is partly responsible for resuspension of sediments. Away from the direct influence of the surf
zone, the clearest bottom waters are found over submarine banks and
the most turbid bottom waters are found in the swales or valleys
landward of the submarine banks (Fig. 2). This suggests that the
bottom turbid layer may develop into a low-density bottom current
near the middle or outer part of the shelf, flowing around the topographic highs. Near-bottom velocity profiles measured on the continental shelf west of Yaquina Bay show a velocity maximum of 15-20 cm/
sec at 150 to 200 cm above the bottom.

Seasonal Fluctuations in the Mud Layer

     Although the mud facies is absent off central Oregon (e.g.,
Yaquina Bay area, Fig. 5), bottom photographs reveal that a thin
blanket of fine-grained material covers the underlying sand facies
during periods of moderate wave conditions. With the arrival of the
long-period waves, the bottom is stirred and the fine materials are
resuspended and transported seaward as a bottom turbid layer. The
bottom is fairly rough in this region, consisting of relict sands
and gravel. This creates a turbulent boundary which enhances the
resuspension of the fine sediments. Because of the low sediment
supply to the shelf in this area and the frequent passage of long-
period surface waves which are capable of rippling the bottom, a
permanent mud facies never develops.

     The mud facies develops when a high sediment supply coincides
with moderate wave conditions. For example, the Columbia River has
its peak discharge during May and June and the fine-grained sediments accumulate on the shelf under waves that stir the shelf deposits to water depths of only 90 m or less on the average. This
increase in sediment supply tends to concentrate the suspended material near the bed (33) and may create a smooth-bottom cohesive
material that is more resistant to erosion during storm conditions.
Despite the nearly ideal conditions, the amount of mud deposited
(6 cm/1000 yrs) by the Columbia River on the adjacent continental
shelf is relatively minor compared to what one would expect from an
annual suspended sediment load of 11 million cubic meters.

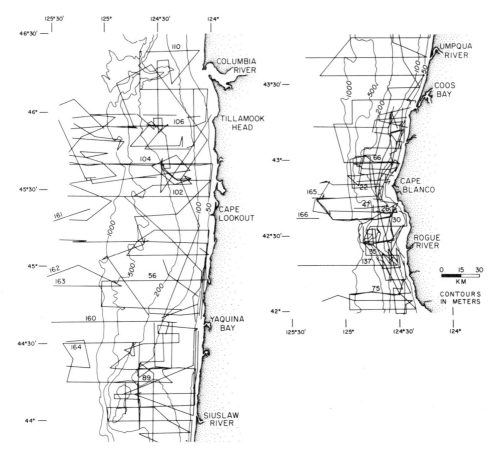

Figure 12. *Seismic reflection profile trackline across the Oregon margin (shelf and slope). Illustrated profiles in Figs. 13 and 14 are numbered (from Kulm and Fowler, 27).*

CONTINENTAL SHELF STRUCTURE

General Framework

A generalized geologic map (Fig. 11) was constructed for the Oregon continental shelf with the aid of more than 8000 km of seismic reflection profiles (Fig. 12) and more than 100 samples. Regional correlation was aided by continuous tracing of the acoustic units laterally from control points with heavy reliance upon the ubiquitous angular unconformities for stratigraphic control.

Some of the most strictly folded and faulted areas of the Oregon margin occur beneath submarine banks near the outer edge of the shelf (Figs. 13 and 14). The principal ones are Nehalem Bank (SP-106), Heceta-Stonewall Bank complex (SP-89), and Coquille Bank (SP-66). Pleistocene to pre-late Miocene rocks are exposed on these banks (Fig. 11); the oldest known exposed sediment occurs in the vicinity of Heceta Bank on the upthrown side of faults.

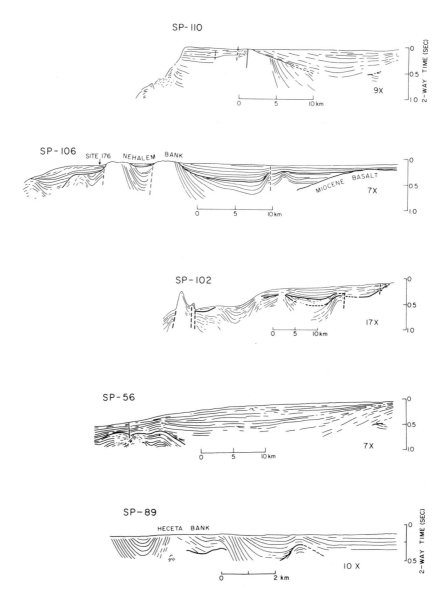

Figure 13. *Seismic reflection profiles of the northern Oregon shelf (see Fig. 12 for location) (from Kulm and Fowler, 27).*

Heceta and Coquille Banks are characterized by positive free-air gravity anomalies which suggest a slight mass excess beneath them. In contrast, the area around Nehalem Bank (about 46°N) shows a negative anomaly and implies a fairly thick low-density sedimentary section (13).

A series of interconnecting shallow synclines occurs between the shore and the outer banks off southern Washington and northern and central Oregon (Fig. 11). The largest of these synclines is

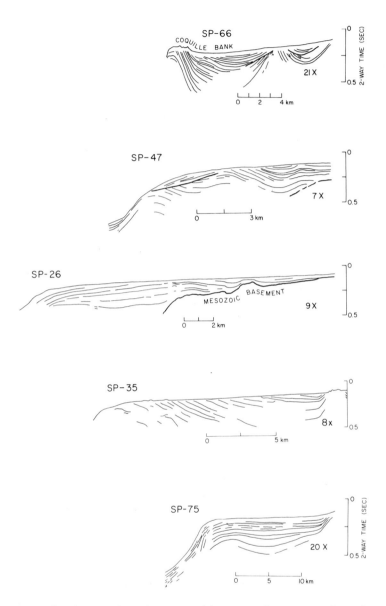

Figure 14. *Seismic reflection profiles of the central and southern Oregon shelf (see Fig. 12 for location) (from Kulm and Fowler, 27).*

basinal and is centered off Cascade Head (Figs. 11 and 13). Along the seaward limb, sediments in the syncline have ponded behind and overlapped a series of subsurface folds (SP-56, -66). Cascade bench (Fig. 2) and other similar benches have been produced by this process of sediment entrapment. The most recent development of these synclines involves Pleistocene sediments that almost always overlie older units unconformably. However, the synclines are generally best

developed offshore from thick late Paleogene and early Neogene sedimentary sections on land that trend offshore.

    Off Washington and north and central Oregon large negative free-
air anomalies coincide with the inner shelf synclines. From reflection profiles and gravity data, it seems clear that much thicker and
older sedimentary basins underlie the shallow Pleistocene basins on
the inner shelf such as those off Cascade Head. Braislin *et al*. (2,
their Fig. 2) show a known sedimentary section of 12,000 feet and
postulate a section in excess of 15,000 feet thick over this area.

    An acoustic basement dips seaward beneath the inner part of the
continental shelf along profile SP-106 (Fig. 13). It is marked by
large positive magnetic anomalies (17, their Fig. 4) and is most
likely the seaward extension of the Miocene volcanic rocks found
onshore (48).

    Between Coos Bay and Cape Sebastian, the inner shelf is under-
lain by intensely folded and faulted sedimentary units (Fig. 11) (31).
Erosion has exposed Mesozoic and younger strata. Structural trends
are variable and limited to a few kilometers; they probably reflect
several stages of deformation associated with rocks of different
ages. Most of these older structures have been truncated (e.g.,
Fig. 14, SP-35) and covered by a thin veneer of Pleistocene sediments.

    From the Rogue River to the California border, the shelf is
underlain by a shallow basin consisting of undeformed sediments which
unconformably overlie older units (Fig. 14, SP-75). This basin apparently trends perpendicular to the shoreline and is characterized by
a large negative free-air gravity anomaly.

    Although the character of the outer shelf is in large part struc-
turally controlled, it also has been shaped by considerable erosion
and sediment progradation (Figs. 13 and 14). A Pleistocene sediment
wedge unconformity overlies older truncated strata which now occur
at water depths as deep as 300 m (Fig. 14, SP-47). Subaerial erosion
during the several Pleistocene lowerings of sea level has truncated
the folded structures over much of the shelf (e.g., Figs. 13 and 14,
SP-106, -89, -35). This suggests in turn that either the earlier sea
level lowerings were substantially lower than the negative 125 m indicated for the late Wisconsin regression (12) or there has been
subsidence during the Pleistocene along portions of the outer shelf
as suggested by Kulm, von Huene *et al*., (29) from deep-sea drilling.

Uplift of the Continental Shelf and Slope

    Assemblages of benthic foraminifera from the sedimentary rocks
collected on the Oregon continental slope and shelf generally are
indicative of deeper environments of deposition than presently found
at the collection sites. The difference between the paleodepth and
the present-day water depth gives the minimum amount of relative
uplift or downwarp for the sedimentary section.

    Uplift of as much as 1,000 meters has been reported for the
central Oregon shelf and upper slope (9, 18, 19). The largest
amounts of uplift (900-1000 meters) on the Oregon continental shelf
involve the late Miocene and early Pliocene rocks in the vicinity

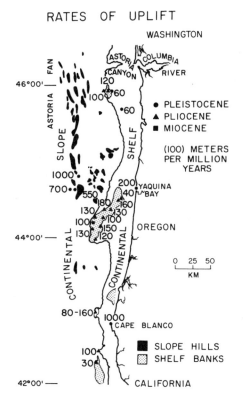

Figure 15. *Minimum rates of uplift on the Oregon continental margin and coastal terraces. Pleistocene is number with solid dot, Pliocene is solid triangle and Miocene is solid square (from Kulm and Fowler, 27).*

of Heceta Bank (Fig. 15). Early to middle Pliocene strata on Nehalem and Coquille Banks have been uplifted as much as 500 to 600 meters. Pleistocene deposits on the submarine banks have an estimated uplift ranging from 0 to 100 m.

The Elk River terrace is the lowest and youngest emergent marine terrace along the Oregon coast (1). An age of 45,000 years is given for the youngest and lowest marine terrace on Cape Blanco (23) which rises 60 m above sea level.

Average rates of uplift range from 100 to 1000 m/m.y. over the Oregon margin with the highest rates associated with coastal terraces at Cape Blanco and with the lower continental slope deposits off central Oregon. Although the largest amounts of uplift of shelf and upper continental slope rocks involve the late Miocene to early Pliocene strata, the rates are only 100 m/m.y. and probably are not more than 200 m/m.y. at a maximum. Rates of uplift over the entire shelf and upper slope are fairly constant suggesting a uniform rate of uplift during late Cenozoic time. The rate of uplift in the coastal region decreases to the north and south from Cape Blanco.

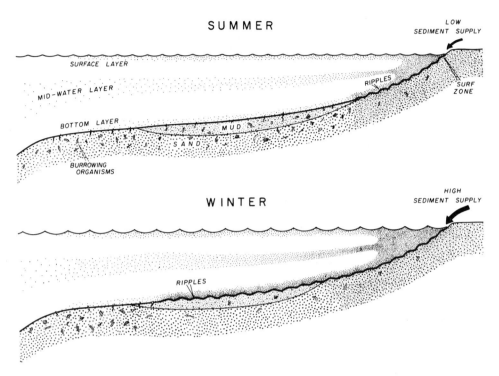

Figure 16. *Seasonal regime for the Oregon continental shelf. Sediment rippling and turbid layer transport are enhanced during the winter due to high sediment input from coastal streams and long period waves that ripple the bottom sediments in water depths of at least 125 m. Turbid layer transport, especially in the surface and mid-water layers, is greatest during the late spring off the Columbia River (after Kulm et al., 28).*

SUMMARY AND CONCLUSIONS

Sedimentation in the Oregon coastal region and on the adjacent continental shelf is controlled by several factors, some related and others independent. The most important factors are: (1) river discharge, coastal erosion, and sediment supply; (2) estuarine circulation system; (3) wave dimensions and direction; (4) subsurface and bottom currents; (5) density stratification of water column; and (6) burrowing benthic organisms.

River discharge which is related to seasonal rainfall in the drainage basin and snow melt, estuarine basin configuration, and tidal range determine the nature of the estuarine circulation system. When a partly mixed and two-layered system develops, the estuaries act as a sediment trap for much of the fluvial sand transported in traction along the bottom and some silts and clays that flocculate and settle in the vicinity of the salt wedge. In addition, marine

sands are transported from the beaches to the estuary by the landward
flowing salt wedge. During periods of unusually high river discharge,
such as that associated with the Columbia River in the early spring,
large volumes of very fine sand (124 to 62 microns) and finer material
may be carried in turbulent suspension to the surf zone and inner continental shelf. Sand-sized material is also supplied to the shelf by
erosion of the coastline during the severe winter storms. Erosion of
the coastline is strongly influenced by its geology and the waves that
attack the sea cliffs and sand buffer zones.

The very fine sand and finer material either is held in suspension or resuspended by surface waves vigorously stirring the bottom
to depths of at least 125 m and possibly 200 m. Long-period winter
waves are capable of stirring the bottom to water depths of 200 m
whereas the shorter period summer waves stir the bottom to depths of
about 90 m (Fig. 16). Although near-bottom currents with velocities
ranging up to 70 cm/sec provide the energy for sand erosion and transport on the inner shelf and mid-shelf region (47), the exact nature
of the sand dispersal system is not known. Fine-grained sediments,
terrigenous silt and clay, and biogenic debris, are carried across
the shelf in a well-defined surface turbid layer and a mid-water turbid layer associated with the seasonal thermocline and the permanent
pycnocline, respectively (Fig. 16). Both the coastal drainages and
the turbulent surf zone, which receives its sediment from the coastline and drainages, supply terrigenous sediment to each layer. A
substantial biogenic input (diatoms, radiolarians, foraminifers) is
made to the surface and mid-water layers as the result of coastal
upwelling during May to September. Some settling of fine-grained
material to the waters below probably occurs during lateral transport.

A bottom turbid layer develops on the shelf in response to the
rippling and resuspension of sediments by surface waves, from unidirectional currents, and from the influx of material from the surf
zone and possibly from the mid-water layer above (Fig. 16). The
bottom layer is prominent on the shelf to the depth of sediment
rippling and is concentrated in the submarine swales and small valleys
near the mid- and outer shelf.

The sedimentary facies on the Oregon continental shelf are related in part to the Holocene transgression of the sea, that deposited
the basal sand facies (Fig. 6), and to those modern-day processes
associated with the present hydraulic regime on the shelf. Benthic
organisms mixed the slowly accumulating mud facies into the underlying
sand facies creating the mixed sand and mud facies (Fig. 6). In the
areas of highest sediment supply (Columbia, Umpqua, and Rogue Rivers),
the mud accumulated more rapidly with the zone of benthic-reworking
shifting upward in the sediment column away from the basal sand. A
mud facies is developing in the areas adjacent to rivers with highest
discharge and sediment supply. In areas of low sediment supply (shelf
adjacent to the central Oregon coast), little mud accumulates on the
rough relict sediment surface. The mixed mud and sand facies generally surrounds the mud facies, which thins towards its periphery.
Wave stirring controls the landward extent of the mud facies and shelf
edge turbulence and sediment bypassing probably limit its seaward

extent.   The mud facies is developing slowly because surface waves
periodically resuspend the silts and clays and transport them seaward in a low density bottom turbid layer.

The structure of the Oregon shelf shows that it has been folded
and faulted, especially in the vicinity of submarine banks that occur
near the outer edge of the shelf (Figs. 11, 13, 14).   Older sedimentary rocks (largely Pliocene to Miocene) have been uplifted from deeper parts of the continental slope and now form the banks and other
rock exposures on the shelf. Rather rapid structural adjustments have
taken place on the shelf during the past several million years.

REFERENCES

1. Baldwin, E. M.   1964.   Geology of Oregon. Edwards Brothers, Inc.,
    Ann Arbor, Michigan.   136 pp.
2. Braislin, D. B., D. D. Hastings, and P. D. Snavely, Jr.   1971.
    Petroleum potential of western Oregon and Washington and
    adjacent continental margin.   Am. Assoc. Petrol. Geol. Mem.
    15.   1:229-238.
3. Burt, W. V., and W. B. McAlister.   1959.   Recent studies in the
    hydrography of Oregon estuaries.   Res. Briefs, Fish Commis-
    sion of Oregon.   7:14-27.
4. Bushnell, D. C.   1964.   Continental Shelf Sediments in the Vi-
    cinity of Newport, Oregon.   M.S. Thesis.   Oregon State
    University, Corvallis.   107 pp.
5. Byrne, J. V.   1962.   Geomorphology of the continental terrace
    off the central coast of Oregon.   Ore Bin 24:65-74.
6. Byrne, J. V.   1963a.   Coastal erosion, Northern Oregon.   *In:*
    Clements, T. (ed.) Essays in Marine Geology in Honor of
    K. O. Emery.   Univ. Southern California Press, Los Angeles,
    pp. 11-33.
7. Byrne, J. V.   1963b.   Geomorphology of the continental terrace
    of the northern coast of Oregon. Ore Bin 25:201-209.
8. Byrne, J. V.   1964.   An erosional classification for the north-
    ern Oregon coast.   Ann. Assoc. Am. Geographers.   54:329-335.
9. Byrne, J. V., G. A. Fowler, and N. M. Maloney.   1966.   Uplift
    of the continental margin and possible continental accretion
    off Oregon.   Science 154:1654-1656.
10. Carlson, P. R. 1968.   Marine Geology of Astoria Submarine Canyon.
    Ph.D. Thesis.   Oregon State University, Corvallis.   259 pp.
11. Chambers, D. M.   1969.   Holocene Sedimentation and Potential
    Placer Deposits on the Continental Shelf off the Rogue River,
    Oregon.   M.S. Thesis.   Oregon State University, Corvallis.
    102 pp.
12. Curray, J. R. 1965.   Late Quaternary history, continental shelves
    of the United States.   *In:* Wright, H. E., and D. C. Frey
    (eds.) The Quaternary of the United States.   Princeton
    University Press, Princeton, New Jersey.   723 pp.
13. Dehlinger, P., R. W. Couch, and M. Gemperle.   1968.   Continental
    and oceanic structure from the Oregon Coast westward across

the Juan de Fuca Ridge. Can. J. Earth Sci. 5:1079-1090.

14. Drake, D. E. 1971. Suspended sediment and thermal stratifi-
    cation in Santa Barbara Channel, California. Deep-Sea Res.
    18:763-769.

15. Duxbury, A. C. 1965. The union of the Columbia River and the
    Pacific Ocean--general features. *In:* Transactions, Ocean
    Science and Ocean Engineering, Joint Conference Marine Tech-
    nological Society and American Society of Limnology and
    Oceanography, Washington, D.C., pp. 914-922.

16. Emery, K. O. 1938. Rapid method of mechanical analysis of
    sand. J. Sediment. Petrol. 8:105-111.

17. Emilia, D. A., J. W. Berg, Jr., and W. E. Bales. 1968. Mag-
    netic anomalies off the northwest coast of the United States.
    Geol. Soc. Am. Bull. 79:1053-1061.

18. Fowler, G. A. 1966. Notes on late Tertiary foraminifera from
    off the central coast of Oregon. Ore Bin 28:53-60.

19. Fowler, G. A., and G. E. Muehlberg. 1969. Tertiary foramini-
    feral paleoecology and biostratigraphy of part of the Ore-
    gon continental margin (abs.). Am. Assoc. Petrol. Geol.
    Bull. 53:467.

20. Harlett, J. C. 1972. Sediment Transport on the Northern Ore-
    gon Continental Shelf. Ph.D. Thesis. Oregon State Univer-
    sity, Corvallis. 120 pp.

21. Harlett, J. C., and L. D. Kulm. 1973. Suspended sediment
    transport on the northern Oregon continental shelf. Geol.
    Soc. Am. Bull. 84:3815-3826.

22. Highsmith, R. M., Jr. 1962. Water. *In:* Highsmith, R. M.,
    Jr. (ed.) Atlas of the Pacific Northwest Resources and
    Development. Oregon State University Press, Corvallis,
    pp. 38-42.

23. Janda, R. J. 1969. Age and correlation of marine terraces
    near Cape Blanco, Oregon (abs.). *In:* Program of the 65th
    Annual Meeting, Cordilleran Section, Geological Society of
    America, pp. 29-30.

24. Komar, P. D., R. H. Neudeck, and L. D. Kulm. 1972. Observa-
    tions and significance of deep-water oscillatory ripple
    marks on the Oregon continental shelf. *In:* Swift, D.,
    D. B. Duane, and O. H. Pilkey (eds.) Shelf Sediment Trans-
    port: Process and Pattern. Dowden, Hutchinson & Ross,
    Stroudsburg, Pennsylvania, pp. 601-619.

25. Kulm, L. D. 1965. Sediments of Yaquina Bay, Oregon. Ph.D.
    Thesis. Oregon State University, Corvallis. 184 pp.

26. Kulm, L. D., and J. V. Byrne. 1966. Sedimentary response to
    hydrography in an Oregon estuary. Mar. Geol. 4:85-118.

27. Kulm, L. D., and G. A. Fowler. 1974. Oregon continental margin
    structure and stratigraphy: A test of the imbricate thrust
    model. *In:* Burk, C. A., and C. L. Drake (eds.) The Geol-
    ogy of Continental Margins. Springer-Verlag, New York,
    pp. 261-283.

28. Kulm, L. D., R. C. Roush, J. C. Harlett, R. H. Neudeck, D. M.
    Chambers, and E. J. Runge. 1975. Oregon continental shelf

sedimentation: Interrelationships of facies distribution
and sedimentary processes. J. Geol. 83:145-175.

29. Kulm, L. D., R. von Huene, *et al.* 1973. Initial Reports of the
Deep Sea Drilling Project, v. XVIII. U. S. Government
Printing Office, Washington, D.C. 1077 pp.

30. Lockett, J. B. 1965. Some Indications of Sediment Transport
and Diffusion in the Vicinity of the Columbia Estuary and
Entrance, Oregon and Washington. International Association
Hydraulic Research, Portland, U. S. Army Engineer Div.,
North Pacific. 7 pp.

31. Mackay, A. J. 1969. Continuous Seismic Profiling Investiga-
tions of the Southern Oregon Continental Shelf between Coos
Bay and Cape Blanco. M.S. Thesis. Oregon State University,
Corvallis. 118 pp.

32. Maloney, N. J. 1965. Geology of the Continental Terrace off
the Central Coast of Oregon. Ph.D. Thesis. Oregon State
University, Corvallis. 233 pp.

33. McCave, I. N. 1972. Transport and escape of fine-grained sediment from shelf areas. *In:* Swift, D., D. B. Duane, and
O. H. Pilkey (eds.) Shelf Sediment Transport: Process and
Pattern. Dowden, Hutchinson, & Ross, Stroudsburg, Pennsylvania, pp. 601-619.

34. National Marine Consultants, Inc. 1961. Waves Statistics for
Three Deep Water Stations along the Oregon-Washington Coast.
U.S. Army Engineers Dist., National Marine Consultants, Inc.,
Santa Barbara, California. 17 pp.

35. Nelson, C. H. 1968. Marine Geology of Astoria Deep-Sea Fan.
Ph.D. Thesis. Oregon State University, Corvallis. 287 pp.

36. Nelson, C. H., L. D. Kulm, P. R. Carlson, and J. R. Duncan.
1968. Mazama ash in the northeastern Pacific. Science
161:46-49.

37. North, W. B. 1964. Coastal Landslides in Northern Oregon.
Master's Thesis. Oregon State University, Corvallis. 85 pp.

38. Pak, H. 1970. The Columbia River as a Source of Marine Light
Scattering Particles. Ph.D. Thesis. Oregon State University, Corvallis. 110 pp.

39. Pak, H., G. B. Beardsley, Jr., and R. L. Smith. 1970. An
optical and hydrographic study of a temperature inversion
off Oregon during upwelling. J. Geophys. Res. 75:629-638.

40. Postma, H. 1967. Sediment transport and sedimentation in the
estuarine environment. *In:* Lauf, G. H. (ed.) Estuaries.
American Association for Advancement of Science, Washington,
D.C., Pub. 83, pp. 158-179.

41. Rogers, L. C. 1966. Blue Water 2 lives up to promise. Oil
and Gas J. 64(33):73-75.

42. Roush, R. C. 1970. Sediment Textures and Internal Structures:
a Comparison between Central Oregon Continental Shelf Sediments and Adjacent Coastal Sediments. M.S. Thesis. Oregon
State University, Corvallis. 75 pp.

43. Runge, E. J. 1966. Continental Shelf Sediments, Columbia River
to Cape Blanco, Oregon. Ph.D. Thesis. Oregon State Univer-

sity, Corvallis.   143 pp.

44. Scheidegger, K. F., L. D. Kulm, and E. J. Runge. 1971. Sediment sources and dispersal patterns of Oregon continental shelf sands. J. Sediment. Petrol. 41:1112-1120.

45. Shepard, F. P. 1963. Submarine Geology. Harper & Row, New York. 557 pp.

46. Silver, E. A. 1971. Small plate tectonics in the northeastern Pacific. Geol. Soc. Am. Bull. 82:3491-3496.

47. Smith, J. D., and T. S. Hopkins. 1972. Sediment transport on the continental shelf off of Washington and Oregon in light of recent current measurements. *In:* Swift, D., D. B. Duane, and O. H. Pilkey (eds.) Shelf Sediment Transport: Process and Pattern. Dowden, Hutchinson, & Ross, Stroudsburg, Pennsylvania, pp. 143-180.

48. Snavely, P. D., Jr., N. S. MacLeod, and H. C. Wagner. 1973. Miocene tholeiitic basalts of coastal Oregon and Washington and their relation to coeval basalts of the Columbia Plateau. Geol. Soc. Am. Bull. 84:387-424.

49. Spigai, J. J. 1971. Marine Geology of the Continental Margin off Southern Oregon. Ph.D. Thesis. Oregon State University, Corvallis. 214 pp.

50. U.S. Army Engineers. 1962. Reservoir Regulations Manual-- Bonneville Dam, Columbia River, Washington and Oregon. Portland, Oregon. 12 pp.

51. White, S. M. 1970. Mineralogy and geochemistry of continental shelf sediments off the Washington-Oregon coast. J. Sediment. Petrol. 40:38-54.

52. Williams, H., F. J. Turner, and C. J. Gilbert. 1954. Petrography: an Introduction to the Study of Rocks in Thin Sections. Freeman, San Francisco, California. 406 pp.

-3-

Physical Characteristics of Pacific
Northwestern Coastal Waters[1]

ADRIANA HUYER AND ROBERT L. SMITH

The coastal waters of the Pacific Northwest have been
studied extensively in the last two decades so that much is known
about their physical characteristics, especially the circulation
and stratification. The physical characteristics determine to a
large extent the natural productivity of the ecosystem, and may determine whether or not it can be enhanced. This chapter will
describe the main physical characteristics (the circulation, stratification, tides, and waves) and the ways these might be limiting.
It will also outline the kinds of information still needed to clarify the problems associated with increasing the marine plant biomass
of the Pacific Northwest.

CIRCULATION

Much of the knowledge about the circulation of coastal waters
is based on direct current observations, made with instruments

[1]We wish to acknowledge the contributions of several programs
to our knowledge of Oregon coastal waters: the Newport Hydro Line
occupied regularly from 1958 through 1971, and funded largely by the
Office of Naval Research; the pioneering coastal current measuring
program from 1965 to 1969, funded mainly by the National Science
Foundation; the first and second Coastal Upwelling Experiments, CUE-I
and -II, which were funded by the Office for the International Decade
for Ocean Exploration, National Science Foundation, and most recently
the program to study the winter-spring transition (WSIP), funded by
the National Science Foundation, NSF, DES 74-22290.
We also wish to acknowledge the aid of H. Pak, H. Pittock,
P.D. Komar, L.F. Small, and R.J. Zaneveld for providing information
on the physical characteristics with which we were less familiar.

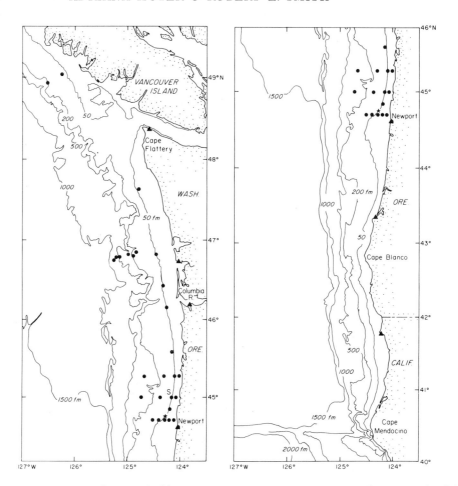

Figure 1. *Locations of direct current measurements (dots) and tide gages (triangles) along the Pacific Northwest Coast. The star at 44°45'N indicates the position of Poinsettia, a mooring maintained for over a year. The mooring at Sunflower (S) was maintained for seven months.*

moored on a taut wire with subsurface flotation. The current meters record speed, direction, and temperature at regular intervals, usually 10 to 20 minutes. Locations where current observations have been made repeatedly, or for periods longer than a month, are shown in Fig. 1. By now, there are sufficient data to resolve signals with periods of several hours, a day, several days, and even the seasonal variation.

    At one location, Poinsettia, 44°45'N, 124°17.5'W, current meters were deployed in December 1972 to monitor the shelf currents. The mooring was serviced at intervals of a few months, and was maintained until May 1974. To study the low frequency currents, the short gaps due to servicing were filled in with artificial data and the result-ing long record was filtered statistically to remove periods of a

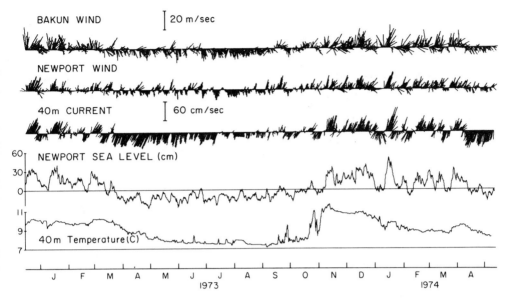

Figure 2. *Time series of wind (computed and observed), the 40 m current at Poinsettia, sea level at Newport, Oregon, and temperature at 40 m, Poinsettia, from December 1972 to May 1974. The observed parameters have been statistically filtered to remove tidal and shorter observations. Bakun's wind is computed from the large scale 6-hourly synoptic pressure charts. The length and direction of each vector represents the speed and direction toward which the flow occurs, with vertical upwards representing the true month.*

day or less. The low-passed time series of the current and temperature at 40 m is shown in Fig. 2, with low-passed time series of the wind and sea level at Newport, and with the wind at 45°00'N, 125°00'W computed from the six-hourly synoptic pressure charts (2). This record demonstrates some important features of the coastal currents:

(a) the currents are usually nearly parallel to the local isobath (in this case along 20° T); the "alongshore" (really along-isobath) component is usually stronger than the "onshore" (cross-isobath) component.

(b) the currents are highly variable in both direction and speed on a time scale of several days.

(c) these fluctuations in the currents are directly correlated to fluctuations in the sea level and the alongshore component of the wind.

(d) there seems to be a significant seasonal variation in the alongshore component of the current. In fall and winter, the mean is northward, but the direction of the current is highly variable. In spring and summer, the mean is strongly southward, and the flow is hardly ever northward.

(e) the seasonal cycle of the currents seems to be asymmetrical, with extremes in winter and spring. The sea level exhibits a

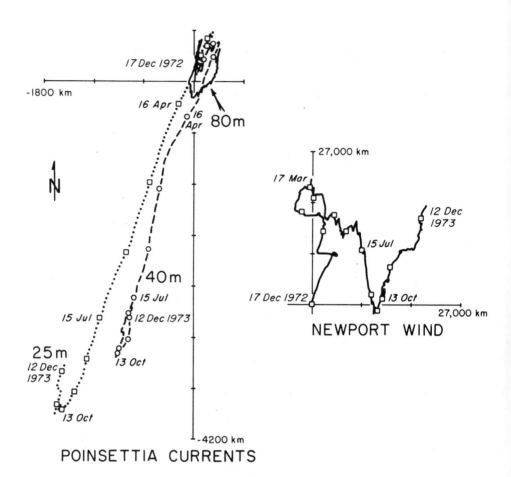

Figure 3. *Progressive vector diagrams of the wind at Newport, Oregon, and of the current at 25 m, 40 m, 80 m at Poinsettia, for 363 days beginning December 17, 1972.*

similar asymmetrical cycle, but the seasonal cycle in the wind is symmetrical, with the strongest northward mean in December or January, and the strongest southward mean in July or August.

Fig. 2 also shows that the wind measured at Newport is similar to the larger scale wind computed from the 6-hourly synoptic pressure charts, on both the seasonal and the multiple-day time-scales. The temperature-time series show little variation at periods of several days, but reveal a strong seasonal signal: the temperature is highest in winter and lowest in summer!

The Mean Circulation

Integrating a year's data from Poinsettia gives some indication of the mean circulation over the middle of the continental shelf off central Oregon. Fig. 3 shows progressive vector diagrams (PVD's) of

the current at three depths, and of the wind. A PVD is a trace of
the endpoint of a vector whose components are the time integrals of
the eastward and northward velocity; if the velocity were independent
of location, this trace would represent a trajectory. The most striking feature of Fig. 3 is the extreme difference between the wind and
current: the current PVD's form almost a straight line which is
nearly parallel to the local isobath, while the wind favors no particular direction (except that it is hardly ever westward). The currents
at 25 and 40 m exhibit a net southward flow during the year, with an
average of about 9 km/day (10 cm/sec) at 25 m, and about 7 km/day
(8.5 cm/sec) at 40 m. The mean Newport wind is northeastward at about
74 km/day (.86 m/sec). The eastward and northward components of the
mean wind stress are 0.17 and 0.20 dynes/cm^2.

Over the inner part of the shelf, the mean current is much weaker.
Although direct current measurements are not available year round,
there are several sets of observations near the 50 m isobath: at
44°40'N from July to October 1972 (45); at 45°16'N during July and
August 1973 (46); and at 45°N from February to May 1975 (13).   These
indicate that the mean flow over the inner part of the continental
shelf is weakly southward, or perhaps even northward.

Less is known about the mean circulation over the outer shelf
and further offshore, but it is generally believed to be southward
at the surface. Geostrophic computations along 44°40'N (25) indicate
that there is mean southward surface flow beyond the shelf, but that
it is weaker than the flow over the mid-shelf. At a depth of 200-
300 m, there is a mean northward undercurrent (12).   Its influence
on water properties (and therefore probably its velocity) decreases
with distance from shore (12, 49).   Its presence very near the continental slope has been verified off Washington in fall (6, 48) and in
summer (20), and off Oregon in spring and early summer (29).

Seasonal Variation

Huyer, Pillsbury, and Smith (26) computed the monthly means of
all the direct current data available for the Oregon continental
shelf including Pointsettia, up to the end of 1973. They found a
strong and regular seasonal variation in the alongshore flow. Another
mooring, Sunflower, was subsequently maintained over the 100 m isobath at 45°N from February through August 1975 (13). Monthly means
of the alongshore flow at both Poinsettia and Sunflower are shown in
Fig. 4. Both records show a similar seasonal cycle: northward
currents at all depths in winter, with little or no mean shear; southward currents at all depths in spring, with strong vertical shear in
the lower half of the water column; and southward surface currents
over a northward undercurrent in summer, with a weaker mean vertical
shear. No long continuous current records are available for the
southern Washington shelf, but there are a series of shorter records
from different seasons near the 70 m isobath at 46°25'N (53).   The
currents there exhibit a similar seasonal cycle except that the
current is more strongly northward in winter and more weakly southward in the spring. This might be because it is effectively nearer

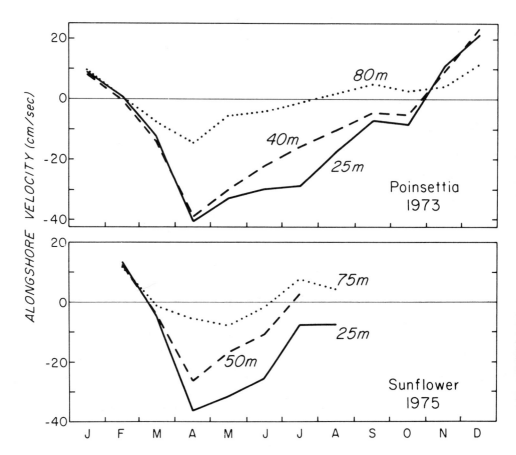

Figure 4. *Monthly mean values of the alongshore component of the currents off the Oregon coast at Poinsettia, 1973, and at Sunflower, 1975. To obtain the alongshore component, the coordinate axes were rotated by -20° at Poinsettia and -13° at Sunflower. Positive values are poleward; negative values indicate equatorward currents.*

shore (in shallower water), but could also be because it is further north.

Events: Fluctuations with Periods of Several Days

The data from Poinsettia (Fig. 2) showed clearly that in the alongshore component of the current there are variations (termed "events") which are associated with similar variations in the wind and sea level. The events that occur over the shelf in summer have been studied extensively in recent years (10, 27, 36, 37, 54). At present, we know that:

(a) the alongshore current fluctuations are nearly independent of depth in both magnitude and phase, at least below 20 m.

(b) the current fluctuations are highly coherent with those in the sea level, and usually coherent with those in the local wind.

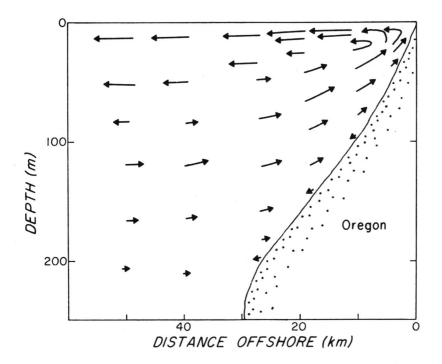

Figure 5. *A diagram of the onshore-offshore circulation that occurs during an upwelling event in summer, when the wind stress has remained favorable for upwelling for several days, from Huyer (24). Offshore velocities in the surface layer are about 20 cm/sec; the strongest onshore velocities are 5 - 10 cm/sec.*

(c) current fluctuations over the continental shelf are highly coherent over alongshore separations of up to 200 km.

(d) the amplitude of the alongshore current fluctuations decreases rapidly with distance from shore.

(e) the current fluctuations generally rotate clockwise.

(f) the orientation of the fluctuations is determined largely by the local bathymetry.

(g) at the lowest frequencies, the current fluctuations behave as a northward propagating continental shelf wave that is continually forced by the wind stress. At other frequencies, the current fluctuations seems to behave as if they are locally driven by the wind stress.

The onshore component of the current also varies with a period of several days. This variability was clearly observed in the summer of 1973 when current meters were moored in the near surface layer as well as at greater depths over the 100 m isobath (14, 24). When the

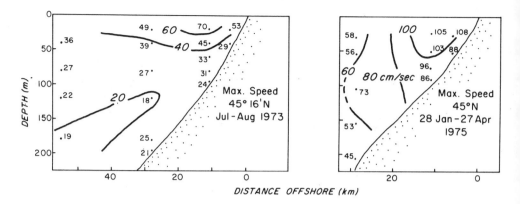

Figure 6. *The vertical-offshore distribution of the maximum current speeds observed during July and August 1973, and during February, March, and April 1975 off Newport, Oregon.*

wind stress is southward, there is strong offshore flow at the surface, which decreases rapidly to zero at about 15 or 20 m. The compensating onshore flow occurs over the remainder of the water column, but is strongest just below the layer of offshore flow. The vertical velocity associated with this circulation decreases rapidly with distance from shore. Fig. 5 shows a diagram of the onshore-offshore circulation during an upwelling event in summer, i.e., when the wind has been southward and strong for several days.

Tidal and Inertial Currents

    The currents oscillate with periods of approximately a day as well as at the longer periods described above. The principal periods are those associated with the diurnal tide (about 25 hrs), the semi-diurnal tide (12.4 hrs), and the inertial period (17.4 hrs off Oregon). The diurnal tidal currents are closely associated with the diurnal oscillation in sea level, but the semi-diurnal currents can be associated with internal waves as well as with the semi-diurnal changes in sea level. Inertial currents are generated impulsively, often by a rapid change in the wind stress (47); in a pure inertial current the centrifugal force is balanced by the Coriolis force.
    Inertial oscillations off central Oregon often have amplitudes exceeding 10 cm/sec (31, 35, 44). Kundu (35), using summer data from one location over the 100 m isobath, showed that the amplitude of the inertial motion decayed slowly with depth; the energy propagated downward from the surface as we would expect if the oscillations were generated by the wind. Wang and Mooers (58) found that inertial energy propagated downward and offshore along an inclined pycnocline (a layer of maximum vertical density gradient) during an upwelling event. There is little spatial correlation between inertial currents as measured by current meters at the same depth (40 m) off Oregon in summer (32).

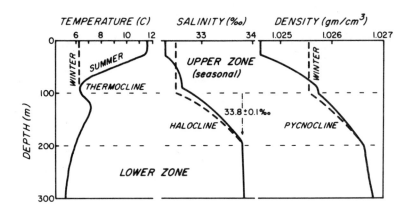

Figure 7. *Typical vertical profiles of the temperature, salinity, and density in the eastern subarctic Pacific, after Tully (57).*

The energy associated with internal semi-diurnal tides is highly variable at the 100 m isobath (19, 44, 58). The kinetic energy of the internal tide is high when the stratification is strong, and weaker when the stratification is reduced. When the stratification decreases abruptly, as during an upwelling event, the semi-diurnal velocity shear may become unstable. When this happens, the tidal currents are dissipated and there is increased mixing (58).

Extreme Currents

Though a discussion of the circulation in terms of the dominant periods during which it varies is of scientific interest, it is of little value in determining whether a particular engineering approach might succeed. The current at any given time is a composite of all of these signals, and maximum speeds can far exceed the amplitude of any of the dominant signals. Fig. 6 shows the maximum hourly speeds as observed during two separate experiments, one in July and August 1973, and the other from late January through April 1975. These values are taken from the data reports (13, 46) that also include time series plots of the hourly data. Even in summer the maximum speed exceeds one knot (50 cm/sec); in winter, the maximum speed exceeds two knots. Note that these speeds were observed at depths greater than 20 m; surface currents are likely to be stronger.

STRATIFICATION

Sea water density is usually described in terms of sigma-t, the specific gravity anomaly multiplied by a thousand (e.g., water with a sigma-t value of 25.6 has a specific gravity of 1.0256, and a density of 1.0256 gm/cm^3 at atmospheric pressure. Sigma-t, a function only of the temperature and salinity, increases with increasing

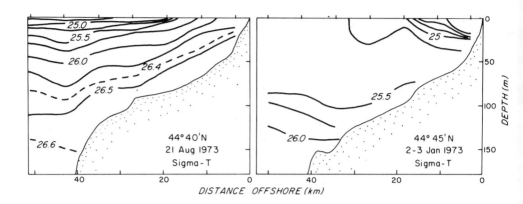

Figure 8. *Vertical sections of sigma-t observed over the central Oregon shelf in January and August 1973. (26)*

salinity and decreasing temperature.

The general structure of the vertical stratification of the waters off the Pacific Northwest is shown in Fig. 7. The deep winter mixed layer is caused by convective cooling and wind mixing; its depth is limited by the strong halocline. Subsequent warming and additional runoff lead to the seasonal thermocline and seasonal halo-cline. A remnant of cold water often persists through the summer as a temperature minimum at the top of the halocline (51, 57).

These water characteristics are modified locally by mixing, changing surface temperature, upwelling, precipitation, and by runoff (especially the Columbia River effluent) (5, 41). Affected by these processes, the density stratification varies with periods of a year, several days, and several hours.

The seasonal density variation over the continental shelf has been described for two locations: off southern Washington (55) and off central Oregon (25). In both locations, isopycnals (lines of equal density) generally slope downward toward shore in winter, and upward toward shore in summer (Fig. 8). Over the entire water col-umn, the water over the inner half of the shelf is denser in summer than in winter. The lowest salinities and densities occur in the Columbia River plume, which lies northward along the Washington coast in winter and southwestward along the Oregon coast in summer (3). The highest temperatures, up to nearly 18°C, also occur in the Colum-bia River plume. Off central Oregon, the surface temperature over the continental shelf varies from about 10°C in winter to about 14°C in early summer, decreasing in summer (because of upwelling) to about 10°C, increasing in early fall to a maximum of about 15°C, before decreasing to the winter value. Near the bottom, the temperature decreases from its winter value of about 10°C to a minimum of about 6.5°C in summer. Surface salinity decreases in spring from about 32.5 °/oo in winter, and increases to about 33 °/oo in summer. At the bottom, the salinity increases from about 32.8 °/oo in winter to

about 33.8 $^o/oo$ in summer. At the surface, sigma-t varies from about
25.0 in winter and summer, to about 23.0 in spring and fall (25). In
summer off Oregon, the depth of the surface mixed layer seldom exceeds
10 m (16); even off Washington, it seldom exceeds 25 m (21). In
winter, the surface mixed layer often exceeds 50 m, except very near
shore where strong stratification is maintained by runoff.

The temperature-salinity characteristics of the water also vary
seasonally. In summer, because of the southward advection, the water
is more strongly subarctic. There is a temperature minimum at the
top of the halocline and temperatures in the halocline are relatively
cool. Where the southward advection is strongest, the temperature
minimum is most extreme (28). In winter, when there is northward
advection over the shelf, the halocline temperatures are warmer, indicating more of the water is of southern origin. There is also a
seasonal difference in the zooplankton populations (43): species
with southern affinities are dominant in winter and species with
northern affinities are dominant in summer.

The stratification over the shelf varies significantly in response to wind fluctuations with periods of several days. In summer,
the largest density changes occur in the upper 20 m and in a coastal
zone about 10 to 15 km wide (15, 30). The density distribution
further offshore and at greater depths is nearly constant. During
these events, the surface temperature may change by as much as $5^{\circ}C$
(22) as recently upwelled water replaces Columbia River plume water.
In winter, there are significant density changes down to 100 m or
more.

The time series of temperature from many of the current meters
(45, 46) indicate that there is a significant semi-diurnal variation
in the stratification, particularly at stations further from shore.
Repeated STD casts (17) show that the vertical displacement of isopycnals is usually about 10 m over the mid-shelf.

Curtin and Mooers (9) observed internal waves with a period of
about an hour in August 1972. These were manifested at the surface
as a series of slicks which propagated toward shore. Below the surface, the depth of an acoustic scattering layer oscillated up and
down with an amplitude of 10 m. Hayes and Halpern (19) showed that
the internal wave energy increased during an upwelling event. The
internal waves may break over the shelf, and may be important in the
mixing of the various types of water present.

Since there are several kinds of water along the Northwest coast,
it can be expected that there will sometimes be strong lateral gradients of density, salinity, temperature, and other water properties.
Strong horizontal gradients are commonly called "fronts." Two main
density fronts are present in summer: one is at the boundary of the
Columbia River plume, and one occurs where the permanent pycnocline
reaches the surface during upwelling. During the upwelling season,
these fronts are often nearly coincident and form a particularly
strong front. This major front is associated with the enhanced southward surface flow during the upwelling season (38). Temperature
fronts are often associated with changes in color (42).

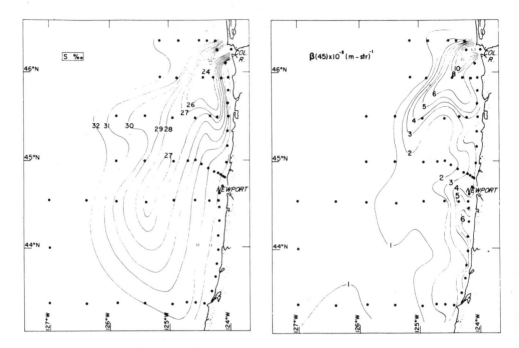

Figure 9. *The distribution of 3 m of salinity and of a light-scattering parameter, β(45), off the Oregon coast, June 20 to July 3, 1968. The scattering parameter can be treated as a linear function of the suspended material (40).*

TURBIDITY

The turbidity of the coastal waters is caused by organic and inorganic suspended particles. The inorganic particles enter the system through river runoff and through wave action in the surf zone or along the bottom.

Away from river outflow and the surf zone, the turbidity of the surface layers depends only on amount of chlorophyll present (52). In areas where rivers are a major source of organic materials, the extinction coefficient is largest at the river mouth and it decreases with distance from the mouth. Pak, Beardsley, and Park (40) demonstrated (Fig. 9) the turbidity of Columbia River plume water.

The depth of the euphotic zone, at which the intensity of light is at least 1 percent of its surface value, is between 30 and 60 m throughout the year at one location, 25 miles off Newport, Oregon (8). Nearer shore the euphotic zone is shallower, but it usually exceeds 10 m (Small, personal communication). In summer, the surface mixed layer is very shallow with a strong seasonal pycnocline beginning at a depth of 10 to 20 m. The strong stratification would tend to keep the phytoplankton within the euphotic zone where they can fully utilize the light. In winter, the mixed layer depth often exceeds the euphotic depth, except very near shore where the stratification is strong.

Analysis of vertical sections of optical parameters (18, 33) show that particles are concentrated in the surface layer, a bottom nepheloid layer, and a layer at intermediate depth. Between these layers the water is quite clear. Much of the suspended material in the bottom layer probably originates in the surf zone, but some of it is stirred up from the bottom by waves or currents (18, 56). The material in the intermediate layer may originate mainly in the surf zone; its depth coincides with the pycnocline (18).

TIDES

The tides along the Northwest coast are mainly semi-diurnal (i.e., sea level falls and rises about twice a day). There is also a significant diurnal component, so that one of these highs is greater than the other. Together, the first semi-diurnal (M_2) and the first diurnal (K_1) tidal constituents account for about 70 percent of the tidal range. The amplitudes of the principal tidal components at Newport are given in Table I.

Table I. *Principal tidal harmonic components at Newport, Oregon.*

Name	Symbol	Period	cm
Principal lunar semi-diurnal	M_2	12.42	88.1
Luni-solar diurnal	K_1	23.93	43.1
Principal lunar diurnal	O_1	25.82	26.4
Principal solar semi-diurnal	S_2	12.00	23.1
Larger lunar eliptic	N_2	11.97	18.1

Both the tidal range and the tidal extremes vary with position along the Northwest coast (Fig. 10). The tidal range is greater to the north, and tidal extremes occur later. The main semi-diurnal component (M_2) progresses more rapidly northward than the main diurnal component (K_1); the speeds are about 214 and 140 M/sec, respectively (39).

Observed sea level differs from the predicted levels (1) because it also varies with atmospheric pressure, changes in water density, currents, and wind stress. The "inverted barometer" response is a decrease in sea level of 1 cm for every 1 mb increase in atmospheric pressure; hence, it has an amplitude of 10 to 20 cm, depending on season. The seasonal variation of sea level associated with seasonal changes in the current and density distribution has an amplitude of about 15 cm (50). The sea level changes associated with several-day fluctuations in the current have amplitudes of 15 to 20 cm (Fig. 2). Storm surges, which usually last less than a day, may be as high as 1.4 m (23). In a typical year, the extremes of sea level

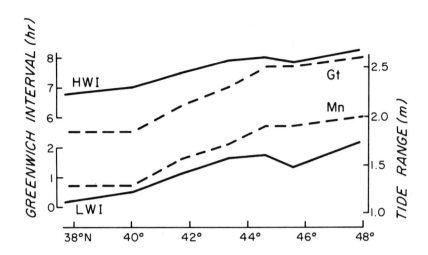

Figure 10. *Variation of tidal range and phase along the Northwest coast. Gt and Mn are two measures of the tidal range: Gt is the mean difference between highest and lowest sea level occurring each day; Mn is the difference between the mean high water and mean low water levels. HWI and LWI are measures of the tidal phase: they are the time intervals between lunar passage of the Greenwich meridian and the local high water (HWI) or local low water (LWI).*

are 1.8 m above and 2.2 m below the mean sea level.

WAVES

Wind waves and swell are monitored at Newport, Oregon, by means of a seismometer which measures microseisms produced by the waves (4, 11). The output was calibrated by a series of visual observations of wave height and period. The system is set to obtain a 10-minute record every 6 hours. For each record, the height and periods of the highest third of the waves are averaged to yield the significant wave height and period. The monitoring has been almost continuous since 1972, and the wave climate has been documented. The waves are much higher in winter than in summer. In July and August, the significant wave height is not expected to exceed 8 feet (2.5 m); in November, December, or January, the significant wave height is likely to exceed 16 ft (5 m) on one day of each month (7). Komar, Quinn, Creech, Rea, and Lizarraga-Areiniega (34) have used the significant wave heights to compute breaker heights at Newport. Fig. 11 shows the minimum, maximum, and mean breaker heights for ten-day periods from July 1972 to July 1973. Note that the maximum and minimum values do not apply to individual waves. They refer to the extreme values of the significant wave heights from the 10-minute records.

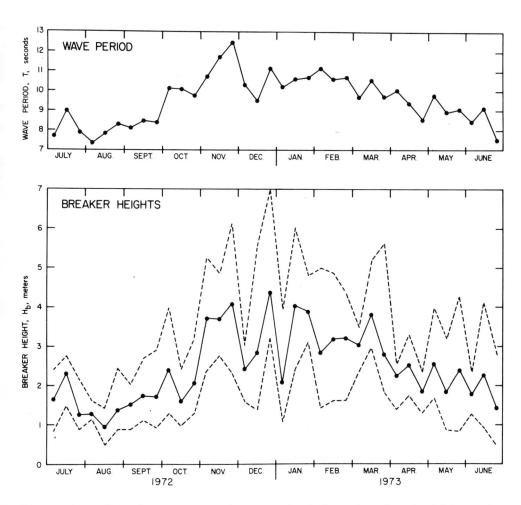

Figure 11. *Ten-day mean, maximum, and minimum breaker heights and periods computed from the significant wave heights as measured by the seismometer at Newport, Oregon. Significant wave height is the mean height of the highest one-third of the waves (34).*

SUMMARY AND CONCLUSIONS

Some important questions about the physical characteristics must be answered before attempting to increase the marine plant biomass of the Pacific Northwest. Maximum current speeds may be prohibitive to major aquaculture projects except in sheltered areas, in the lee of capes or banks, where the currents are weaker. Turbidity is high in areas where there is river outflow. There is insufficient information regarding light levels at many locations and at different seasons which are favorable for increased productivity. The wave climate at Newport is now quite well known but more research is needed off southern Oregon.

REFERENCES

1. Anon. 1976. Tide Tables 1977: West Coast of North and South America, including the Hawaiian Islands. U.S. Dept. of Commerce, National Oceanic and Atmospheric Administration (National Ocean Survey). 222 pp.
2. Bakun, A. 1975. Daily and Weekly Upwelling Indices, West Coast of North America, 1967–1973. NOAA Tech. Rep. NMFS SSRF-693. 114 pp.
3. Barnes, C. A., A. C. Duxbury, and B. A. Morse. 1972. Circulation and selected properties of the Columbia River effluent at sea. *In:* Pruter, A. T., and Alverson, D. L. (eds.) The Columbia River Estuary and Adjacent Ocean Waters. University of Washington Press, Seattle, pp. 41–80.
4. Bodvarsson, G. 1975. Ocean Wave-generated Microseisms at the Oregon Coast. M.S. Thesis. Oregon State University, Corvallis. 83 pp.
5. Bourke, R. H., and J. G. Pattullo. 1974. Seasonal variation of the water mass along the Oregon-northern California coast. Limnol. Oceanogr. 19:190–198.
6. Cannon, G. A., N. P. Larid, and T. V. Ryan. 1975. Flow along the continental slope off Washington, autumn 1971. J. Mar. Res. 33(Suppl.):97–107.
7. Creech, C. 1977. Five-year Climatology (1972–1976) of Near-shore Ocean Waves off Yaquina Bay, Oregon. Technical Report, OSU Sea Grant Program, Corvallis, Oregon.
8. Curl, H. C., Jr., and L. F. Small. 1965. Variations in photosynthetic assimilation ratios in natural, marine phytoplankton communities. Limnol. Oceanogr. 10:(Suppl. R67-R73).
9. Curtin, T. B., and C. N. K. Mooers. 1975. Observation and interpretation of a high-frequency internal wave packet and surface slick pattern. J. Geophys. Res. 80:882–894.
10. Cutchin, D. L., and R. L. Smith. 1973. Continental shelf waves: low-frequency variations in sea level and currents over the Oregon continental shelf. J. Phys. Oceanogr. 3:73–82.
11. Enfield, D. B. 1974. Prediction of Hazardous Columbia River Bar Conditions. Ph.D. Thesis. Oregon State University, Corvallis. 205 pp.
12. Favorite, F., A. J. Dodimead, and K. Nasu. 1976. Oceanography of the Subarctic Pacific Region, 1960–1971. Int. North Pac. Fish. Comm. Bull. No. 33. 187 pp.
13. Gilbert, W. E., A. Huyer, E. D. Barton, and R. L. Smith. 1976. Physical Oceanographic Observations off the Oregon Coast, 1975: WISP and UP-75. Ref. 76-4, Oregon State University, School of Oceanography, Corvallis. 189 pp.
14. Halpern, D. 1976. Structure of a coastal upwelling event observed off Oregon during July 1973. Deep-Sea Res. 23:495–508.
15. Halpern, D. 1974. Variations in the density field during coastal upwelling. Tethys 6:363–374.

16. Halpern, D., and J. R. Holbrook. 1972. STD Measurements Off
     the Oregon Coast, July/August 1972. International Decade
     of Ocean Exploration. Coastal Upwelling Ecosystems Analy-
     sis. Data Rep. No. 4. Ref. M72-82, University of Wash-
     ington Department of Oceanography, Seattle. 381 pp.
17. Halpern, D., J. R. Holbrook, and R. M. Reynolds. 1973. Physi-
     cal Oceanographic Observations Made by the Pacific Oceano-
     graphic Laboratory off the Oregon Coast during July and
     August 1972. International Decade of Ocean Exploration.
     Coastal Upwelling Ecosystems Analysis. Tech. Rep. 3.
     Ref. M73-46, University of Washington Department of Ocean-
     ography, Seattle. 205 pp.
18. Harlett, J. C. 1972. Sediment Transport on the Northern Ore-
     gon Continental Shelf. Ph.D. Thesis. Oregon State Univer-
     sity, Corvallis. 120 pp.
19. Hayes, S., and D. Halpern. 1976. Observations of internal
     waves and coastal upwelling off the Oregon coast. J. Mar.
     Res. 34:247-267.
20. Hickey, B. 1976. The relationship between currents on the
     Washington continental shelf and slope during the summer
     season. *In:* Book of Abstracts: Joint Oceanographic
     Assembly, Edinburgh, 1976. Food and Agricultural Organi-
     zation of the United Nations, Rome, p. 176.
21. Holbrook, J. R. 1975. STD Measurements off Washington and
     Vancouver Island during September 1973. U.S. Dept. of
     Commerce. NOAA Tech. Mem. ERL PMEL - 5. 88 pp.
22. Holladay, C. G., and J. J. O'Brien. 1975. Mesoscale varia-
     bility of sea surface temperatures. J. Phys. Oceanogr.
     5:761-772.
23. Humphrey, J. H., and D. E. Dorratcague. 1976. Numerical sim-
     ulation of storm surges on the Pacific Northwest coast.
     In: Conference on Coastal Meteorology. American Meteor-
     ological Society, Boston, Massachusetts. 127 pp.
24. Huyer, A. 1976. A comparison of upwelling events in two loca-
     tions: Oregon and northwest Africa. J. Mar. Res. 34:531-
     546.
25. Huyer, A. 1977. Seasonal variation in temperature, salinity
     and density over the continental shelf off Oregon. Limnol.
     Oceanogr. 22:442-453.
26. Huyer, A., R. D. Pillsbury, and R. L. Smith. 1975a. Seasonal
     variation of the alongshore velocity field over the contin-
     ental shelf off Oregon. Limnol. Oceanogr. 20:90-95.
27. Huyer, A., B. M. Hickey, J. D. Smith, R. L. Smith, and R. D.
     Pillsbury. 1975b. Alongshore coherence at low frequencies
     in currents observed over the continental shelf off Oregon
     and Washington. J. Geophys. Res. 80:3495-3505.
28. Huyer, A., and R. L. Smith. 1974. A subsurface ribbon of cool
     water over the continental shelf off Oregon. J. Phys.
     Oceanogr. 4:381-391.
29. Huyer, A., and R. L. Smith. 1976. Observations of a poleward
     undercurrent over the continental slope off Oregon, May -

June 1975. Abstract only. Trans., Am. Geophys. Union 57:263.

30. Huyer, A., R. L. Smith, and R. D. Pillsbury. 1974. Observations in a coastal upwelling region during a period of variable winds (Oregon coast, July 1972). Tethys 6:391-404.

31. Johnson, W. R., J. C. Van Leer, and C. N. K. Mooers. 1976. A cyclesonde view of coastal upwelling. J. Phys. Oceanogr. 6:556-574.

32. Kindle, J. C. 1974. The horizontal coherence of inertial oscillations in a coastal region. Geophys. Res. Lett. 1:127-130.

33. Kitchen, J. C., D. Menzies, H. Pak, and J. R. Zaneveld. 1975. Particle size distributions in a region of coastal upwelling analyzed by characteristic vectors. Limnol. Oceanogr. 20: 775-783.

34. Komar, P. D., W. Quinn, C. Creech, C. C. Rea, and J. R. Lizarraga-Areiniega. 1976. Wave conditions and beach erosion on the Oregon coast. Ore Bin 38:103-112.

35. Kundu, P. K. 1976. An analysis of inertial oscillations observed near Oregon coast. J. Phys. Oceanogr. 6:879-893.

36. Kundu, P. K., and J. S. Allen. 1976. Some three-dimensional characteristics of low-frequency current fluctuations near the Oregon coast. J. Phys. Oceanogr. 6:181-199.

37. Kundu, P. K., J. S. Allen, and R. L. Smith. 1975. Modal decomposition of the velocity field near the Oregon coast. J. Phys. Oceanogr. 5:683-704.

38. Mooers, C. N. K., C. A. Collins, and R. L. Smith. 1976. The dynamic structure of the frontal zone in the coastal upwelling region off Oregon. J. Phys. Oceanogr. 6:3-21.

39. Munk, W., F. Snodgrass, and M. Wimbush. 1970. Tides offshore: transition from California coastal to deep-sea waters. Geophys. Fluid Dynamics 1:161-235.

40. Pak, H., G. F. Beardsley, Jr., and P. K. Park. 1970. The Columbia River as a source of marine light-scattering particles. J. Geophys. Res. 75:4570-4578.

41. Pattullo, J., and W. Denner. 1965. Processes affecting seawater characteristics along the Oregon coast. Limnol. Oceanogr. 10:443-450.

42. Pearcy, W. G., and D. F. Keene. 1974. Remote sensing of water color and sea surface temperatures off the Oregon coast. Limnol. Oceanogr. 19:573-583.

43. Peterson, W. T., and C. B. Miller. 1977. The seasonal cycle of zooplankton abundance and species composition along the central Oregon coast. Fish. Bull. (in press)

44. Pillsbury, R. D. 1972. A Description of Hydrography, Winds and Currents during the Upwelling Season near Newport, Oregon. Ph.D. Thesis. Oregon State University, Corvallis. 163 pp.

45. Pillsbury, R. D., J. S. Bottero, R. E. Still, and W. E. Gilbert. 1974a. A compilation of Observations from Moored Current Meters: Vol. VI; Oregon Continental Shelf, April - October

1972. Ref. 74-2, Oregon State University, School of Oceanography, Corvallis. 230 pp.

46. Pillsbury, R. D., J. S. Bottero, R. E. Still, and W. E. Gilbert. 1974b. A Compilation of Observations from Moored Current Meters: Vol. VII; Oregon Continental Shelf, July - August 1973. Ref. 74-7, Oregon State University, School of Oceanography, Corvallis. 87 pp.

47. Pollard, R. T. 1970. On the generation by winds of inertial waves in the ocean. Deep-Sea Res. 17:795-812.

48. Reed, R. K., and D. Halpern. 1976. Observations of the California undercurrent off Washington and Vancouver Island. Limnol. Oceanogr. 21:389-398.

49. Reid, J. L. 1965. Intermediate waters of the Pacific Ocean. Johns Hopkins, Baltimore, Maryland. 85 pp.

50. Reid, J. L, and A. W. Mantyla. 1976. The effect of geostrophic flow upon coastal sea elevations in the northern North Pacific Ocean. J. Geophys. Res. 81:3100-3110.

51. Roden, G. I. 1964. Shallow temperature inversions in the Pacific Ocean. J. Geophys. Res. 69:2899-2914.

52. Small, L. F., and H. Curl, Jr. 1968. The relative contribution of particulate chlorophyll and river tripton to the extinction of light off the coast of Oregon. Limnol. Oceanogr. 13:84-91.

53. Smith, J. D., B. Hickey, and J. Beck. 1976. Observations from Moored Current Meters on the Washington Continental Shelf from February 1971 to January 1974. Spec. Rep. 65. University of Washington Department of Oceanography, Seattle.

54. Smith, R. L. 1974. A description of current, wind, and sea level variations during coastal upwelling off the Oregon coast, July - August 1972. J. Geophys. Res. 79:435-443.

55. Stefansson, U., and F. A. Richards. 1964. Distributions of dissolved oxygen, density, and nutrients off the Washington and Oregon coasts. Deep-Sea Res. 11:355-380.

56. Sternberg, R. W., and L. H. Larsen. 1975. Threshold of sediment movement by open ocean waves: observations. Deep-Sea Res. 22:299-309.

57. Tully, J. P. 1964. Oceanographic regions and processes in the seasonal zone of the North Pacific Ocean. *In:* Yoshida, K. (ed.) Studies on Oceanography, University of Tokyo Press, Tokyo, Japan, pp. 68-84.

58. Wang, D. P., and C. N. K. Mooers. 1977. Evidence for interior dissipation and mixing during a coastal upwelling event off Oregon. J. Mar. Res. (in press)

-4-

Chemical Characteristics of Pacific
Northwestern Coastal Waters—Nutrients, Salinities,
Seasonal Fluctuations[1]

ELLIOT L. ATLAS, LOUIS I. GORDON,
AND RICHARD D. TOMLINSON

The Pacific Northwest coastal waters comprise an extremely dynamic oceanographic system. The physical and chemical properties fluctuate seasonally, weekly, and even daily as a result of currents, winds, and biological, biochemical, and chemical processes.
Several years of observations off the Oregon Coast will be used here
to describe the average chemical characteristics of these waters and
to examine the seasonal fluctuations and some short time-scale phenomena which occur. Surface and near-coastal waters will be emphasized.

The chemical characteristics of the coastal waters are strongly
affected by the interplay of three important oceanographic regimes:
the Northeast Pacific Ocean, coastal upwelling, and the Columbia
River plume. Each of these regimes is chemically unique, especially
in its nutrient characteristics (Table I). From the standpoint of
this chapter, the Northeast Pacific Ocean provides a relatively
steady background of low-nutrient water compared to the more dramatic
changes in horizontal and vertical distributions associated with
coastal upwelling and the Columbia River.

Coastal upwelling is a strongly seasonal phenomenon that brings
cold, nutrient-rich waters to the surface along the coast. In addition to the seasonal upwelling cycle, there are short individual upwelling events which occur on an intermittent basis. The high pro-

_____

[1]This research was funded in part by the Office of Naval Research
under Contract N00014-67-A-0369-0007, Project NR083-102 and by the
National Science Foundation Grant NSF-GA-12113. Numerous contracts
and grants from ONR and NSF have funded both our work and that of
Drs. Burt, Patullo, Park, Smith, and Huyer. We are grateful for this
support.

Table I. *Annual average nutrient levels in Pacific Northwest coastal waters*[1]

	$NO_3 + NO_2$ (μM)	PO_4 (μM)	SiO_4 (μM)
Northeast Pacific Surface Water[2]	5	0.7	10
Columbia River Water[3]	10	0.5	150
Upwelling Water[4]	35	2.4	45

[1] The symbols NO_3, NO_2, PO_4, and SiO_4 signify total reactive nitrate, nitrite, phosphate, and silicate, respectively, as measured by the analytical methods employed. The concentration units (μM) are micromolar, taken to be identical to microgram atoms per liter of nitrate-nitrogen, nitrite-nitrogen, etc.

[2] From (12)

[3] From (14)

[4] From (1)

ductivity of upwelling areas is well-known (15) and is due mainly to the supply of nutrients to the photic zone.

The Columbia River also adds nutrients to the surface waters of the Northeast Pacific, but over a much more extensive area than just the upwelling zones. The Columbia River discharge can be traced for several hundred miles from the river mouth (3). Seasonal variations in winds, currents, and river flow strongly modify the direction and areal extent of the river effluent. Winter flow is northward close inshore while summer flow is on the average southwestward.

Thus, superimposed on the "background" Northeast Pacific water are the seasonal fluctuations and intermittent pulses of water from coastal upwelling and the Columbia River. The result is a complex ocean area; its chemical nature and changes are the subject of this paper.

Most of the data discussed here were taken off the Oregon coast during a series of hydrographic or special cruises conducted by Oregon State University from 1959 to 1972 (4, 5, 6, 17, 18, 19, 20, 21, 22, 23, 24). Some of these cruises have been described by Atlas (1), Ball (2), Cissell (7), Gordon (10), Hager (11), Hager and Bourke (12), Kantz (13), and Tomlinson *et al.* (17).

AVERAGE CHEMICAL CHARACTERISTICS AND THEIR SEASONAL VARIATIONS

Chemical data from a single hydrographic station 25 miles off the Oregon coast (NH-25, 44°39'N, 124°39'W) can be used to illustrate seasonal variations and mean chemical characteristics of coastal waters. Individual monthly averages, rather than individual measurements, were used to compute the long-term means. This averaging procedure gives equal weight to each year's data. Extremes in the individual measurements and the standard deviation of the monthly averages for surface water (0 m) are given in Table II.

The averages show a fairly coherent picture of seasonal fluctuations (Fig. 1 a-g) and provide an indication of the vertical distribution of chemical properties in the water column. Typically, salinity and nutrient concentrations increase with depth off the Oregon coast, and temperature and oxygen decrease with depth. The vertical gradients in all properties are maximal in the upper 50 m during the spring and summer. The upper 50 m are quite well mixed during the winter.

The impact of the Columbia River on the seasonal cycle is evident primarily in the salinity distribution (Fig. 1 b). Low salinity surface water flows southward during the summer and may be detected near shore. Upwelling has a significant influence on all chemical and physical properties, especially during the spring and summer months.

The increased light and the availability of nutrients during the early spring months stimulates plant growth. The increasing biological activity quickly reduces nitrate concentrations to near zero and gradually removes phosphate from the water (Fig. 1 d,e). Nutrient supplies to most of the water column are then increased during the summer by upwelling waters. Temperature and salinity are most affected near the surface; they show only minor variations at depth. Nutrient and oxygen concentrations, however, are significantly altered down to 100 to 200 m (Fig. 1 c-f).

On the average, nitrate tends to be removed before phosphate, which suggests that nitrogen may be the important limiting nutrient in these coastal waters. This is also indicated by the nutrient enrichment experiments of Glooshenko and Curl (9) and Frey (8). Another way to examine nutrient limitation is the N:P ratio in the water (Fig. 1 g). "Average phytoplankton" assimilate N and P in the ratio of 15-16:1. Lower ratios indicate nitrogen limitation. Virtually the entire water column shows nitrogen limitation during the year. We calculated N:P based only upon nitrate + nitrite nitrogen rather than all available nitrogen forms. There are few published ammonia data available for these waters. Ammonia levels are usually very low (~1 µM) and would have little effect on the N:P ratio.

The extreme values for each month over the six- to thirteen-year sampling period (Table II) present an interesting perspective on the variability of the chemical constituents. For example, both the maximum *and* minimum surface oxygen concentrations for the entire sampling period were observed during July. In general, the largest variations in chemical properties occur during the summer months. There is much less year-to-year variation during the winter. During the winter, storms mix the upper waters well and there is diminished biological activity. Virtually the same levels of nutrients, temperature, and salinity can be expected every year. During the summer,

Table II. Maxima, minima, and standard deviation of surface water properties 25 miles off Oregon (44°39'N, 124°39'W)[1]

		Jan	Feb	Mar	Apr	May	Jun	Jul	Aug	Sep	Oct	Nov	Dec	Overall
T (°C)	max	10.82	10.88	10.34	10.42	13.00	16.60	16.21	15.83	16.44	15.27	12.99	11.86	16.60
	min	9.60	9.01	8.28	9.22	10.00	8.73	11.70	10.80	12.68	12.36	10.72	10.18	8.28
	s.d.	0.49	0.62	0.72	0.50	1.14	1.56	1.82	2.17	1.34	0.99	0.64	0.56	–
S (‰)	max	32.79	32.69	32.63	32.58	32.03	32.81	32.54	32.51	32.41	32.60	32.78	32.57	32.79
	min	32.27	31.11	31.16	30.94	30.43	22.63	30.17	31.47	31.34	31.46	31.98	32.30	22.63
	s.d	0.16	0.52	0.50	0.58	0.55	2.64	0.79	0.33	0.40	0.38	0.21	0.21	–
O_2 (ml/l)	max	6.53	6.74	7.62	7.08	7.24	7.11	7.54	6.75	6.84	6.79	6.43	6.26	7.54
	min	6.17	6.35	6.51	6.73	6.36	6.06	5.87	6.18	5.89	6.14	6.04	6.11	5.87
	s.d.	0.13	0.14	0.40	0.13	0.30	0.35	0.54	0.34	0.35	0.24	0.13	0.06	–
PO_4 (μM)	max	0.90	1.15	0.86	0.73	0.52	0.62	1.00	0.51	0.94	0.73	0.67	0.79	1.15
	min	0.54	0.76	0.25	0.22	0.18	0.11	0.23	0.37	0.18	0.12	0.48	0.53	0.11
	s.d.	0.15	0.21	0.31	0.23	0.17	0.14	0.32	0.06	0.36	0.25	0.08	0.18	–

Table II – cont'd.

		Jan	Feb	Mar	Apr	May	Jun	Jul	Aug	Sep	Oct	Nov	Dec	Overall
NO_3	max	6.0	4.4	6.8	1.1	0.7	4.9	6.8	1.8	1.7	0.2	0.9	4.4	6.8
$+ NO_2$	min	2.9	6.0	0.1	0	0.3	0.1	0.5	0.1	0	0	1.9	0	0
(μM)	s.d.	1.60	0.9	3.4	0.5	0.3	0.7	3.2	0.9	0.8	0	0.5	3.1	–
SiO_4	max	12	7	13	6	18	16	14	ND[2]	9	4	3	4	18
	min	6	6	13	3	5	3	1	ND[2]	1	1	1	4	1
(μM)	s.d.	2.5	0.7	–	1.5	6	5	6	ND[2]	0.7	2	1	–	–

[1] For explanation of symbols and units see footnote 1 to Table 1.

[2] ND = No data

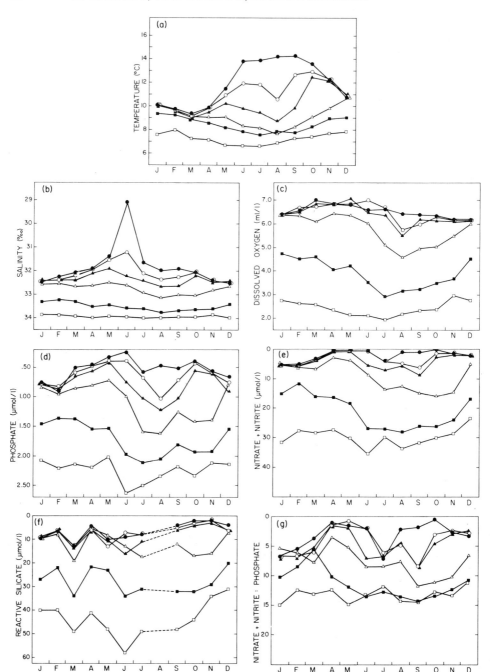

Figure 1. *Seasonal changes in the characteristics of sea water in the various depths at station NH-25, 46 km off the Oregon coast. Closed circles, 0 m; open circles, 10 m; closed triangles, 20 m; open triangles, 50 m; closed squares, 100 m; open squares, 200 m. See text for data sources.*

however, upwelling, intense biological activity, and the Columbia
River plume can produce significant variations in a relatively short
time during any month. Oxygen-depleted upwelled water can become
supersaturated water in a matter of days as a result of an intense
phytoplankton bloom. (A more detailed example of the short-term
variations is given later.) Thus, even though a coherent and logical
seasonal sequence may be evident in the monthly average data, the
likelihood of finding the "average" levels at any particular station
change during the year. During the summer, or upwelling months, when
the chemical and physical properties are most variable, an "average"
distribution is only a transient, unlikely condition.

SEASONAL EFFECTS ON AREAL DISTRIBUTIONS

The distribution of nutrients at the surface is affected by
river input, upwelling, and biological activity. This was shown
above for a single station and it is generally true for all coastal
waters between northern California and the Strait of Juan de Fuca,
Washington. Representative surface nutrient concentrations are
shown in Figs. 2-5 (16).
    The winter contours (Fig.2) show strong nitrogen input from the
Columbia River and Strait of Juan de Fuca. Much of the river discharge stays near the coast and moves northward. The remainder of
the coastal waters show $NO_3 + NO_2 \approx 5$ µM and $PO_4 \sim 1$ µM.
    With spring the nutrient supply from the Columbia River and the
Strait of Juan de Fuca decreases, and biological production removes
nutrients from large areas of the coastal waters. Significant areas
are totally depleted in nitrates, though low levels of phosphate
remain (0.2-0.5 µM).
    A sharp gradient appears during the summer (Fig. 4) which sepa-
rates high-nutrient upwelled water (>2 µM P, >8 µM N) from nutrient-
depleted offshore waters (0.2 µM P, 0 µM NO_3). During this period,
the Columbia River does not show much nutrient contrast with the
Northeast Pacific regional surface waters. The flow of the Columbia
River effluent during the summer is generally south. A clear representation of the river plume can be seen in Fig. 6 (6).
    The fall conditions (Fig. 5) indicate continued upwelling in a
small area north of Cape Blanco (~ 124°W, 41°N) and a somewhat higher
flow of nutrients from the Strait of Juan de Fuca and the Columbia
River.
    The changes in nutrient levels are most significant very near
shore (within 40 km). In any season, there is relatively little
latitudinal variation in nutrients; the largest gradients occur
perpendicular to the coast. Local upwelling areas, such as off
Cape Blanco, may produce small regions of anomalous chemical characteristics.
    We very briefly discuss in succeeding sections some additional
time and space development features of Oregon coastal water nutrient
chemistry, as well as point out the possible roles of yet other
nutrient materials.

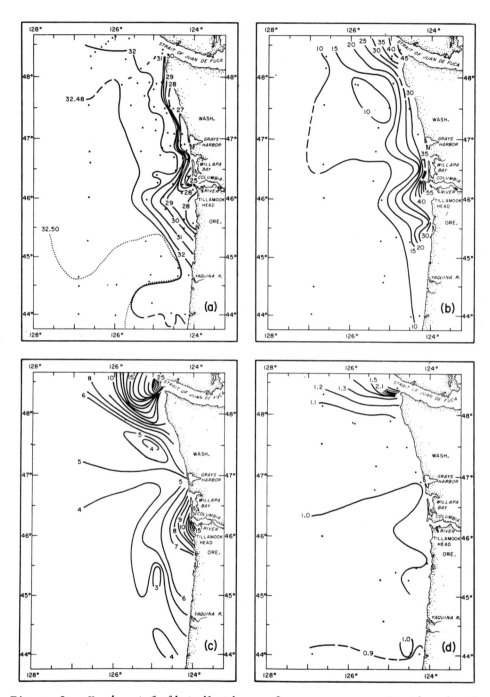

Figure 2.   *Horizontal distributions of average concentration in the upper 10 m, January to February 1962, of a) Salinity, b) Silicates, c) Nitrates, d) Phosphates.   All nutrient concentrations are in µg-at./liter.   (From 16)*

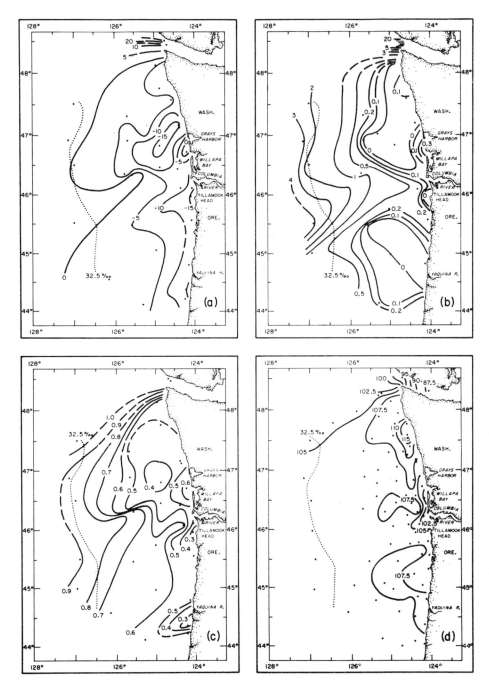

Figure 3.   *Horizontal distributions, 0-10 m averages, in spring, off the Washington and Oregon coasts of a) Silicate anomalies, b) Nitrates, c) Phosphates, d) Oxygen, percent saturation.   Nutrient concentrations and anomalies are in µg-at./liter.   (From 16)*

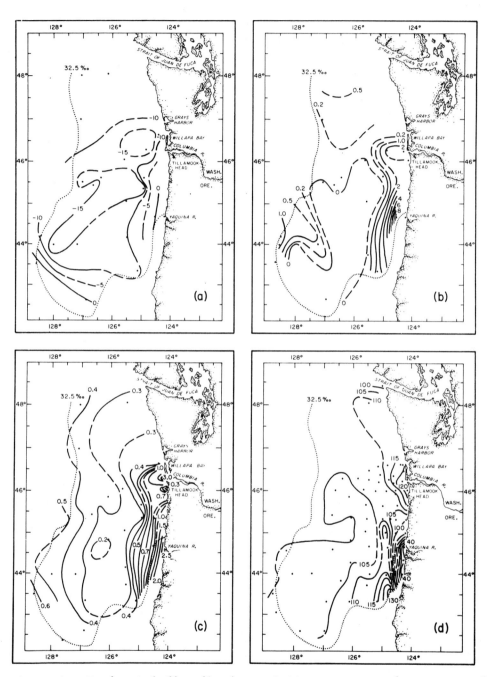

Figure 4. *Horizontal distributions, 0-10 m averages, in summer, off the Washington and Oregon coasts of a) Silicate anomalies, b) Nitrates, c) Phosphates, d) Oxygen, percent saturation. Nutrient concentrations and anomalies are in μg-at./liter. (From 16)*

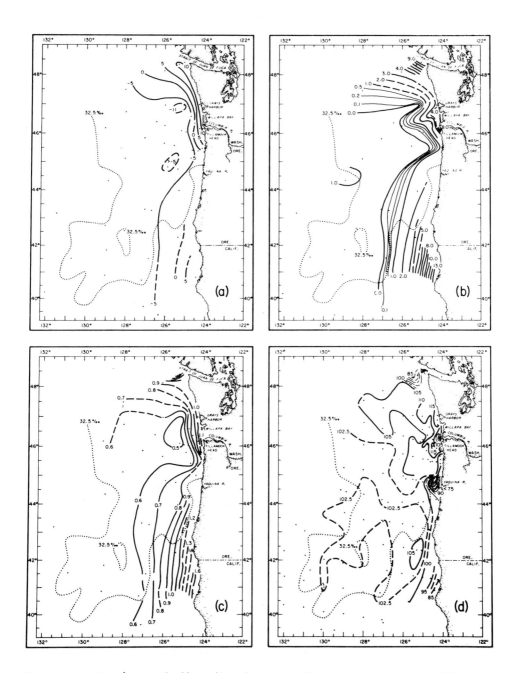

Figure 5. *Horizontal distributions, 0-10 m averages, in fall, off the Washington and Oregon coasts of a) Silicate anomalies, b) Nitrates, c) Phosphates, d) Oxygen, percent saturation. Nutrient concentrations and anomalies are in μg-at./liter.   (From 16)*

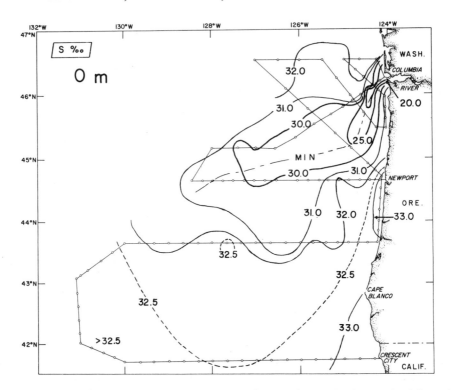

Figure 6. *Typical summer extent and intensity of the Columbia River plume as shown by surface salinity in July 1967.* (7)

EFFECTS OF AN UPWELLING EVENT ON CHEMICAL DISTRIBUTIONS

During the spring and summer months, the near-shore chemistry and biology are influenced by upwelling which occurs in response to the northerly winds dominant at that time of year. However, the winds and the upwelling are not constant. Rather, the "upwelling season" occurs as a series of pulses or events having a time scale of days to weeks (Huyer and Smith, this volume).

A cruise during June 1970 spanned such an event (1); we will briefly describe it here and illustrate its effect on nutrient distributions. The wind pattern during the cruise showed consistent north winds (Fig. 7). There was a gradual increase in the intensity of the winds until the end of the cruise when the winds suddenly reversed and diminished in speed. A single hydrographic line (NH) was occupied three times during the cruise, and the resultant salinity and nitrate data chronicle the progress of this upwelling event (Figs. 8, 9).

A lens of low salinity water observed about 15 km offshore at the beginning of the cruise reemphasizes the effect of the Columbia River. Nitrate was totally absent at the surface. Several days of northerly winds then drastically changed the vertical distributions. Waters in the upper 25 m and **very near shore** were the most altered. Nitrate concentrations in previously depleted surface water 10 km

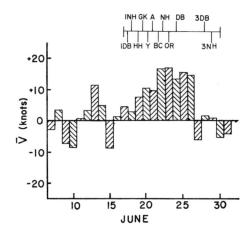

Figure 7. *Daily averages of the northerly component of wind velocity measured at the south jetty, Newport, Oregon, during June 1970, and the times of occupation of hydrographic lines during cruise Y7006A (1). (V̄ is positive for a north wind.) The symbols along the bar at the top of the figure represent: DB, Depoe Bay line; NH, Newport hydrographic line; HH, Heceta Head line; GK, Gwynn Knolls line; Y, Yachats line; A, Alsea line; BC, Beaver Creek line; and OR, Otter Rock line.*

offshore increased to 25–30 μM.

    The hydrographic and chemical structures of the deeper water had changed during the final sampling period. A mass of more saline, nutrient-rich water had intruded over the slope area. Surface nutrient concentrations inshore (0–10 km) had begun to decrease indicating some biological uptake. The time from the first to the last sampling of the line was only 11 days.

    The surface distributions of temperature, sigma-t, oxygen, silicates, and phosphate showed large variations corresponding to those discussed for salinity and nitrate (Fig. 10) during the course of this upwelling event. In the distance-time field shown in Fig. 10, a contour sloping upward to the right indicates offshore movement of the given level of that variable. Thus, for example, the simpler patterns shown by temperature and density indicate the progressive movement offshore of cold, high density water. The less conservative nutrients and oxygen reflect the additional and simultaneous effect of phytoplankton production. Hence, their contours are more stationary (or horizontal as drawn in Fig. 10) with time. The largest changes and the highest nutrient concentrations were recorded within 2 km of the coastline.

PATCHINESS IN CHEMICAL DISTRIBUTIONS

    The variability associated with upwelling and the distribution of biological populations in turn produce patchiness in nutrient chemical distributions. These facts, together with practical rates

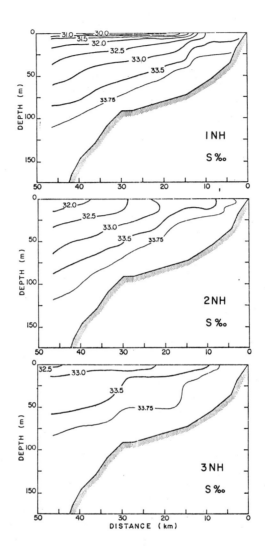

Figure 8. *Vertical distributions of salinity during three occupations of the Newport, Oregon, hydrographic line during June 1970.*

of oceanographic data acquisition, make synoptic data extremely difficult to obtain. The horizontal distributions presented earlier, while accurate with respect to large-scale features, are too broad to define some of the small-scale features which may be significant when estimating potential biomass production off the coast. Surface nutrient data taken during a period of several days in July 1972 show some of the small-scale features of the nutrient distributions (Figs. 11, 12). These data show characteristic scale lengths for variations in phosphate concentrations of factors 2, 3, and 4 over horizontal distances of 2 to 30 km. Such longshore variability on

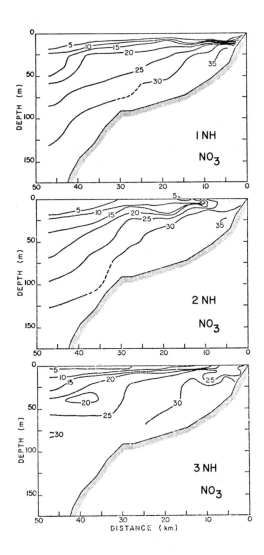

Figure 9. *Vertical distributions of nitrate (µM) on the Newport,*
Oregon, hydrographic line during June 1970.

this spatial scale is quite common along the central Oregon coast.
    Cycles with periods of 12 to 24 hours are observable in the
fluctuations in chemical properties in coastal waters. Tidal and
(possibly) biological processes cause these. The effects of such
tidal fluctuations are apparent in data collected at two coastal
drift stations (Fig. 13) sampled during a 30-hour period in May
1972 during cruise Y7205 (16). A conservative variable (salinity)
and a nonconservative variable (phosphate) showed similar fluctuations
(Fig. 14), the variability at the inshore station being significantly
higher than at the offshore station. The close correlation of the

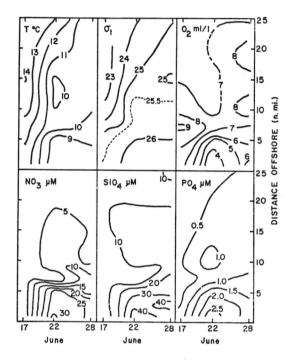

Figure 10. *Temporal changes of several variables at the surface along the Newport, Oregon, hydrographic line during June 1970.*

conservative salinity and the nutrient, phosphate, indicates that at this time and location the physical processes (semi-diurnal tides) dominated the variability in the region. Typical ranges of these variables over the 30-hour period were: offshore, $\Delta S = 0.1‰$, $\Delta PO_4 = 0.1$ µM; inshore, $\Delta S = 0.1‰$, $\Delta PO_4 = 0.4$ µM.

ESTIMATES OF NUTRIENT SUPPLIES TO THE COASTAL EUPHOTIC ZONE

A crucial factor in biomass production is the supply and avail-ability of nutrients. Can the nutrient supply satisfy growth require-ments of a large biomass? This is difficult to answer accurately because in addition to supply rates it requires a detailed knowledge of rates of plant uptake, nutrient regeneration, etc. Nevertheless, we have attempted two simple, first-order estimates of nutrient sup-plies to Oregon coastal waters.

The first of the calculations considers longshore nutrient flow through a box 1 m^2 x 50 m deep, using nitrate because it is probably the limiting nutrient for ultimate biomass growth (shown for the case of phytoplankton in 8, 9). From the data in Fig. 1e, the yearly av-erage nitrate concentration can be estimated to be 0.26 moles N in the upper 50 m^3 at NH-25. Upwelling months averaged separately give the same value. Further, assuming that the average longshore current is vertically uniform to 50 m and flows at 20 cm/sec (Huyer and Smith

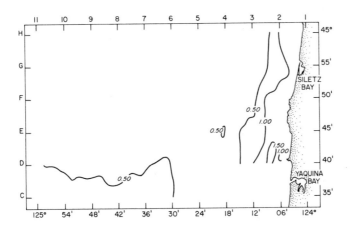

Figure 11. *Small-scale patchiness in surface phosphate concentrations in waters off the Oregon coast, July 23, 1972. (17)*

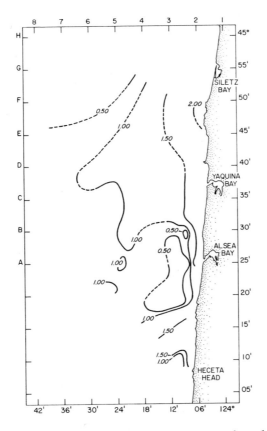

Figure 12. *Small-scale patchiness in surface phosphate concentrations in waters off the Oregon coast, July 19-22, 1972. (17)*

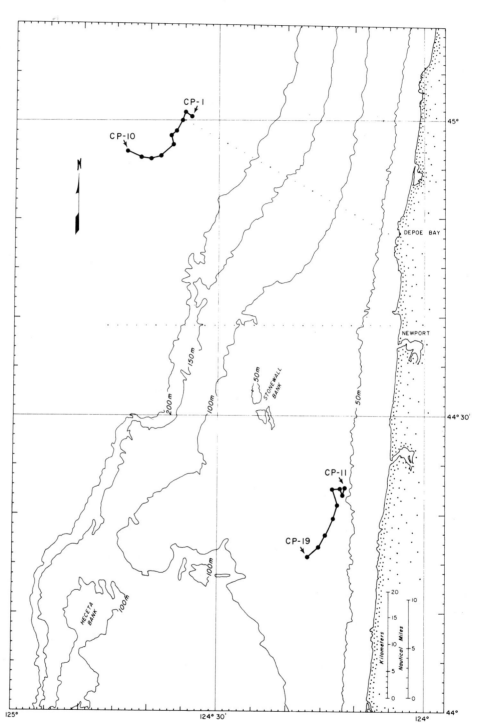

Figure 13. *Locations of two, thirty-hour drift stations occupied in May 1972 off the Oregon coast. (17)*

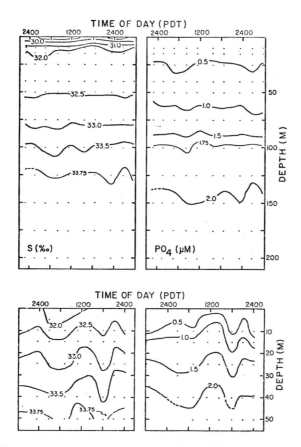

Figure 14. *Time changes in vertical distributions of salinity and phosphate for drift stations CP1-10 (upper) and CP11-19 (lower) during cruise Y7205-C off the Oregon coast in May 1972.*

this volume), then 63 kg of N flow through the hypothetical 50 m^3 box per day. (The onshore-offshore component of flow is much less.) If kelp are typically 1% N and can totally utilize the available nitrate, it follows that daily kelp production would be on the order of 6300 kg per m^2 of ocean surface at NH-25. Closer inshore the potential kelp growth would be larger because of the higher nitrate content of the water column. Practically speaking, however, this limit could not be reached because of a number of factors. Only the most upstream box in the longshore sense could support this growth, the ones downstream being depleted in N by the growth in their upstream neighbor boxes. Other factors are discussed elsewhere in this volume.

    This calculation can be compared to an earlier estimate of the maximum onshore nutrient transport during the upwelling event described above. From calculated mass transport, Atlas (1) estimated that 2200 moles/day of N were transported onshore along one meter of coastline. This corresponds to an onshore transport of 31 kg N/day/m, approximately half the longshore transport calculated above. However, this onshore transport represents "new" nutrients added to the surface waters by upwelling *during an event*. A daily average influx taken over the upwelling season would be less by a factor of four or more. Such upwelling events as used for this calculation take place during perhaps a quarter of the upwelling season.

The figure 63 kg N/m^2/day also represents an average, steady-state longshore flow and is probably a more accurate estimate of the nutrient flow through the coastal waters during the growth of *natural* populations. Thus, the major flux of nutrients through the coastal water presently is associated with the longshore flow, although for mariculture the more significant amounts to consider would be the "new" nutrients added via upwelling. It should be emphasized that these are rough, first-order calculations and that significant variations could be expected for any particular set of local conditions.

MICRONUTRIENTS

Some trace metals are essential for phytoplankton growth. One key element affecting productivity in Pacific Northwest coastal waters is iron. Glooshenko and Curl (9) demonstrated that in nutrient depleted surface waters additions of Fe caused a greater increase in photosynthesis than did additions of N or P. They concluded that iron is the limiting nutrient. Interestingly, addition of N, P, and Fe to newly upwelled waters did not increase photosynthesis. This observation suggested the possibility of yet other limiting nutrients. Some recent work by Frey (8) indicates that both vitamins and trace metals may be decisive in limiting growth *rate* of phytoplankton biomass. There is insufficient data on the total levels *and* chemical speciation of trace metals in the ocean, as well as in laboratory nutrition experiments (North, Chapter 12), to determine precisely which trace metals might be limiting productivity.

SUMMARY AND CONCLUSIONS

There are important influences on the distribution and seasonal variation of nutrient concentrations in Oregon coastal waters. These are the regional Northeast Pacific water, Columbia River discharge, coastal upwelling, and uptake of nutrients by plants. Seasonal levels and variations of nutrients have been tabulated and various influences are clearly reflected by them. For instance, the Columbia River and upwelling introduce characteristic fingerprints onto the regional Northeast Pacific waters; e.g., the river introduces fresh, high silicate moderate nitrogen, and comparable phosphate water into the Northeast Pacific, while the coastal upwelling introduces high salinity, nitrogen, phosphate, and silicate water. However, in Oregon waters both these processes are generally important only during the spring, summer and early fall. From relatively low N:P ratios during summer, we infer that nitrogen may at times be a limiting nutrient, while previous phytoplankton work indicates that this may not always be the case.

An 11-day upwelling event showed how a mass of nutrient-rich upwelling water moves onshore. The patchiness in nutrient distributions reflected all the influences summarized above and had characteristic horizontal scale lengths of a few km for two- to fourfold variations in phosphate concentrations. Two 30-hour experiments demonstrated that another process, semidiurnal tidal motion, was also important in onshore-offshore nutrient transport.

Although very crude and indicative of maximal fluxes only, two calculations of nutrient transport indicate two points. First, the present longshore advective flow of nitrogen typically might be 63 kg/m/day through an east-west section. Second, the onshore transport during an upwelling event only half as great might be more significant to a large biomass growing on the shelf, because it represents "new" nitrogen being introduced to the euphotic zone. Finally, evidence is cited that while nitrogen, iron, and maybe phosphorous might be limiting for growth of the largest possible biomass, trace nutrients, metals, and vitamins might limit growth *rates*.

ACKNOWLEDGMENTS

This work reflects a tremendous amount of work over many years by many scores of individuals. The officers and crews of the OSU research vessels enabled us to gather the necessary hydrographic and chemical data. The hydrographic work was energetically directed, in the earlier years, by Wayne Burt and the late June Patullo, and in more recent years by Robert Smith and Jane Huyer. The earlier chemical observations owe their existence to the enthusiasm and direction of Kilho Park. A host of fellow students, hydrographic technicians, and analytical chemists made most of the actual measurements, and much of the data were processed by computer specialists, chiefly William Gilbert, David Standley, and Lynn Jones. Lynda Barstow, Joseph Jennings, and Cheryl Schurg helped prepare the manuscript.

REFERENCES

1. Atlas, E. L.   1973.   Changes in Chemical Distributions and Relationships during an Upwelling Event off the Oregon Coast. Master's Thesis. Oregon State University, Corvallis. 100 pp.
2. Ball, D. S.   1970.   Seasonal Distribution of Nutrients off the Coast of Oregon, 1968. Master's Thesis. Oregon State University, Corvallis.   71 pp.
3. Barnes, C. A., A. C. Duxbury, and B. A. Morse. 1972. Circulation and selected properties of the Columbia River effluent at sea. *In:* Pruter, A. T., and Alverson, S. L. (eds.) The Columbia River Estuary and Adjacent Ocean Waters, Bioenvironmental Studies. University of Washington Press, Seattle, pp. 41-80.
4. Barstow, D., W. Gilbert, K. Park, R. Still, and B. Wyatt. 1968. Hydrographic Data from Oregon Waters, 1966. School of Oceanography, Oregon State University, Corvallis. Data Rep. 33. 114 pp.
5. Barstow, D., W. Gilbert, and B. Wyatt. 1969. Hydrographic Data

from Oregon Waters, 1967. School of Oceanography, Oregon
State University, Corvallis. Data Rep. 35. Reference 69-3.
82 pp.

6. Barstow, D., W. Gilbert, and B. Wyatt. 1969. Hydrographic Data
from Oregon Waters, 1968. School of Oceanography, Oregon
State University, Corvallis. Data Rep. 36. Reference 69-6.
92 pp.

7. Cissell, M. C. 1969. Chemical Features of the Columbia River
Plume off Oregon. Master's Thesis. Oregon State University
Corvallis. 45 pp.

8. Frey, B. 1977. Effects of Micro-nutrients and Major Nutrients
on the Growth and Species Composition of Natural Phytoplankton Populations. Ph.D. Dissertation. Oregon State University, Corvallis. 70 pp.

9. Glooshenko, W. A., and H. Curl, Jr. 1971. Influence of nutrient enrichment on photosynthesis and assimilation ratios in
natural North Pacific phytoplankton communities. J. Fish.
Res. Bd. Can. 28(5):790-793.

10. Gordon, L. I. 1973. A Study of Carbon Dioxide Partial Pressures in Surface Waters of the Pacific Ocean. Ph.D. Dissertation. Oregon State University, Corvallis. 216 pp.

11. Hager, S. W. 1969. Processes Determining Silicate Concentrations in the Northeastern Pacific Ocean. Master's Thesis.
Oregon State University, Corvallis. 58 pp.

12. Hager, S. W., and R. H. Bourke. 1971. Oxygen and nutrients.
In: Oceanography of the Nearshore Coastal Waters of the
Pacific Relating to Possible Pollution. EPA Water Pollu-
tion Control Research Series. I(13):139-142.

13. Kantz, K. W. 1972. Chemistry and Hydrography of Oregon Coastal
Waters and the Willamette and Columbia Rivers: March and
June, 1971. Master's Thesis. Oregon State University,
Corvallis. 70 pp.

14. Park, K. P., M. Catalfomo, G. R. Webster, and B. H. Reid. 1970.
Nutrients and carbon dioxide in the Columbia River. Limnol.
Oceanogr. 15:70-79.

15. Ryther, J. H. 1969. Photosynthesis and fish production in the
sea. Science 166:72-76.

16. Stefansson, J., and F. A. Richards. 1963. Processes contributing to the nutrient distributions off the Columbia River
and Strait of Juan de Fuca. Limnol. Oceanogr. 8:394-410.

17. Tomlinson, R. D., L. Barstow, D. R. Standley, S. Williams, L. I.
Gordon, and P. K. Park. 1973. Chemical Data from Oregon
Waters, 1972. School of Oceanography, Oregon State Univer-
sity, Corvallis. Data Rep. 56. References 73-15. 119 pp.

18. Wyatt, B., and N. Kujala. 1961. Physical Oceanography Data Off
shore from Newport and Astoria, Oregon, for July 1959 to Jun
1960. School of Oceanography, Oregon State University,
Corvallis. Data Rep. 5. Reference 61-3. 17 pp.

19. Wyatt, B., and N. Kujala. 1962. Hydrographic Data from Oregon
Coastal Waters, June 1960 through May 1961. School of
Oceanography, Oregon State University. Data Rep. 7.

Reference 62-6.   80 pp.

20. Wyatt, B., and W. Gilbert.   1967.   Hydrographic Data from Oregon
     Waters, 1962 through 1964.   School of Oceanography, Oregon
     State University, Corvallis.   Data Rep. 24.   Reference 67-1.
     178 pp.

21. Wyatt, B., R. Still, D. Barstow, and W. Gilbert.   1967.   Hydro-
     graphic Data from Oregon Waters, 1965.   School of Oceano-
     graphy, Oregon State University, Corvallis.   Data Rep. 27.
     Reference 67-28.   59 pp.

22. Wyatt, B., W. Gilbert, L. Gordon, and D. Barstow.   1960.   Hydro-
     graphic Data from Oregon Waters, 1969.   School of Oceano-
     graphy, Oregon State University, Corvallis.   Data Rep. 42.
     Reference 70-12.   163 pp.

23. Wyatt, B., R. Tomlinson, W. Gilbert, L. Gordon, and D. Barstow.
     1971.   Hydrographic Data from Oregon Waters, 1970.   School
     of Oceanography, Oregon State University, Corvallis.   Data
     Rep. 49.   Reference 71-23.   135 pp.

24. Wyatt, B., R. Tomlinson, W. Gilbert, L. Gordon, and D. Barstow.
     1972.   Hydrographic data from Oregon waters, 1971.   School
     of Oceanography, Oregon State University, Corvallis.   Data
     Rep. 53.   Reference 72-14.   78 pp.

- 5 -

Persistent Blooms of Surf Diatoms
along the Northwest Coast[1]

JOYCE LEWIN

The marine waters of the Pacific Northwest Coast harbor a
great variety of unicellular algal species: those that always exist
as single cells as well as those that always have their cells joined
to form filaments or colonies. Diatoms are the most important
representatives of both phytoplankton and benthic microalgal communities, followed by dinoflagellates and other small flagellate species.
In a recent study of the diatom flora of the Strait of Georgia/Juan
de Fuca Strait systems, 219 different diatom taxa were identified,
and the community structure was observed to change radically with
different seasons (9).

During late autumn and winter, phytoplankton cell numbers
remain low due to light limitation. With increasing light in spring,
blooms of various diatom species take place, so that cell numbers of
diatoms are high in coastal waters during spring and early summer.
The spring bloom in Puget Sound is a sequence of blooms of various
diatom species; blooms of dinoflagellates generally occur in the
latter part of the summer. Where the water has become stratified
due to warming at the surface (oceanic surface water and water in
enclosed bodies such as bays or fjords), nitrate may become depleted
as a result of these phytoplankton blooms.

[1]This research was supported through funds provided by Grant
(OCE) 76-20155 from the National Science Foundation; by ERDA Contract
E76S062225TA26#5 from the Energy Research and Development Administra-
tion (reference RLO 2225-T26-47); and by the Washington Sea Grant
Program, which is maintained by the National Oceanographic and Atmos-
pheric Administration of the U.S. Department of Commerce. Contribu-
tion No. 956 from the Department of Oceanography, University of
Washington, Seattle, Washington 98195.

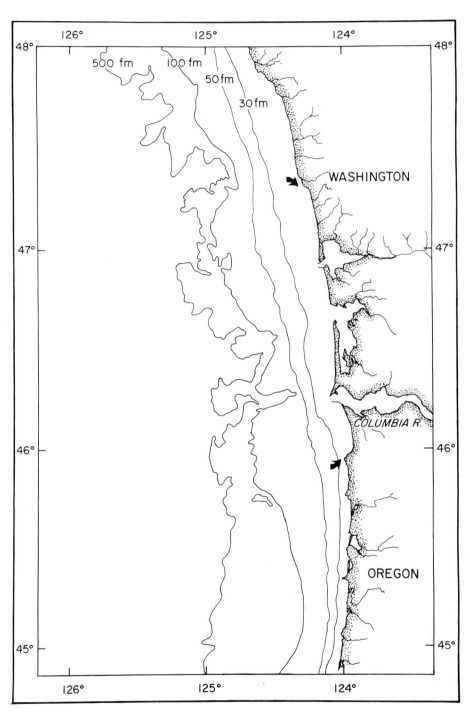

Figure 1. *Coastal areas of Washington and Oregon. Beaches where blooms of surf diatoms are most extensive are delimited by arrows.*

DIATOM BLOOMS ALONG SANDY BEACHES

Rocky areas on the coastline of Washington and Oregon form a favorable habitat for a rich seaweed flora (Phinney, this volume); however, sections with long stretches of wide gently sloping beaches, fine sand, and pounding surf are devoid of seaweed. Instead, a different sort of algal growth occurs in the form of surf diatoms. These diatom blooms are most abundant and most obvious along the beaches between Tillamook Head in Oregon and Point Grenville in Washington (about 80 miles of sandy beach *in toto,* interrupted by the Columbia River mouth and entrances to Willapa Bay and Gray's Harbor, Fig. 1). They are also present along many other beach areas to the south and north of this section of the coast. The algal growth is so dense that the breaking waves are brown and thick deposits of diatoms may be left on the beach when the tide recedes (Figs. 2, 3).

These diatom blooms were studied in the 1920s and '30s, when it was thought that a clue to the origin of petroleum might be found in a study of such massive amounts of algal cells; early investigators referred to the blooms as "epidemics" (2, 13).

The blooms of surf diatoms have two unique features: (1) there are only two species of significance comprising the blooms, and (2) they are persistent in that they are almost always present and most abundant during autumn, winter, and early spring.

SPECIES COMPRISING THE BLOOMS

The two species comprising the blooms along the Washington and Oregon coasts are *Chaetoceros armatum* T. West (a centric diatom species) and *Asterionella socialis* Lewin and Norris (a pennate diatom species). Both species exist as colonial forms (Fig. 4); the cells of *C. armatum* are joined tightly together in chains or filaments enclosed in a clay coat, while the cells of *A. socialis* are held together by mucilage pads forming a scruffy looking colony.

CELL CONCENTRATIONS

Populations of the two species in surf samples have been monitored at Copalis Beach for the past 6 years (Fig. 5a, b).

Cell numbers of *C. armatum* are always fewer in nighttime samples than in daytime samples (6, 8) and, therefore, are plotted separately in the figure (5a). This is because the cell chains rise to the surface in the early morning and float as large patches of a stabilized foam (Fig. 2), in the daytime. Wave action then transports the floating material toward the beach; hence, there are higher cell concentrations in daytime samples. In late afternoon, the floating patches disappear and cell chains disperse, resulting in lower cell counts in nighttime samples.

As seen from data in Fig. 5, best growth and maximum development of each of the species takes place during autumn, winter, and spring of most years. In these respects, *C. armatum* and *A. socialis* differ from other phytoplankton species and from seaweed species along this coast; the latter forms either disappear or grow very

Figure 2. *The breakers at Copalis Beach, Washington. An extensive stable foam of diatom cells (Chaetoceros armatum) floats on the surface of the water.*

Figure 3. *Diatom deposits (Chaetoceros armatum) stranded on Copalis, Beach, Washington, by the receding tide.*

Figure 4. *Photomicrograph of surf diatom species Chaetoceros armatum and Asterionella socialis. (x 375)*

slowly during winter months. The ability to do well under adverse physical conditions is due to special physiological adaptations of the two surf species (7).

Cell concentrations of both species are less dense during summer months, and in some years *A. socialis* has disappeared completely in mid- or late summer (Fig. 5b). Even though cell numbers would indicate that *C. armatum* was not as severely affected as was *A. socialis*, evidence from physiological and morphological studies indicates that cells of *Chaetoceros* are under a definite physiological stress in summer; for example, nitrate reductase activity is depressed (3), the C:N ratio of the cells is considerably higher (4), and the generation time is longer (8).

ENVIRONMENTAL FACTORS ASSOCIATED WITH THE BLOOMS

Other chapters in this volume deal with the geology of the Northwest Coast and the physical and chemical oceanography of coastal waters. Much of the material is directly relevant to biological and ecological studies along coastal beaches. Chemical changes taking place beyond the breakers and in the river estuaries will have an effect on biological reactions in the surf. Winds (direction, magnitude, duration), upwelling, and river discharge (affected by rainfall and snow melt) interact to influence the surf-zone ecosystem. The shallowness of the water (gentle slope of the coast, particularly in Washington cf. Fig. 1), results in a high degree of inhomogeneity in the environment. Patches of water of varying properties may form and remain along the beach for considerable periods of time.

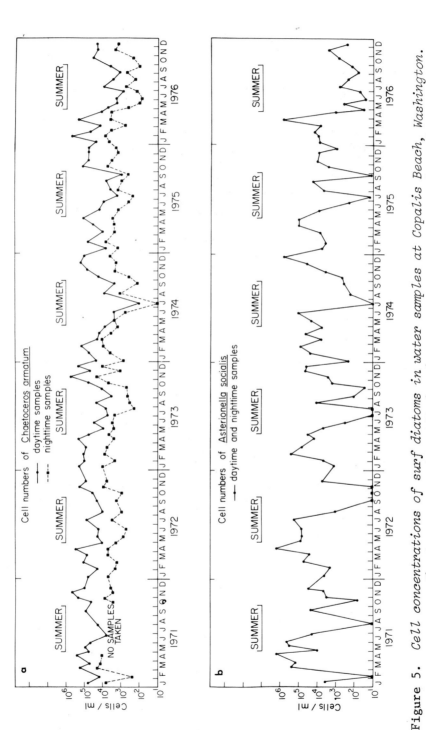

Figure 5. Cell concentrations of surf diatoms in water samples at Copalis Beach, Washington.

    a. *Chaetoceros armatum*. Each point represents average numbers from several samples collected at 2 stations, either in daytime or nighttime.

    b. *Asterionella socialis*. Each point represents average numbers from several samples at 2 stations during a 24-hour sampling period.

Salinity and Temperature

During 6 years of monitoring the surf environment, salinity concentrations in surf samples ranged between 16.8 °/oo and 35.7 °/oo, while temperatures ranged between 3.5°C and 21.2°C. During most of the year, salinity concentrations lie between 20 °/oo and 30 °/oo; higher salinity water above 30 °/oo only reaches the surf region in summer during periods of upwelling.

Nutrients

Concentrations of phosphate show seasonal variations but do not appear to become limiting during summer (6). Silicate is always in abundant supply (6); the primary source of silicate is rivers which drain from an area of volcanic deposition. During most of the year (autumn, winter, spring), silicate concentration in the surf is inversely correlated with salinity (i.e., lower salinity water contains the highest levels of silicate); on the other hand, during periods of upwelling in the summer months, deeper oceanic water of high salinity also contains high concentrations of silicate.

Nitrate becomes depleted in surf samples by late April of each year (Fig. 6a-f) due to spring phytoplankton blooms in the upper water layers of inshore water lying beyond the breaker zone (cf. 1, 10, 11). In April and May of 1974, a bloom of *Thalassiosira pacifica*, a centric diatom species, was found covering many square kilometers. This species was responsible for depletion of nitrate in the upper water layers beyond the breakers (Fig. 7); the two surf diatom species were either absent or present in very low numbers in water samples from beyond the breaker zone (Lewin, unpubl.).

Reintroduction of nitrate into the surf over the summer depends on the extent of upwelling (which in turn depends on the magnitude and duration of northerly winds during summer) and on the extent of river drainage (low salinity water lying along the coast may mask upwelling). Any nitrate that reaches the surf during summertime is in association with higher salinity water (30 °/oo and greater) (Fig. 8). The low nitrate levels during a 6-month period in 1971 (Fig. 6a) were due to extraordinarily weak upwelling; higher levels of nitrate in the summer of 1973 (Fig. 6c) resulted from strong upwelling, combined with record low rates of freshwater discharge from coastal rivers.

Since nitrate is often in short supply in the surf, the availability of other forms of nitrogen is important. Ammonium levels are high in surf samples (Fig. 9) and, during the critical summer period, usually exceed nitrate concentrations.

INTERDEPENDENCE OF RAZOR CLAMS AND SURF DIATOMS

Beaches with the greatest abundance of surf diatoms are the most highly productive razor clam beaches on the Pacific coast (5, 12). The success of the razor clams can be attributed to the abundant and almost continuously available algal food supply; gut contents of razor clams from these beaches contain digested material of the two

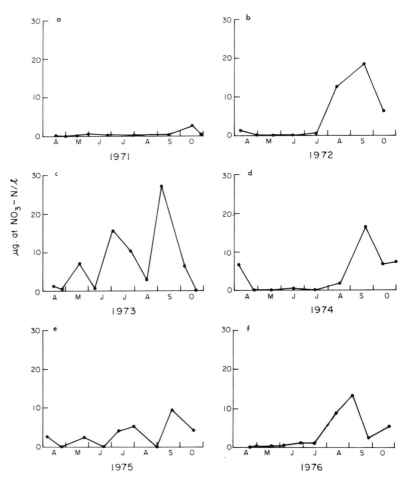

Figure 6. *Concentrations of NO_3-N (µg at/liter) in surf samples at Copalis Beach, Washington, from mid-April to mid-October each year from 1971 to 1976. Each point represents an average concentration from many duplicate samples collected during a 24-hour sampling period.*

surf diatom species almost exclusively.

In turn, the razor clams recycle nutrients used by the diatoms (Fig. 10). In particular, the major source of regenerated nitrogen (ammonium) in the surf environment comes from metabolic processes of razor clams. The clam population is large; between 6 and 15 million large clams are removed from coastal beaches of Washington each year by sports fishermen. Since these are taken from the part of the beach accessible on low tides, it is thought that a much large population resides out-of-reach of diggers. The metabolic activities of the razor clam population will have a definite influence on the surf environment, and excretion of ammonium is one of these influences.

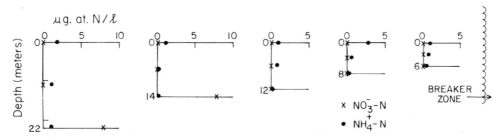

Figure 7. *Profiles of nitrogen concentrations (both NO_3^- and NH_4^+) beyond the breaker zone at Copalis Beach, Washington, on April 27, 1974.*

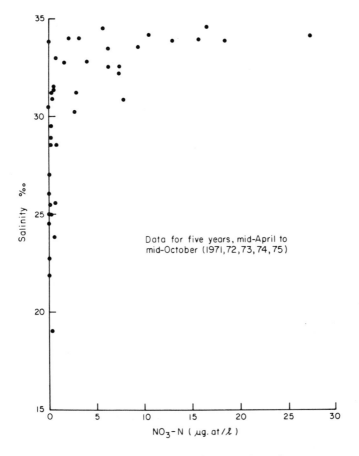

Figure 8. *Nitrate concentrations and salinity in water samples at Copalis Beach, Washington, between mid-April and mid-October. Each point represents an average from many samples collected over a 24-hour period.*

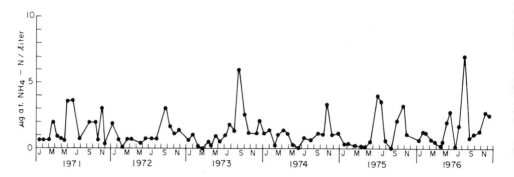

Figure 9. *Concentrations of ammonium (NH$_4^+$) in surf samples at Copali: Beach, Washington. Each point represents an average from many sample: collected over a 24-hour period.*

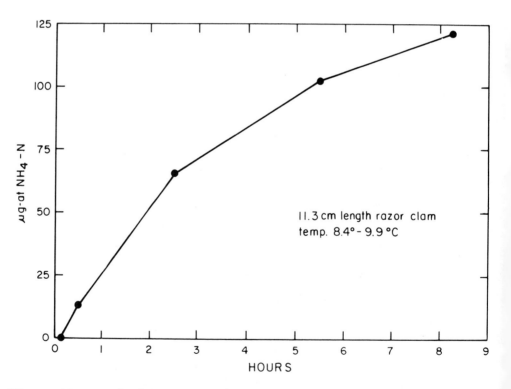

Figure 10. *Typical curve showing excretion of ammonium vs. time by a razor clam (11.3 cm shell length) in a field experiment at Copalis Beach, Washington.*

SUMMARY AND CONCLUSIONS

Much of the information that has resulted from this study about the physical and chemical environment along the beach will be applicable to other parts of the Northwest coast; persons concerned with the feasibility of seaweed farming operations will be interested in these data. In particular, the presence of nitrogen-depleted water along the shore during summer will be a factor to be considered.

The surf diatoms constitute a high algal biomass with a rapid growth rate, and this source of plant material might be used as a living food supply for culture of bivalve molluscs in aquaculture projects along the coast. Or, perhaps a controlled farming operation of razor clams might be contemplated utilizing the surf diatoms as the food supply. A model now being developed of the population dynamics of the surf diatom species will be useful in making future predictions as to the feasibility of such undertakings.

If the predictions made by Edward Hall (this volume) should come to pass, these surf species might be considered for algal biomass to be utilized as sources of petrochemicals and liquid hydrocarbon fuels.

REFERENCES

1. Anderson, G.C. 1964. The seasonal and geographic distribution of primary productivity off the Washington and Oregon coasts. Limnol. Oceanogr. 9:284-302.
2. Becking, L.B., C.F. Tolman, H.C. McMillan, J. Field, and T. Hashimoto. 1927. Preliminary statement regarding diatom "epidemics" at Copalis Beach, Washington, and an analysis of diatom oil. Econ. Geol. 22:356-368.
3. Collos, Y., and J. Lewin. 1974. Blooms of surf-zone diatoms along the coast of the Olympic Peninsula, Washington. IV. Nitrate reductase activity in natural populations and laboratory cultures of *Chaetoceros armatum* and *Asterionella socialis*. Mar. Biol. 25:213-221.
4. Collos, Y., and J. Lewin. 1976. Blooms of surf-zone diatoms along the coast of the Olympic Peninsula, Washington. VII. Variations of the carbon-to-nitrogen ratio in field samples and laboratory cultures of *Chaetoceros armatum*. Limnol. Oceanogr. 21:219-225.
5. Hirschhorn, G. 1962. Growth and mortality rates of the razor clam (*Siliqua patula*) on Clatsop Beaches, Oregon. Fish Commission of Oregon, Portland, Oregon. Contrib. No. 27.
6. Lewin, J., T. Hruby, and D. Mackas. 1975. Blooms of surf-zone diatoms along the coast of the Olympic Peninsula, Washington. V. Environmental conditions associated with the blooms (1971 and 1972). Estuarine Coastal Mar. Sci. 3:229-241.
7. Lewin, J., and D. Mackas. 1972. Blooms of surf-zone diatoms along the coast of the Olympic Peninsula, Washington. I. Physiological investigations of *Chaetoceros armatum* and

Asterionella socialis in laboratory cultures. Mar. Biol.
16:171–181.

8. Lewin, J., and V.N.R. Rao. 1975. Blooms of surf-zone diatoms
    along the coast of the Olympic Peninsula, Washington. VI.
    Daily periodicity phenomena associated with *Chaetoceros
    armatum* in its natural habitat. J. Phycol. 11:330–338.

9. Shim, J.H. 1976. Distribution and Taxonomy of Planktonic
    Marine Diatoms in the Strait of Georgia, B.C. Ph.D. Thesis.
    University of British Columbia, Vancouver.

10. Stefánsson, U., and F.A. Richards. 1963. Processes contribu-
    ting to the nutrient distributions off the Columbia River
    and Strait of Juan de Fuca. Limnol. Oceanogr. 8:394–410.

11. Stefánsson, U., and F.A. Richards. 1964. Distributions of dis-
    solved oxygen, density, and nutrient off the Washington and
    Oregon coasts. Deep-Sea Res. 11:355–380.

12. Tegelberg, H.C. 1964. Growth and ring formation of Washington
    razor clams. Washington Dept. Fisheries, Fisheries Res.
    Papers 2:69–103.

13. Thayer, L.A. 1935. Some Experiments on the Biogenetic Origin
    of Petroleum. Ph.D. Thesis. Stanford University, Palo Alto,
    California.

-6-
The Macrophytic Marine Algae of Oregon
HARRY K. PHINNEY

The species composition of the marine algal flora of Oregon appears to conform to the patterns discussed by Setchell (9, 10), Scagel (7, 8), Druhl (3, 4), and others. Briefly stated, the temperate segment of the flora of the eastern margin of the North Pacific basin ranges from Point Conception, California, about 34° 25' on the south, to Sitka, Alaska, 57° 10' on the north. The rocky headlands of the Oregon Coast lie mostly between the 42nd and the 46th parallel and are almost in the center of this range of temperate marine species. The list of species known in Oregon contains 67 species of the Chlorophyta, 87 species of the Phaeophyta, and 207 species of the Rhodophyta making a total of 361 species of macrophytic marine algae. There are, in addition, 13 species of the Chlorophyta, 14 species of the Phaeophyta, and 66 species of the Rhodophyta known in Washington and in California, but which have not been found in Oregon. Thus, an additional 93 species (20%) might be added to the list.

The species list was studied to determine the probability that a significant portion of these 93 species with apparently discontinuous distributions have simply been overlooked by collectors in Oregon. According to Abbott and Hollenberg (1), of the 13 species of Chlorophyta in this group, one is subtidal, two are minute, and four are found only in sheltered locations. Of the species in the Phaeophyta, six are subtidal, two are small, and one is limited to sheltered areas. Assessment of the species of the Rhodophyta in this category shows: 21 are subtidal, 12 are small, three are rare; and two are found only in sheltered areas. Thus, there would seem to be some probability that 54 of these 93 species eventually may be found in Oregon. Subjective estimation suggests that the probability of locating most of these species is poor. That the minute species have been overlooked is quite probable. However, subtidal collections from Oregon comprise species common in the lowest intertidal areas without revealing forms uniquely subtidal.

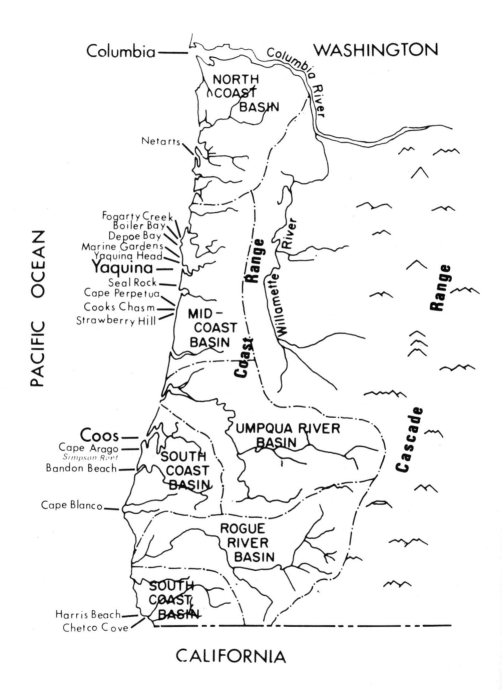

Figure 1.   *Rocky shores and headlands that are major collecting areas for macrophytic marine algae on the Oregon Coast.*

A logical disposition of the remaining 39 species is not simple.
Descriptions of their habits give no indication why they are not
found in Oregon. This suggests strongly that many of these entities
have discontinuous distributions and may never be found. This presumption invites speculation concerning discontinuities. The most
obvious characteristic common to the collecting areas along the
Oregon Coast is the degree of exposure. Although technically
described as protected rocky outer coast (P.R.O.C.), the offshore
reef provides minimal protection and along the entire coast the only
truly protected areas lie in estuaries. There are no lagoons or rocky
shored bays that are not part of estuaries. Although the estuarine
intertidal environment is secure from the great physical violence
of surf action, the estuaries of Oregon are small in relation to the
volume of fresh water passing through them during the months of peak
run-off. This flow of fresh water causes extremely low salinities
throughout the estuaries at low tide. Salinities of less than 10 $^o/oo$
occur at the sea water intake of the Oregon State Marine Science
Center at Newport on Yaquina Bay at low tide during freshets. Even
in Coos Bay, the second largest estuary in Oregon and the largest
bay, similar salinities have been recorded (5) near the entrance at
Charleston.

The high turbidities that characterize Oregon's inshore coastal
waters must also be considered. During heavy run-off, the estuaries
receive fluviatile sediments that are apparently mostly deposited
within the estuary. However, extremely high loads of sediments are
characteristic of moving inshore waters here at all seasons of the
year. The flow of sediments onto and off the shore has given rise
to the poetic description of the beaches as "rivers of sand."

Although the building of the extensive sandy beaches between
the rocky headlands may be the most obvious result of the onshore
transport of sand, the pattern of onshore deposition includes rocky
shores even to the rocky headlands. Locations of major collecting
sites are given in Fig. 1. There are several rocky beaches along
the central Oregon Coast where the onshore transport of sand has
resulted in the upper intertidal to supratidal levels becoming extensive sand beaches. This situation exists at the northern end of
the reef at Boiler Bay, at the Marine Gardens reef just north of
the Devil's Punch Bowl near Otter Rock, at Seal Rock, at the reef
at Cape Perpetua south of Cook's Chasm, and at Strawberry Hill at
the southern end of Neptune State Park. The rocks at Depoe Bay,
Yaquina Head, Yachats, and at Cove Creek on Cape Perpetua show only
relatively minor onshore movement and that seldom extends above mean
low tide level.

The volume of sand deposited over a reef during the summer beach
building episode is frequently impressive. During the summers 1974
to 1976, the deposit on the boulder field at the southern end of the
Marine Gardens reef laid down between mid-July and mid-September
averaged 8 to 10 ft in depth in a band from the ±1.0 ft line to the
12 ft line (above mean low water). At this same time in 1974 and
1975 the boulders at the bottom of the Devil's Punch Bowl were buried
approximately 12 ft deep. The deposit of sand at Marine Gardens and

Seal Rock averaged 8 ft or less by October 1976, but this deposit was still in place in mid-December, when only one fall storm of any magnitude had been experienced.

Although the amount of sand being transported into other rocky areas of the central Oregon Coast is seldom as spectacular as the previous examples, there is evidence that even in less conspicuous cases there may be serious impact. At Simpson Reef on Cape Arago, an extensive boulder reef with many tide pools below mean low water, the transport and deposition of a coarse shell sand had filled all of the smaller and most of the larger pools in the outer third of the reef by the end of July 1976. Reports from students at the Institute of Marine Biology at Charleston indicate this has been an annual event for at least six years.

In the past 30 years, the pattern of deposition has deviated. There is no threat of permanent burial of any of the rocky reefs. In the near future the pattern of deposition will change again and other areas will receive the deposits. The present areas will be scoured quite clean in a season or two.

The effect of transport and deposition lies in the relatively short-term stability of the affected rocky intertidal areas. It also appears that the inshore subtidal areas are subject to massive molar effects generated by the tremendous "bedload" of sediments transported across them. One is incited to speculate upon the possible connection between the changing patterns of transport and deposition of sand and the sometimes fluctuating size and location of beds of *Nereocystis leutkeana* (bull whip kelp) and the fact that the only established bed of *Macrocystis integrifolia* (California giant kelp) in Oregon is on Simpson Reef, Cape Arago.

A phenomenon obviously associated with sand transport is the characteristically high turbidity of the inshore waters. With the establishment of the "winter" pattern of upper air circulation and the concurrent monsoon, the coast of Oregon is subject to a steady succession of storm fronts accompanied by strong winds (to 110 mph at coastal stations) from the southwestern quadrant. The waves from these frequent local storms, which mostly move onto the coast of Oregon from the Gulf of Alaska, are interspersed by waves generated by severe disturbances (typhoons, etc.) initiated in the western Pacific basin by winds of even greater magnitude. During this season, the sand deposited in the intertidal area during summer is scoured out and transported into deeper subtidal areas. The turbidity of the inshore water at this time is so great that divers have reported that they literally cannot see their hands before their faces immediately beneath the surface.

Even during the relatively storm-free season from mid-March through mid-September, divers report that only 2 of 3 diving days afford suitable visibility (i.e., 3 ft) as they approach the bottom in the shallow waters (40 to 70 ft) of a reef 2 to 2½ miles offshore. It must be recalled that this is when the sand is being transported across the lower subtidal and deposited on the upper subtidal and intertidal areas. However, not all poor visibility in summer is the result of suspended sediment. During persistent upwelling, dense phytoplankton blooms also contribute to the problem.

Intertidal areas subjected to the effects of massive sand transport and deposition of sand exhibit a macrophytic flora much reduced
in numbers of species, and those present are frequently depauperate
or show damage from molar action   (Table I).

Table I. *Species of algae characteristic of areas subject to transport and deposition of sand on the Oregon Coast.*

CHLOROPHYTA	RHODOPHYTA
Cladophora columbiana	*Ahnfeltia gigartinoides*
Cladophora flexuosa	*Ahnfeltia plicata*
Cladophora hutchinsiae	*Bangia fusco-purpurea*
Codium setchellii	*Botryoglossum farlowianum*
Enteromorpha sp.	*Cryptosiphonia woodii*
Urospora penicilliformis	*Dilsea californica*
	Gelidium pusillum
	Gigartina papillata
PHAEOPHYTA	*Gracilaria sjoestedtii*
	Grateloupia doryphora
	Gymnogongrus linearis
Alaria marginata	*Halosaccion glandiforme*
Egregia menziesii	*Plocamium oregonum*
Laminaria dentigera	*Porphyra lanceolata*
Laminaria sinclairii	*Prionitis filiformis*
Petalonia fascia	*Prionitis lanceolata*
Ralfsia pacifica	*Prionitis linearis*
Ralfsia fungiformis	*Schizymenia pacifica*
Sphacelaria didichotoma	*Stenogramme interrupta*
Sphacelaria racemosa	

The number of algal species in Oregon that may hold some promise
for harvest from native populations or that might be suitable subjects
for large-scale cultivation is limited. Among the Chlorophyta, only
species of *Ulva* and *Enteromorpha* appear to produce organic materials
at a rate high enough to warrant consideration. The possibility that
biologically active compounds of potential value are produced at some
stage or stages of the life cycle of such plants needs further investigation. However, members of the Phaeophyta have been widely utilized,
particularly in temperate and colder waters. Because of their bulk,
*Macrocystis integrifolia, Nereocystis luetkeana, Alaria marginata,
Egregia menziesii, Laminaria saccharina, Sargassum muticum, and*
possibly *Hedophyllum sessile* deserve consideration.

The natural population of *Macrocystis* on the Oregon Coast is too
limited to be of interest. The only feasible source would be cultivated plants. Although the native population of *Nereocystis* forms
numerous beds along the Oregon Coast, these tend to be small, vary

in size and position, and frequently a bed will disappear for a few
years and eventually, gradually be reestablished. Most of these beds
are close inshore and commercial harvest has never been attractive.

Both *Alaria marginata* and *Egregia menziesii* are essentially
ubiquitous in rocky areas and have been harvested locally as a source
of fertilizer. *A. marginata* was formerly a source of potassium salts
in the kelp industry (6), but there has never been extensive commercial
harvest of either species in Oregon. Both species have a reasonably
high alginic acid content (up to 20% by dry weight) and warrant consideration for cultivation.

Hedophyllum is a marginal prospect in view of the relatively
slight bulk of the plants. The rate of growth does not seem sufficient
to recommend it as a candidate for cultivation. However, *Hedophyllum*
produces large amounts of hydrocolloid and other materials of potential
commercial value, and Levring, Hoppe, and Schmid (6) indicate that the
development of cultivated *Hedophyllum* species has been studied.

Two forms have been omitted from the preceding discussion
because they appear to be of particular promise. *Laminaria saccharina*
can grow to a length of 2 meters in 8 to 10 weeks. However, once the
plant has attained its maximum size for a given location, thallus
growth slows to a point approximately equalling the rate of tissue
loss by erosion from the blade tip. Some conditions might be arranged
in cultivation to prevent or at least minimize this tissue loss.

The second algal species that appears to have some promise of
success in cultivation is *Sargassum muticum*. The success of this
introduced species in competition with the native laminarians and
its rapid spread signal this as an extremely aggressive plant. The
spread of *S. muticum* from Coos Bay in 1947 to Humboldt County, California, by 1965 (2) and then to San Diego County by 1971, its
ability to grow both intertidally and subtidally, and its occurrence
both in estuaries and on the outer coast support the characterization
of this species as aggressive.

In 1948 *S. muticum* was known only in estuaries. In the summer
of 1973, Dr. Peter Rothlisberg found an attached plant of *S. muticum*
at Seal Rock. The following spring it was learned that a population
had been well-established on Simpson Reef at North Cove, Cape Arago,
Coos County, Oregon, for at least three years. At the present time
this is the only known, permanently established population on the
outer coast in Oregon. At this site *S. muticum* occurs in the upper,
mid-intertidal surrounded by *Phyllospadix scouleri*.

Native populations of the bulkier rhodophytes certainly are
spread more thinly than the laminarians. Successful exploitation
would most certainly depend upon cultured plants. Even so, there
are a few local species that might be candidates for investigation
(Table II).

Because of the lack of information on the occurrence of biologically active materials in marine plants, it is probable that the
species suggested for further study omits some of value. Consideration for cultivation in Oregon should be given to any and all
species whose distributions lie north of Point Conception, California.

In order to record the species which are known for Oregon, a

Table II. *Native species of algae in the Rhodophyta of potential interest for cultivation off the Oregon Coast.*

Species	Compound Produced
*Ahnfeltia gigartinoides	agaroid
Ahnfeltia plicata	agaroid
*Gelidium purpurascens	agar and agarose
Gigartina exasperata	
Gigartina papillata	carrageenan, vitamin C
*Gracilaria sjoestedtii	agaroid
*Gymnogongrus linearis	carrageenan
Iridaea cordata	iridophycan
Iridaea flaccida	iridophycan
*Neoagardhiella baileyi	iota-carrageenan
*Rhodomela larix	bromine (enriched to 3%)

*Potential candidates for cultivation

checklist prepared in collaboration with John B. Kramer is included
in Table III. This checklist was conceived to include only the
multicellular and colonial marine algae large enough to be conspicuous to the unaided eye in the field. The unicellular algae, motile
algae, the diatoms, and the blue-green algae have been omitted.

The list has been constructed to provide the maximum information in a brief form. The names of entities are listed alphabeti-
cally. The great majority of entities named have had their occurrence in Oregon recorded repeatedly. In a few cases a brief note
indicates that a species has not been reported since an early record.
The names of entities whose presently known distribution includes
records in both Washington and California but for which there are
no records in Oregon are preceded by an asterisk.

Names used in commonly available checklists and manuals that
have been reduced to synonymy are listed enclosed in brackets both
in proper alphabetical order and also listed with the presently
accepted epithet. A question mark preceding a name indicates a
doubtful record. Names of species now believed to be stages in
heteromorphic life cycles as indicated by recent cultural studies
are enclosed in quotation marks.

Table III. *An abbreviated checklist of the marine macroalgae of Oregon.*[1]

CHLOROPHYTA

BLIDINGIA

  B. minima var. minima (Nägeli ex Kützing)
    Kylin [Enteromorpha minima]
  B. minima var. subsalsa (Kjellman) Scagel
    [Enteromorpha micrococca f. subsalsa]
    [Enteromorpha minima var. subsalsa]
  B. minima var. vexata (Setch. & Gard.)
    J. Norris [Ulva vexata]
    [Enteromorpha vexata]

BOLBOCOLEON

  *B. piliferum Pringsheim

BRYOPSIS

  B. corticulans Setchell
  B. hypnoides Lamouroux

CHAETOMORPHA

  C. aerea (Dillwyn) Kützing
    Included in Chaetomorpha linum,
    A. & H. (1)
  *C. californica Collins
  C. linum (Müller) Kützing
    Includes C. aerea, A. & H. (1)
  C. tortuosa (Dillwyn) Kützing
    [Rhizoclonium tortuosum], synonym of
    R. riparium, A. & H. (1)

CHLORANGIUM

  [C. marinum Cienkowski]
    Prasinocladus marinus

CHLOROCHYTRIUM

  "C. inclusum" Kjellman
    Supposed sporophyte of Spongomorpha
    sp.
  C. porphyrae Setch. & Gard.

CLADOPHORA

  *C. albida (Hudson) Kützing
  C. columbiana Collins
    [C. hemispherica]
    [C. trichotoma]
  *C. flexuosa (Griffiths) Harvey
    Included in C. sericea, A. & H. (1)
  C. gracilis (Griffiths) Kützing
  C. glaucescens (Griffiths) Harvey

    Included in C. sericea, A. & H. (1)
  C. graminea Collins
  [C. hemisphaerica Gardner apud Collins]
    C. columbiana
  C. hutchinsiae (Dillwyn) Kützing
  C. membranacea nov. var.
  C. microcladioides Collins
  C. pulverulenta (Mertens) nov. comb.
  C. sericea (Hudson) Kützing
    Includes C. flexuosa & C. glaucescens,
    A. & H. (1)
  *C. stimpsonii Harvey
  [C. trichotoma (C. Agardh) Kützing]
    C. columbiana

CODIOLUM

  C. gregarium A. Brown
    Probably the sporophyte of Urospora sp.
  C. petrocelidis Kuckuck
    Probably the sporophyte of Spongo-
    morpha coalita

CODIUM

  C. fragile (Suringar) Hariot
  C. setchellii Gardner

COLLINSIELLA

  C. tuberculata Setch. & Gard.
    Probably a stage of Enteromorpha sp.
    and/or Monostroma sp.

DERBESIA

  D. marina (Lyngbye) Solier
    Diploid stage of Halicystis ovalis

ENDOPHYTON

  E. ramosum Gardner

ENTEROMORPHA

  E. ahlneriana Bliding
  [E. angusta (Setch. & Gard.) Doty]
    Ulva angusta
  E. clathrata (Roth) Greville var. clathrata
  *E. clathrata var. crinita (Roth) Hauck
    [E. crinita]
  E. compressa (L.) Greville
  E. flexuosa (Roth) J. Agardh
    [E. prolifera var. flexuosa]
  [E. grevillei Thuret]
    Monostroma grevillei

---

  *The names of entities whose presently known distribution includes records in both
Washington and California but for which there are no records in Oregon.
  [1]This checklist has been compiled from herbarium specimens and the literature by
Harry K. Phinney and John B. Kramer.

ENTEROMORPHA (continued)

    E. intestinalis var.*cylindracea* J. Agardh
    E. intestinalis (L.) Link var.*intestinalis*
    **E. intestinalis* f. *clavata* J. Agardh
    **E. intestinalis* var. *maxima* J. Agardh
    E. linza (L.) J. Agardh
       [*Ulva linza*]
    **E. marginata* J. Agardh
    [*E. micrococca* f. *subsalsa* Kjellman]
       Blidingia minima var. *subsalsa*
    [*E. minima* Nägeli]
       Blidingia minima var. *minima*
    [*E. minima* var. *subsalsa* (Kjellman) Doty]
       Blidingia minima var. *subsalsa*
    **E. prolifera* (Müller) J. Agardh var.
       prolifera
    [*E. prolifera* var. *flexuosa* (Wolfen) Doty]
       E. flexuosa
    **E. torta* (Mertens) Reinbold
    E. tubulosa Kützing
       Included in *E. flexuosa* A. & H. (1)
    [*E. vexata* (Setch. & Gard.) Doty]
       Blidingia minima var. *vexata*

ENTOCLADIA

    E. codicola Setch. & Gard.
    E. viridis Reinke

GAYELLA

    [*G. constricta* Setch. & Gard.]
       Prasiola meridionalis

GOMONTIA

    "*G. polyrhiza*" (Lagerheim) Bornet &
    Flahualt
       Probably a stage of *Monostroma*, *Ulva*
       or other genera

HALICYSTIS

    [*H. ovalis* (Lyngbye) Areschoug]
       The haploid stage of *Derbesia marina*

HORMIDIUM

    H. rivulare Kützing
       A freshwater species reported by Doty,
       1947, in Winchester Bay

LOLA

    L. lubrica (Setch. & Gard.) A. & G. Hamel

MONOSTROMA

    M. fuscum (Postels & Ruprecht) Wittrock
    f. *fuscum*
    **M. grevillei* (Thuret) Wittrock
       [*Enteromorpha grevillei*]
    M. oxyspermum (Kützing) Doty

       [*Ulva oxyspermum*]
       [*Ulvaria oxyspermum*]
    M. zostericola Tilden

PERCURSARIA

    **P. dawsonii* Hollenberg & Abbott
    P. percursa (C. Agardh) Rosenvinge

PRASINOCLADUS

    P. ascus Proskauer
    P. lubricus Kuckuck
       Included in *P. marinus*, A. & H. (1)
    P. marinus (Cienkowski) Waern
       [*Chlorangium marinum*]
       Includes *P. lubricus*, A. & H. (1)

PRASIOLA

    P. meridionalis Setch. & Gard.
       [*Gayella constricta*]
       [*Rosenvingiella constricta*]

PSEUDODICTYON

    P. geniculatum Gardner

PSEUDULVELLA

    **P. applanata* Setch. & Gard.
    P. consociata Setch. & Gard.

RHIZOCLONIUM

    R. implexum (Dillwyn) Kützing
    R. riparium (Roth) Harvey
       Includes [*R. tortuosum*],A. & H. (1)
    [*R. tortuosum* (Dillwyn) Kützing]
       Chaetomorpha tortuosa
       Included in *R. riparium*, A. & H. (1)

ROSENVINGIELLA

    [*R. constricta* (Setch. & Gard.) Silva]
       Prasiola meridionalis

SPONGOMORPHA

    S. arcta (Dillwyn) Kützing
    S. coalita (Ruprecht) Collins
    S. mertensii (Ruprecht) Setch. & Gard.
    S. saxatilis (Ruprecht) Collins
    S. spinescens Kützing

TRENTEPOHLIA

    T. sp. (a fog belt aerophilous alga)

ULOTHRIX

    U. flacca (Dillwyn) Thuret *in* Le Jolis
    [*U. implexa* Kützing]
       U. pseudoflacca

ULOTHRIX (continued)

    U. laetevirens (Kützing) Collins
    U. pseudoflacca Wille
        [*U. implexa*] *in* Smith, 1944

ULVA

    U. angusta Setch. & Gard.
        [*Enteromorpha angusta*]
    U. californica Wille
    U. expansa (Setchell) Setch. & Gard.
    U. fenestrata Postels & Ruprecht
    U. lactuca L.
    [*U. linza* L.]
        Enteromorpha linza
    U. lobata (Kützing) Setch. & Gard.
    [*U. oxyspermum* Kützing]
        Monostroma oxyspermum
    U. rigida C. Agardh
    U. stenophylla Setch. & Gard.
    U. taeniata (Setchell) Setch. & Gard.
    [*U. vexata* Setch. & Gard.]
        Blidingia minima var. *vexata*

ULVARIA

    [*U. oxyspermum* Kützing]
        Monostroma oxyspermum

ULVELLA

    *U. setchellii Dangeard

UROSPORA

    U. doliifera (Setch. & Gard.) Doty
    U. grandis Kylin
    [*U. mirabilis* Areschoug var. *mirabilis*]
        U. penicilliformis
    U. penicilliformis (Roth) Areschoug
        [*U. mirabilis* var. *mirabilis*]
    U. wormskjoldii (Mertens) Rosenvinge

PHAEOPHYTA

AGARUM

    *A. fimbriatum Harvey

ALARIA

    A. marginata Postels & Ruprecht
    A. nana Schrader

ANALIPUS

    A. japonicus (Harvey) Wynne
        [*Heterochordaria abietina*]

COILODESME

    C. bulligera Strömfelt

C. californica (Ruprecht) Kjellman

COLPOMENIA

    C. bullosa (Saunders) Yamada
        [*Scytosiphon bullosus*]
    C. peregrina (Sauvageau) Hamel
        [*Colpomenia sinuosa*]*sensu* Smith, 1944
    C. tuberculata Saunders, PP2752

COMPSONEMA

    C. intricatum Setch. & Gard., J. F. Chad-
    wick 789
    [*C. secundum* Setch. & Gard.]
        Hecatonema streblonematoides
    C. sessile Setch. & Gard.

COSTARIA

    C. costata (Turner) Saunders
    C. mertensii J. Agardh

CYLINDROCARPUS

    C. rugosus Okamura
        [*Petreospongium rugosum*]

CYSTOSEIRA

    C. geminata C. Agardh
        Reports are probably based on early
        occurrences of *Sargassum muticum*.
    C. osmundacea (Turner) C. Agardh

DESMARESTIA

    [*D. herbacea* Lamouroux]
        D. ligulata var. *ligulata*
    D. intermedia Postels & Ruprecht
    *D. kurilensis Yamada
    D. latifrons Kützing
    D. ligulata (Lightfoot) Lamouroux var.
    ligulata
        [*D. herbacea*]
        [*D. munda*]
    [*D. munda* Setch. & Gard.]
        D. ligulata var. *ligulata*
    *D. viridis (Müller) Lamouroux

DICTYONEUROPSIS

    D. reticulata (Saunders) Smith

DICTYONEURUM

    D. californicum Ruprecht

DICTYOSIPHON

    D. chordaria Areschoug *in* Saunders

DICTYOTA

    *D. binghamiae J. Agardh

DICTYOTA (continued)

D. *flabellata* (Collins) Setch. & Gard.

ECTOCARPUS

E. *acutus* Setch. & Gard. var. *acutus*
E. *acutus* var. *haplogloiae* Doty
E. *confervoides* (Roth) Le Jolis f.
    confervoides
E. *confervoides* var. *pygmaeus* (Areschoug)
  Kjellman
    Included in *E. parvus*, A. & H. (1)
E. *corticulatus* Saunders
    [*E. corticulans*] *in* Smith, 1944
[E. *cylindricus* Saunders]
    Feldmannia cylindrica
E. *dimorphus* Silva
    Included in *E. parvus*, A. & H. (1)
[E. *granulosus* (J. E. Smith) C. Agardh]
    Giffordia granulosa
E. *mucronatus* Saunders
    Included in *Feldmannia irregularis*,
    A. & H. (1)
[E. *oviger* Harvey]
    Giffordia oviger
    Included in *Giffordia granulosa*,
    A. & H. (1)
*E. *parvus* (Saunders) Hollenberg
    Includes *E. confervoides* var.
    pygmaeus & *E. dimorphus*, A. & H.(1)
E. *pygmaeus* Areschoug *apud* Kjellman
E. *siliculosus* (Dillwyn) Lyngbye
E. *simulans* Setch. & Gard.
    Includes *E. terminalis*, which
    occurs in Oregon, A. & H. (1)
E. *terminalis* Kützing
    Included in *E. simulans*, A. & H.(1)
E. *tomentosus* (Hudson) Lyngbye
E. *variabilis* (Saunders) G. M. Smith

EGREGIA

E. *menziesii* (Turner) Areschoug ssp.
    menziesii

EISENIA

*E. *arborea* Areschoug

ELACHISTA

E. *fucicola* (Velley) Areschoug

EUDESME

E. *virescens* J. Agardh

FELDMANNIA

F. *cylindrica* (Saunders) Hollenberg &
  Abbott
    [*Ectocarpus cylindricus*]
*F. *irregularis* (Kützing) Hamel

Includes *Ectocarpus mucronatus*,
found in Oregon, A. & H. (1)

FUCUS

F. *distichus* ssp. *edentatus*(De La Pylaie)
  Powell
    [*F. furcatus*]
    [*F. gardneri*]
F. *distichus* ssp. *edentatus* f. *abbrevia-*
tus (Gardner) Hollenberg & Abbott
    F. furcatus f. *abbreviatus*
F. *evanescens* C. Agardh f. *evanescens*
F. *evanescens* f. *oregonensis* Gard.
F. *evanescens* f. *robustus* Setch. & Gard.
[F. *furcatus* C. Agardh]
    F. distichus ssp. *edentatus*
[F. *furcatus* f. *abbreviatus* Gard.]
    F. distichus ssp. *edentatus* f.
    abbreviatus
[F. *gardneri* Silva]
    F. distichus ssp. *edentatus*

GIFFORDIA

G. *granulosa* (J. E. Smith) Hamel
    [*Ectocarpus granulosus*]
    Includes *G. oviger*, A. & H., (1)
G. *granulosoides* Setch. & Gard.
    Included in *G. sandriana*,A.& H.,(1)
G. *oviger* (Harvey) Hollenberg & Abbott
    [*Ectocarpus oviger*]
    Included in *G. granulosa*,A.& H.,(1)
*G. *sandriana* (Zanardini) Hamel
    Includes *G. granulosoides*, found in
    Oregon, A. & H. (1)

HAPLOGLOIA

H. *andersonii* (Farlow) Levering

HECATONEMA

[H. *variabile* Setch. & Gard.]
    H. streblonematoides
H. *primarium* (Setch. & Gard.) Loiseaux
    [*Myrionema foecundum* f. *simplicissi-*
    mum]
    [*Myrionema primarium*]
H. *streblonematoides* (Setch. & Gard.)
  Loiseaux
    [*Compsonema secundum*]
    [*H. variabile*]

HEDOPHYLLUM

H. *sessile* (C. Agardh) Setchell *in*
  Collins, Holden & Setchell
H. *subsessile* (Areschoug) Setchell *in*
  Collins, Holden & Setchell

HETEROCHORDARIA

[H. *abietina* (Ruprecht) Setch. & Gard.]
    Analipus japonicus

ILEA

[*I. fascia* (Müller) Fries]
    Petalonia fascia

LAMINARIA

[*L. andersonii* Eaton ex. Farlow]
    L. dentigera
L. cuneifolia J. Agardh f. *cuneifolia*
L. dentigera Kjellman
    [*L. andersonii*]
    [*L. setchellii*]
L. ephemera Setchell
L. farlowii Setchell
L. saccharina Lamouroux f. *saccharina*
L. saccharina f. *membranacea* J. Agardh
[*L. setchellii* Silva]
    L. dentigera
L. sinclairii (Harvey) Farlow, Anderson
  & Eaton

LEATHESIA

L. difformis (L.) Areschoug
L. nana Setch. & Gard.

LESSONIOPSIS

L. littoralis (Farlow & Setchell) Reinke

MACROCYSTIS

M. integrifolia Bory
? *M. pyrifera* (L.) C. Agardh
    H. B. S. Womersly states that re-
    ports resulted from a misunderstand-
    ing of specific characters.

MELANOSIPHON

M. intestinalis (Saunders) Wynne
    [*Myelophycus intestinalis*]
M. intestinalis f. *tenuis* Setch. & Gard.
    [*Myelophycus intestinalis* f. *tenuis*]

MYELOPHYCUS

[*M. intestinalis* Saunders]
    Melanosiphon intestinalis
[*M. intestinalis* f. *tenuis* Setch.& Gard.]
    Melanosiphon intestinalis f. *tenuis*

MYRIONEMA

M. corunnae var. *sterile* Setch. & Gard.
[*M. foecundum* f. *simplicissimum* Setch. &
  Gard.]
    Hecatonema primarium
M. globosum f. *affine* Setch. & Gard.
M. phyllophilum Setch. & Gard.
[*M. primarium* Setch. & Gard.]
    Hecatonema primarium
M. strangulans Greville

NEREOCYSTIS

N. luetkeana (Mertens) Postels & Ruprecht

PACHYDICTYON

P. coriaceum (Holmes) Okamura

PELVETIA

P. fastigiata (J. Agardh) De Toni
    All reports of this entity north of
    central California are based on col-
    lections of *Pelvetiopsis* made prior
    to Gardner's recognition of the genus.

PELVETIOPSIS

P. limitata (Setchell) Gardner f.
    limitata

PETALONIA

[*P. debilis* (C. Agardh) Derbès & Solier
    debilis]
    P. fascia
P. fascia (Müller) Kuntze
    [*Ilea fascia*]
    [*P. debilis* f. *debilis*]
    The various forms listed by Setchell
    & Gardner may represent a single
    variable species.

PETRODERMA

P. maculiforme (Wollny) Kuckuck

PETROSPONGIUM

[*P. regosum* (Okamura) Setch. & Gard.]
    Cylindrocarpus rugosus

PHAEOSTROPHION

P. irregulare Setch. & Gard.

PILAYELLA

P. gardneri Collins
P. littoralis (L.) Kjellman
P. tenella Setch. & Gard.
P. unilateralis Setch. & Gard.

PLEUROPHYCUS

P. gardneri Setchell & Saunders *in*
  Saunders

POSTELSIA

P. palmaeformis Ruprecht

PTERYGOPHORA

P. californica Ruprecht

PUNCTARIA

P. hesperia Setch. & Gard.

PYLAIELLA

see Pilayella

RALFSIA

R. fungiformis (Gunnerus) Setch. & Gard.
R. pacifica Hollenberg

SARGASSUM

S. muticum (Yendo) Fensholt

SCYTOSIPHON

[S. attenuatus (Foslie) Doty]
    S. dotyi
[S. bullosus Saunders]
    Colpomenia bullosa
[S. complanatus (Rosenvinge) Doty]
    S. dotyi
S. dotyi Wynne
    [S. attenuatus]
    [S. complanatus]
S. lomentaria J. Agardh

SORANTHERA

S. ulvoidea Postels & Ruprecht f.
ulvoidea

SPHACELARIA

S. didichotoma Saunders
S. furcigera Kützing
    Includes S. subfusca in A. & H.,
    (1)
S. racemosa Greville
[S. subfusca Setch. & Gard.]
    Included in S. furcigera in A. & H.,
    (1)

SPONGONEMA

*S. tomentosum (Hudson) Kützing

STICTYOSIPHON

S. tortilis (Ruprecht) Reinke

STREBLONEMA

S. aecidioides De Toni
    [S. aecidioides f. pacificum]
[S. aecidioides f. pacificum Setch. &
Gard.]
    S. aecidioides
S. evagatum Setch. & Gard., PP3262
*S. pacificum Saunders
S. myrionematoides Setch. & Gard.
S. vorax Setch. & Gard.

SYRINGODERMA

*S. abyssicola (Setch. & Gard.) Levring

RHODOPHYTA

ACROCHAETIUM

A. amphiroae (Drew) Papenfuss
*A. arcuatum (Drew) Tseng
*A. barbadense (Vickers) Børgesen
    [A. macounii]
*A. daviesii (Dillwyn) Nägeli
A. desmarestiae Kylin
[A. macounii (Collins) Hamel]
    A. barbadense
*A.microscopicum (Kützing) Nägeli
A. pacificum Kylin
*A. pectinatum (Kylin) Hamel
*A. porphyrae (Drew) Smith
*A. rhizoideum (Drew) Jao
A. subimmersum (Setch. & Gard.)
*A. vagum (Drew) Jao
A. variable (Drew) Smith

AEODES

[A. gardneri Kylin]
    Halymenia coccinea

AGARDHIELLA

[A. coulteri (Harvey) Setchell]
    Neoagardhiella baileyi
[A. tenera (J. Agardh) Schmitz]
    Neoagardhiella baileyi

AGLAOTHAMNION

A. endovagum (Setch. & Gard.) Abbott

AHNFELTIA

[A. concinna J. Agardh]
    A. gigartinoides
    Specimens identified as A.
    an Hawaiian species, are probably
    A. gigartinoides.
A. gigartinoides J. Agardh
    [A. concinna]
A. plicata (Hudson) Fries

AMPLISIPHONIA

A. pacifica Hollenberg

ANISOCLADELLA

A. pacifica Kylin

ANTITHAMNION

[A. baylesiae Gardner]
    Hollenbergia subulata

ANTITHAMNION (continued)

A. *defectum* Kylin
  [*A. pygmaeum*]
[A. *densiusculum* Gardner]
  Hollenbergia subulata
[A. *glanduliferum* Kylin]
  Antithamnionella glandulifera
A. *kylinii* Gardner
[A. *nigricans* Gardner]
  Hollenbergia nigricans
[A. *occidentale* Kylin]
  Scagelia occidentale
[A. *pacificum* (Harvey) Kylin]
  Antithamnionella pacifica var.
  pacifica
[A. *pygmaeum* Gardner]
  A. *defectum*
[A. *subulatum* (Harvey) J. Agardh]
  Hollenbergia subulata
A. *tenuissimum* Gardner
[A. *uncinatum* Gardner]
  Antithamnionella pacifica var.
  uncinata

ANTITHAMNIONELLA

A. *glandulifera* (Kylin) Wollaston
  [*Antithamnion glanduliferum*]
A. *pacifica* var. *pacifica* (Harvey)
  Wollaston
  [*Antithamnion pacificum*]
A. *pacifica* var. *uncinata* (Gardner)
  Wollaston
  [*Antithamnion uncinatum*]

ASTEROCOLAX

A. *gardneri* (Setch.) Feldmann & Feldmann
  [*Polycoryne gardneri*]

BANGIA

B. *fusco-purpurea* (Dillwyn) Lyngbye
  [*B. vermicularis*]
[B. *vermicularis* Harvey]
  B. *fusco-purpurea*

BAYLESIA

[B. *plumosa* Setchell]
  Schimmelmannia plumosa

BESA

*B. *stipitata* Hollenberg & Abbott

BONNEMAISONIA

[B. *californica* Buffham]
  B. *nootkana* (Esper) Silva
*B. *nootkana* (Esper) Silva
  [*B. californica*]
  [*Pikea nootkana* (Esper) Silva]
  Not *Pikea nootkana* (Esper) Doty

BOSSEA

see *Bossiella*

BOSSIELLA

B. *californica* (Decaisne) Silva ssp.
  californica
B. *californica* ssp. *schmittii* (Manza)
  Johansen
  [*Calliarthron schmittii*]
B. *chiloensis* (Decaisne) Johansen
  [*B. corymbifera*]
[B. *corymbifera* (Manza) Silva]
  B. *chiloensis*
[B. *dichotoma* (Manza) Silva]
  B. *orbigniana* ssp. *dichotoma*
[B. *frondescens* (Postels & Ruprecht)
  Dawson]
  Corallina frondescens
[B. *frondifera* Manza]
  B. *plumosa*
B. *orbigniana* (Manza) Silva ssp.
  orbigniana
B. *orbigniana* ssp. *dichotoma* (Manza)
  Johansen
  [*B. dichotoma*]
B. *plumosa* (Manza) Silva
  [*B. frondifera*]

BOTRYOCLADIA

B. *pseudodichotoma* (Farlow) Kylin,OSU 241

BOTRYOGLOSSUM

B. *farlowianum* (J. Agardh) De Toni var.
  farlowianum
B. *farlowiana* var. *anomalum* Hollenberg &
  Abbott
B. *ruprechtiana* (J. Agardh) De Toni
  [*Cryptopleura ruprechtiana*]

BRANCHIOGLOSSUM

*B. *woodii* (J. Agardh) Kylin

CALLIARTHRON

C. *cheilosporiodes* Manza
[C. *pinnulatum* Manza]
  Serraticardia macmillanii
[C. *regenerans* Manza]
  C. *tuberculosum*
[C. *Schmittii* Manza]
  Bossiella californica ssp. *schmitti*
[C. *setchelliae* Manza]
  C. *tuberculosum*
C. *tuberculosum* (Postels & Ruprecht)
  Dawson
  [*C. regenerans*]
  [*C. setchelliae*]

CALLITHAMNION

C. *acutum* Kylin
  [*C. californicum in* Smith, 1944]
*C. *biseriatum* Kylin
[*C. californicum* Gardner]
  C. *acutum*
[*C. laxum* Setch. & Gard. *in* Gardner]
  C. *pikeanum*
[*C. lejolisea* Farlow]
  Ptilothamnionopsis lejolisea
C. *pikeanum* Harvey
  [*C. laxum*]
  [*C. pikeanum* var. *laxum*]
  [*C. pikeanum* var. *pacificum*]
[*C. pikeanum* var. *laxum* (Setch. & Gard.)
  Doty]
  C. *pikeanum*
[*C. pikeanum* var. *pacificum* (Harvey) Setch.
  & Gard. *in* Gardner]
  C. *pikeanum*

CALLOCOLAX

*C. *fungiformis* Kylin
  [*C. globulosis*]
[*C. globulosis* Dawson]
  C. *fungiformis*

CALLOPHYLLIS

C. *crenulata* Setchell
C. *edentata* Kylin
*C. *firma* (Kylin) Norris
C. *flabellulata* Harvey
  [*C. marginifructa*]
C. *heanophylla* Setchell
  One collection reported by Doty, 1947
[*C. marginifructa* Setchell & Swezy *apud*
  Setchell]
  C. *flabellulata*
[*C. megalocarpa* Setchell & Swezy *apud*
  Setchell]
  C. *violacea*
C. *obtusifolia* J. Agardh, H.R. Hayden; 7,
  8a, 10
C. *pinnata* Setchell & Swezy *apud* Setchell
*C. *thompsonii* Setchell
C. *violacea* J. Agardh
  [*C. megalocarpa*]

CERAMIUM

C. *californicum* J. Agardh
C. *codicola* J. Agardh
C. *eatonianum* (Farlow) De Toni
C. *gardneri* Kylin
C. *pacificum* (Collins) Kylin
C. *washingtoniensis* Kylin
*C. *zacae* Setch. & Gard.

CHOREOCOLAX

C. *polysiphoniae* Reinsch

CLATHROMORPHUM

C. *parcum* (Setchell & Foslie) Adey
  [*Lithothamnion parcum*]
  [*Polyporolithon parcum*]

CONCHOCELIS

C. *rosea* Batters
  A filamentous stage of *Porphyra*

CONSTANTINEA

C. *simplex* Setchell

CORALLINA

[*C. chilensis* Decaisne *apud* Harvey]
  C. *officinalis* var. *chilensis*
[*C. densa* (Collins) Doty]
  C. *vancouveriensis*
C. *frondescens* Postels & Ruprecht
  [*Joculator delicatulus*]
  [*Bossiella frondescens*]
[*C. gracilis* var. *densa* Collins]
  C. *vancouveriensis*
C. *officinalis* var. *chilensis* (Harvey)
  Kützing
  [*C. chilensis*]
C. *vancouveriensis* Yendo
  [*C. densa*]
  [*C. gracilis* var. *densa*]

CRUORIA

*C. *profunda* Dawson
C. sp., PP3220

CRUORIOPSIS

*C. *aestuarii* Hollenberg

CRYPTONEMIA

C. *borealis* Kylin
C. *obovata* J. Agardh
C. *ovalifolia* Kylin

CRYPTOPLEURA

[*C. brevis* Gardner]
  C. *violacea*
C. *crispa* Kylin
C. *lobulifera* (J. Agardh) Kylin, OSU1493
[*C. ruprechtiana* (J. Agardh) Kylin]
  Botryoglossum ruprechtiana
C. *violacea* (J. Agardh) Kylin
  [*C. brevis*]

CRYPTOSIPHONIA

C. *woodii* J. Agardh

CUMAGLOIA

C. *andersonii* (Farlow) Setch. & Gard. *in* Gardner

CUMATHAMNION

*C. *sympodophyllum* Wynne & Daniels

DASYOPSIS

[D. *plumosa* (Harvey & Bailey) Schmitz]
    Rhodoptilum plumosum

DELESSERIA

D. *decipiens* J. Agardh

DERMATOLITHON

[D. *ascripticia* (Foslie) Setchell & Mason]
    Tenarea ascripticia
[D. *dispar* (Foslie) Foslie]
    Tenarea dispar

DERMOCORYNUS

*D. *occidentalis* Hollenberg

DILSEA

D. *californica* (J. Agardh) O. Kuntze

ENDOCLADIA

E. *muricata* (Postels & Ruprecht) J. Agardh

ERYTHROCLADIA

E. *irregularis* Rosenvinge
E. *subintegra* Rosenvinge

ERYTHROGLOSSUM

E. *californicum* J. Agardh
    Not reported since Doty, 1947

ERYTHROPHYLLUM

E. *delesserioides* J. Agardh
E. *splendens* Doty
    Not reported from Oregon since Doty,
    1947

ERYTHROTRICHIA

E. *carnea* (Dillwyn) J. Agardh
E. *pulvinata* Gardner
E. *welwitschii* (Ruprecht) Batters

FARLOWIA

*F. *compressa* J. Agardh
F. *conferta* (Setchell) Abbott

[*Leptocladia conferta*]
[*Pikea nootkana* (Esper) Doty]
    NOT *Pikea nootkana* (Esper) Silva
F. *mollis* (Harvey & Bailey) Farlow &
Setchell

FAUCHEA

*F. *fryeana* Setchell
F. *laciniata* J. Agardh

FAUCHEOCOLAX

*F. *attenuata* Setchell

FOSLIELLA

[F. *ascripticia* (Foslie) Smith]
    Tenarea ascripticia
[F. *dispar* (Foslie) Smith]
    Tenarea dispar

FRYEELLA

F. *gardneri* (Setchell) Kylin

GASTROCLONIUM

G. *coulteri* (Harvey) Kylin

GELIDIUM

[G. *caloglossoides* Howe]
    Pterocladia caloglossoides
[G. *cartilagineum* var. *robustum* Gardner]
    G. *robustum*
G. *coulteri* Harvey
*G. *crinale* (Turner) Lamouroux
[G. *pulchrum* Gardner]
    G. *purpurascens*
G. *purpurascens* Gardner
    [G. *pulchrum*]
G. *pusillum* (Stackhouse) Le Jolis
*G. *robustum* (Gardner) Hollenberg & Abbott
    [G. *cartilagineum* var. *robustum*]
G. *sinicola* Gardner
    see *Pterocladia media*

GIGARTINA

G. *agardhii* Setch. & Gard.
    Identification of numerous specimens
    from Oregon given this name has not
    been confirmed. The tetrasporangial
    stage has been reported as *Petrocelis
    franciscana*.
[G. *binghamiae* J. Agardh]
    G. *corymbifera*
[G. *californica* J. Agardh]
    G. *exasperata*
G. *canaliculata* Harvey
G. *corymbifera* (Kützing) J. Agardh
    [G. *binghamiae*]
G. *cristata* (Setchell) Setch. & Gard.

GIGARTINA (continued)

       Included in *G. papillata* A. & H. (1)
  G. exasperata Harvey & Bailey
     [*G. californica*]
*G. harveyana (Kützing) Setch. & Gard.
[*G. mamillosa* (Goodenouth & Woodward)
   J. Agardh]
     G. papillata
  G. papillata (C. Agardh) J. Agardh
     [*G. mamillosa*]
     Includes *G. cristata*, A. & H. (1)
*G. tepida Hollenberg
  G. volans (C. Agardh) J. Agardh *in* C.
  Agardh

GLOIOPELTIS

  G. furcata (Postels & Ruprecht) J. Agardh

GLOIOPHLOEA

  [*G. confusa* Setchell]
     Pseudogloiophloea confusa

GLOIOSIPHONIA

  [*G. californica* (Farlow) J. Agardh]
     G. capillaris
  G. capillaris (Hudson) Berkeley
     [*G. californica*]
  G. verticillaris Farlow

GONIMOPHYLLUM

*G. skottsbergii Setchell

GONIOTRICHOPSIS

*G. sublittoralis Smith

GONIOTRICHUM

*G. alsidii (Zanardini) Howe
     [*G. elegans*]
*G. cornu-cervi (Reinsch) Hauck
[*G. elegans* (Chauvin) Zanardini]
     G. alsidii

GRACILARIA

  G. sjoestedtii Kylin
     [*Gracilariopsis sjoestedtii*]
  G. verrucosa (Hudson) Papenfuss

GRACILARIOPHILA

  G. oryzoides Setchell & Wilson *apud* Wilson

GRACILARIOPSIS

  [*G. sjoestedtii* (Kylin) Dawson]
     Gracilaria sjoestedtii

GRATELOUPIA

  [*G. californica* Kylin]
     G. doryphora
  G. doryphora (Montagne) Howe
     [*G. californica*]
     [*G. maxima*]
  [*G. maxima* (Gardner) Kylin]
     G. doryphora
     Reported by Doty on three specimens
  G. setchellii Kylin

GRIFFITHSIA

  G. pacifica Kylin

GYMNOGONGRUS

  G. leptophyllus J. Agardh
  G. linearis (Turner) J. Agardh
  G. platyphyllus Gardner

HALOSACCION

  H. glandiforme (Gmelin) Ruprecht

HALYMENIA

*H. californica Smith & Hollenberg
  H. coccinea (Harvey) Abbott
     [*Aeodes gardneri*]
  H. schizymenioides Hollenberg & Abbott

HERPOSIPHONIA

  H. grandis Kylin
     Reported by Sanborn & Doty but not
     listed by Doty, 1947
  H. plumula (J. Agardh) Hollenberg var.
  plumula
     [*H. rigida*]
  H. plumula var. *parva* (Hollenberg)
  Hollenberg
     [*H. parva*]
     [*H. pygmaea*]
  [*H. parva* Hollenberg]
     H. plumula var. *parva*
  [*H. pygmaea* Hollenberg *apud* Smith]
     H. plumula var. *parva*
  [*H. rigida* Gardner]
     H. plumula var. *plumula*
*H. verticillata (Harvey) Kylin

HETEROSIPHONIA

*H. japonica Yendo

HILDENBRANDIA

*H. dawsonii (André) Hollenberg
  H. occidentalis Setchell *in* Gardner
  H. prototypus Nardo
     [*H. rosea*]
  [*H. rosea* Kützing]
     H. prototypus

HOLLENBERGIA

*H. nigricans (Gardner) Wollaston
    [Antithamnion nigricans]
H. subulata (Harvey) Wollaston
    [Antithamnion baylesiae]
    [Antithamnion densiusculum]
    [Antithamnion subulatum]

HOLMESIA

H. californica (Dawson) Dawson

HYDROLITHON

H. decipiens (Foslie) Adey
    [Lithophyllum decipiens]

HYMENENA

H. cuneifolia Doty
H. flabelligera (J. Agardh) Kylin
H. kylinii Gardner
H. multiloba (J. Agardh) Kylin
H. setchellii Gardner
H. smithii Kylin

IRIDAEA

I. cordata (Turner) Bory var. cordata
I. cordata var. splendens (Setch. & Gard.)
  Abbott
    [I. coriacea]
    [I. splendens]
    [Iridophycus fulgens]
    [Iridophycus oregona]
    [Iridophycus parvulum]
[I. coriacea (Setch. & Gard.) Scagel]
    I. cordata var. splendens
I. cornucopiae Postels & Ruprecht
    Probably includes specimens previously
    identified in Oregon as Rhodoglossum
    parvum
I. flaccida (Setch. & Gard.) Hollenberg
  & Abbott
I. heterocarpa Postels & Ruprecht
I. lineare (Setch. & Gard.) Kylin
I. punicea Postels & Ruprecht
    [I. whidbeyana]
*I. sanguinea (Setch. & Gard.) Hollenberg
  & Abbott
[I. splendens (Setch. & Gard.) Papenfuss]
    I. cordata var. splendens
[I. whidbeyana (Setch. & Gard.) Scagel]
    I. punicea

IRIDOPHYCUS

see Iridaea
[I. fulgens Setch. & Gard.]
    Iridaea cordata var. splendens
[I. oregona Doty]
    Iridaea cordata var. splendens
[I. parvulum (Kjellman) Setch. & Gard.

    Iridaea cordata var. splendens

JANCZEWSKIA

J. gardneri Setchell & Guernsey apud
  Setchell

JOCULATOR

[J. delicatulus Doty]
    Corallina frondescens

KALLYMENIA

*K. oblongifructa (Setchell) Setchell

KYLINIA

*K. arcuata (Drew) Kylin
? K. gynandra (Rosenvinge) Kylin, PP3132

LAURENCIA

L. spectabilis Postels & Ruprecht var.
  spectabilis

LEPTOCLADIA

[L. conferta Setchell]
    Farlowia conferta

LEPTOFAUCHEA

L. pacifica Dawson

LITHOPHYLLUM

[L. decipiens (Foslie) Foslie]
    Hydrolithon decipiens
? L. grumosum (Foslie) Foslie
*L. lichenare Mason
[L. neofarlowii Setchell & Mason]
    Pseudolithophyllum neofarlowii

LITHOTHAMNION

L. californicum Foslie
[L. conchatum Setchell & Foslie apud
  Foslie]
    Mesophyllum conchatum
[L. lamellatum Setchell & Foslie]
    Mesophyllum lamellatum
*L. microsporum (Foslie) Foslie
L. pacificum (Foslie) Foslie
[L. parcum Setchell & Foslie]
    Clathromorphum parcum
*L. phymatodeum Foslie
[L. reclinatum Foslie]
    Neopolyporolithon reclinatum

LITHOTHRIX

L. aspergillum Gray

LOMENTARIA

*L. hakodatensis Yendo

LOPHOSIPHONIA

[*L. villum* (J. Agardh) Setch & Gard.]
    Polysiphonia scopulorum var. *villum*

MELOBESIA

M. marginata Setchell & Foslie *in* Foslie
M. mediocris (Foslie) Setchell & Mason

MEMBRANELLA

?**M. nitens* Hollenberg & Abbott
    One uncertain report north of Cali-
    fornia, on the Olympic peninsula

MEMBRANOPTERA

M. dimorpha Gardner
M. multiramosa Gardner
M. platyphylla (Setch. & Gard.) Kylin
M. weeksiae Setch. & Gard. *apud* Gardner

MESOPHYLLUM

M. conchatum (Setchell & Foslie) Adey
    [*Lithothamnion conchatum*]
    [*Polyporolithon conchatum*]
M. lamellatum (Setchell & Foslie) Adey
    [*Lithothamnion lamellatum*]

MICROCLADIA

M. borealis Ruprecht
M. coulteri Harvey

MYRIOGRAMME

*M. repens Hollenberg
M. spectabilis (Eaton) Kylin

NEMALION

N. helminthoides (Velley) Batters
    [*N. lubricum*]
[*N. lubricum* Duby]
    N. helminthoides

NEOAGARDHIELLA

N. baileyi (Kützing) Wynne & Taylor
    [*Agardhiella coulteri*]
    [*Agardhiella tenera*]

NEOPOLYPOROLITHON

*N. reclinatum (Foslie) Adey & Johansen
    [*Lithothamnion reclinatum*]
    [*Polyporolithon reclinatum*]

NEOPTILOTA

N. californica (Harvey) Kylin
    [*Ptilota californica*]
N. densa (C. Agardh) Kylin
    [*Ptilota densa*]
N. hypnoides (Harvey) Kylin
    [*Ptilota hypnoides*]

NIENBURGIA

N. andersoniana (J. Agardh) Kylin
    [*N. borealis*]
[*N. borealis* (Kylin) Kylin]
    N. andersoniana

NITOPHYLLUM

N. cincinnatum Abbott

ODONTHALIA

O. floccosa (Esper) Falkenberg
O. lyallii (Harvey) J. Agardh
O. oregona Doty
O. washingtoniensis Kylin

OPUNTIELLA

O. californica (Farlow) Kylin

OZOPHORA

*O. latifolia Abbott

PETROCELIS

P. franciscana Setch. & Gard. *apud* Gardner
    Possibly a sporangial stage in the
    life history of *Gigartina agardhii*

PETROGLOSSUM

P. pacificum Hollenberg

PEYSSONELLIA

P. meridionalis Hollenberg & Abbott
P. pacifica Kylin

PHYCODRYS

P. isabelliae R. Norris & Wynne
P. setchellii Skottsberg
    Not reported since Doty 1947

PIKEA

P. californica Harvey
[*P. nootkana* (Esper) Doty]
    Farlowia conferta
    NOT *P. nootkana* (Esper) Silva
[*P. nootkana* (Esper) Silva]
    Bonnemaisonia nootkana

PIKEA (continued)

  [*P. pinnata* Setchell *in* Collins, Holden &
     Setchell]
         P. robusta
  P. robusta Abbott
         [*P. pinnata*]

PLATYSIPHONIA

  *P. clevelandii (Farlow) Papenfuss
  *P. decumbens Wynne

PLATYTHAMNION

  P. heteromorphum (J. Agardh) J. Agardh
     Reported north of San Francisco only
     by Doty, 1947
  P. pectinatum Kylin
  P. reversum (Setch. & Gard.) Kylin
  P. villosum Kylin

PLEONOSPORIUM

  *P. squarrosum Kylin
  *P. squarrulosum (Harvey) Abbott
  *P. vancouverianum J. Agardh

PLOCAMIOCOLAX

  P. pulvinata Setchell

PLOCAMIUM

  P. cartilagineum (L.) Dixon
         [*P. coccineum* var. *pacificum*]
         [*P. pacificum*]
  [*P. coccineum* var. *pacificum* (Kylin) Dawson]
         P. cartilagineum
  P. oregonum Doty
  [*P. pacificum* Kylin]
         P. cartilagineum
  P. tenue Kylin
         Not reported since Doty, 1947
  P. violaceum Farlow

PLUMARIA

  see *Ptilota*

POLYCORYNE

  [*P. gardneri* Setchell]
     Asterocolax gardneri

POLYNEURA

  P. latissima (Harvey) Kylin

POLYPOROLITHON

  [*P. conchatum* (Setchell & Foslie) Mason]
     Mesophyllum conchatum
  [*P. parcum* (Setchell & Foslie) Mason]

     Clathromorphum parcum
  [*P. reclinatum* (Foslie) Mason]
     Neopolyporolithon reclinatum

POLYSIPHONIA

  [*P. californica* Harvey]
         P. paniculata
  [*P. collinsii* Hollenberg]
         P. hendryi var. *gardneri*
  *P. hendryi var. *deliquescens* (Hollenberg)
     Hollenberg
  P. hendryi var. *gardneri* (Kylin) Hollenberg
         [*P. collinsii*]
  P. hendryi Gardner var. *hendryi*
  P. mollis Hooker & Harvey
         [*P. snyderiae* var. *snyderiae*]
  P. pacifica Hollenberg var. *pacifica*
  *P. pacifica var. *determinata* Hollenberg
  *P. pacifica var. *distans* Hollenberg
  P. pacifica var. *delicatula* Hollenberg,
     PP2647
  *P. pacifica var. *disticha* Hollenberg
  *P. pacifica var. *gracilis* Hollenberg
  P. paniculata Montagne
         [*P. californica*]
         [*P. senticulosa*]
  P. scopulorum var. *villum* (J. Agardh)
     Hollenberg
         [*Lophosiphonia villum*]
  [*P. senticulosa* Harvey]
         P. paniculata
  [*P. snyderiae* Kylin var. *snyderiae*]
         P. mollis

PORPHYRA

  P. lanceolata (Setchell & Hus) G.M. Smith
  P. miniata (C. Agardh) C. Agardh
         [*P. miniata* var. *cuneiformis*]
  [*P. miniata* var. *cuneiformis* Setchell &
     Hus *apud* Hus]
         P. miniata
  [*P. naiadum* Anderson]
     Smithora naiadum
  P. nereocystis Anderson *in* Blankenship &
     Keeler
  P. occidentalis Setchell & Hus *apud* Hus
         [*P. variegata*]
  P. perforata J. Agardh
  P. schizophylla Hollenberg *apud* Smith &
     Hollenberg
  *P. smithii Hollenberg & Abbott
  P. thuretii Setchell & Dawson *apud* Dawson
  [*P. variegata* (Kjellman) Hus]
         P. occidentalis "carposporangial
         stage"

PORPHYRELLA

  P. gardneri Smith & Hollenberg

PRIONITIS

[P. andersonii Eaton in Farlow]
    P. lyallii
P. filiformis Kylin
P. lanceolata Harvey
P. linearis Kylin
P. lyallii Harvey
    [P. andersonii]

PSEUDOGLOIOPHLOEA

P. confusa (Setchell) Levering in
  Svedelius, OSU 1373
    [Gloiophloea confusa]

PSEUDOLITHOPHYLLUM

*P. neofarlowii (Setchell & Mason) Adey
    [Lithophyllum neofarlowii]

PTEROCHONDRIA

P. woodii (Harvey) Hollenberg var. woodii

PTEROCLADIA

P. caloglossoides (Howe) Dawson
    [Gelidium caloglossoides]
P. media Dawson
    Some specimens reported as Gelidium
    sinicola may belong here.

PTEROSIPHONIA

P. bipinnata (Postels & Ruprecht)
  Falkenberg
    [P. bipinnata var. robusta]
    [P. robusta]
[P. bipinnata var. robusta (Gardner) Doty]
    P. bipinnata
P. dendroidea (Montagne) Falkenberg
    [P. gracilis]
[P. gracilis Kylin]
    P. dendroidea
[P. robusta Gardner]
    P. bipinnata

PTILOTA

P. asplenioides (Turner) Doty
[P. californica Ruprecht ex Harvey]
    Neoptilota californica
[P. densa C. Agardh]
    Neoptilota densa
P. filicina (Farlow) J. Agardh
    [P. tenuis]
[P. hypnoides Harvey]
    Neoptilota hypnoides
P. pectinata (Gunner) Kjellman
    Reported by Sanborn & Doty but not
    Doty, 1947
[P. tenuis Kylin]
    P. filicina

PTILOTHAMNIONOPSIS

*P. lejolisea (Farlow) Dixon
    [Callithamnion lejolisea]

PUGETIA

P. fragilissima Kylin

RHODOCHORTON

R. purpureum (Lightfoot) Rosenvinge
    [R. rothii]
[R. rothii (Turton) Nägeli]
    R. purpureum

RHODODERMIS

[R. georgii (Batters) Collins]
    Rhodophysema georgii

RHODOGLOSSUM

R. affine (Harvey) Kylin
*R. californicum (Harvey) Kylin
[R. parvum Smith & Hollenberg]
    R. roseum
R. roseum (Kylin) Smith
    [R. parvum]

RHODOMELA

R. larix (Turner) C. Agardh

RHODOPHYSEMA

*R. elegans var. polystromatica (Batters)
  Dixon
R. georgii Batters
    [Rhododermis georgii]
    No known specimen supports this
    record of N.L. Gardner, "on eel-
    grass....Shore Acres."
*R. minus Hollenberg & Abbott

RHODOPTILUM

*R. plumosum (Harvey & Bailey) Kylin
    [Dasyopsis plumosa]

RHODYMENIA

R. californica Kylin var. californica
R. pacifica Kylin
R. palmata f. mollis Setch. & Gard.
R. pertusa (Postels & Ruprecht) J. Agardh
R. rhizoides Dawson

RHODYMENIOCOLAX

R. botryoides Setchell

SARCODIOTHECA

*S. furcata (Setch. & Gard.) Kylin

SCAGELIA

 S. occidentale (Kylin) Wollaston
  [*Antithamnion occidentale*]

SCHIMMELMANNIA

 S. plumosa (Setchell) Abbott, PP 2067
  [*Baylesia plumosa*]

SCHIZYMENIA

 S. epiphytica (Setchell & Lawson) Smith
  & Hollenberg
 S. pacifica (Kylin) Kylin

SERRATICARDIA

*S. macmillanii (Yendo) Silva
  [*Calliarthron pinnulatum*]

SMITHORA

 S. naiadum (Anderson) Hollenberg
  [*Porphyra naiadum*]

STENOGRAMME

 S. interrupta (C. Agardh) Montagne

TENAREA

 T. ascripticia (Foslie) Adey
  [*Dermatolithon ascripticia*]
  [*Fosliella ascripticia*]
 T. dispar (Foslie) Adey
  [*Dermatolithon dispar*]
  [*Fosliella dispar*]

TIFFANIELLA

*T. snyderiae (Farlow) Abbott

ZANARDINULA

 see *Prionitis*

ADDENDUM

A recent paper by James W. Markham and Julie L. Celestino (Syesis 9:253–266, 1976) reports the following additions to this list, all from Clatsop County, Oregon:

Bryopsis plumosa (Hudson) C. Agardh

[*Enteromorpha crinita* (Roth) J. Agardh]
  E. clathrata var. *crinita* (Roth)
  Hauck

E. torta (Mertens) Reinbold

Laminaria groenlandica Rosenvinge

Pilayella littoralis f. *rupincola*
 (Areschoug) Kjellman

Soranthera ulvoidea f. *difformis* Setch.
 & Gard.

Achrochaetium plumosum (Drew) Smith

A. porphyrae (Drew) Smith

Iridaea sanguinea (Setch. & Gard.) Hollenberg & Abbott

Kallymenia reniformis (Turner) J. Agardh

Kylinia arcuata (Drew) Kylin

Lithothamnion phymatodeum Foslie

Neoptilota asplenioides (Esper) Kylin

Polyneuropsis stolonifera Wynne, McBride
 & West

Polysiphonia hendryi var. *luxurians*
 (Hollenberg) Hollenberg

Polysiphonia pacifica var. *distans*
 Hollenberg

Polysiphonia urceolata (Dillwyn) Greville

Prionitis cornea (Okamura) Dawson

Rhodomela lycopodioides (L.) C. Agardh

Rhodophysemia elegans f. *polystromatica*
 (Batters) Dixon

REFERENCES

1. Abbott, I. A., and G. J. Hollenberg. 1976. Marine Algae of California. Stanford University Press, California. 827 pp., 701 figs.
2. Dawson, E. Y. 1965. Marine Algae in the Vicinity of Humboldt State College, Arcada, California. Humboldt State College Book Store. 76 pp.
3. Druehl, L. D. 1968. Taxonomy and distribution of northeast Pacific species of *Laminaria*. Can. J. Bot. 46:539–547.
4. Druehl, L. D. 1970. The pattern of Laminariales distribution in the northeast Pacific. Phycologia 9(3/4):237–247.
5. Gilbert, W. 1972. Surface Temperature and Salinity Observations at Pacific Northwest Shore Stations During 1972. Unnumbered Document, School of Oceanography, Oregon State University, Corvallis. Offset 18 pp.
6. Levring, T., H. A. Hoppe, and O. J. Schmid. 1969. Marine Algae. Cram, DeGruyter, and Co., Hamburg, Germany. 421 pp., 13 Tables, 118 Figs.
7. Scagel, R. F. 1957. An Annotated List of the Marine Algae of British Columbia and Northern Washington. Nat. Mus. Can. Bull. No. 152. 289 pp.
8. Scagel, R. F. 1963. Distribution of attached marine algae in relation to oceanographic conditions in the northeast Pacific. *In:* Dunbar, M. J. (ed.) Marine Distributions. University of Toronto Press. Royal Society of Canada Spec. Pub. No. 5:37–50.
9. Setchell, W. A. 1893. On the classification and geographical distribution of the Laminariaceae. Trans. Conn. Acad. 9:333–375.
10. Setchell, W. A. 1935. Geographic elements of the marine flora of the north Pacific Ocean. Amer. Nat. 69:560–577.

Growth of Pacific Northwest Marine Algae
in Semi-Closed Culture[1]

J. ROBERT WAALAND

Traditional methods of cultivating benthic seaweeds use artificial substrates in open marine waters. Such methods have been used successfully for growing *Porphyra* (24), *Eucheuma* (7, 34) and *Laminaria* and other kelps (6, 16, 37). The application of artificial substrate culture methods to seaweeds native to the Pacific Northwest has been the subject of recent research (45, 48), and refinements of such a culture method are discussed elsewhere in this volume by Mumford. Artificial substrate cultivation methods usually involve considerable labor, and seaweeds sited in open waters are subject to the vicissitudes of the environment. This chapter will discuss alternative enclosed or semi-closed culture methods that have been under investigation.

The first completely "closed" cultures of algae involved small species that were conveniently suited to laboratory study. Cultivation of microscopic algae on a large scale in closed systems has received much attention (5, 35, 36, 41, 43). Investigators have attempted cultivation of macroscopic marine algae in closed culture (20) or in enclosures situated in open waters (3, 19). The major difficulties encountered in large closed cultures are supplying an adequate growth medium to support a large biomass and preventing growth of unwanted species. In floating enclosures situated in natural waters, damage from storms, waves, grazing, vandalism, and

[1]This research was primarily supported by the Washington Sea Grant Program and Marine Colloids, Inc. I thank E.W. Duffield for his sustained and conscientious assistance with the experiments reported here and wish to acknowledge the indispensable support and cooperation of the Municipality of Metropolitan Seattle (METRO), the Seattle Parks Department, the Department of Natural Resources, Olympia, and the National Marine Fisheries Service Aquaculture Research Station, Manchester, Washington.

fouling by other algae and by animals are significant problems. Some
seaweed culture studies have been conducted in greenhouses equipped
for marine algal culture and in outdoor aquaria or lighted aquaria
(29). Seaweeds grown in such situations frequently become overgrown
by diatoms, other epiphytes, or competing algae. Such experiences
have convinced many phycologists that fouling and competition would
be a serious problem in large-scale cultures unless unialgal cultures
were used and seawater was sterilized or filtered. Some researchers
have exploited the growth of marine microalgae that dominated semi-
closed cultures (13). So it was encouraging when detached plants
of *Chondrus crispus* were grown for nearly a year in a marine greenhouse (27). Some attention to epiphyte and grazer control was necessary, but these experiments were so successful that they stimulated
further research on semi-closed culture of *C. crispus* and other benthic marine algae. Research showed the importance of sufficient
water motion, nutrients, and other factors to seaweed growth. It
also revealed the value of strain selection and vegetative propagation for large-scale cultivation of marine algae (38, 39, 40).
These successful results with *C. crispus,* an important source of
carrageenan, stimulated the research on semi-closed culture of
Pacific Northwest seaweeds which is described in this report.

The basic requirements of an algal aquaculture enterprise have
been listed by Neish (28). Of the many requirements, research on
Pacific Northwest species has been concerned with (1) the culture
environment required for seaweed growth and (2) growth rates and
other biological characteristics of selected species. Many have
been tested (see Table I), but the major emphasis has been on two
carrageenan producers, *Iridaea cordata* and *Gigartina exasperata.*
Since the culture environment that works for these species works
for many others, its features will be discussed before a discussion
of cultivation experiments testing the species-dependent properties
of *I. cordata* and *G. exasperata.*

CULTIVATION FACILITIES

Growth experiments have been conducted at two sites in Puget
Sound. The first experiments (1973 to 1975) were done at the
Manchester Aquaculture Research Station of the National Marine
Fisheries Service on Clam Bay, Kitsap County, Washington. Later
experiments (1976) were conducted at the West Point Seaweed Research
Station, Seattle, King County, Washington, just west of the sewage
treatment facility of the Municipality of Metropolitan Seattle (METRO)
Several tank designs have been used: 1400-liter polyester resin
coated or plastic lined plywood tanks with $1.2m^2$ surface area, 3500-
liter painted or plastic lined plywood tanks with $2.8m^2$ surface area,
and 73-liter plastic tanks with $0.26m^2$ surface area. Water depth in
the large tanks was 1.1m and in the small tanks 0.3m; for some experiments water depth in the large tanks was reduced to 0.45m. One of
the large tanks was fitted with a transparent plexiglass side (Fig. 1)
The plants and the sea water in the tanks were continuously moved by
aeration supplied through a perforated plastic pipe at the middle or

Table I. *Species and growth rates observed in semi-closed cultures at Manchester Aquaculture Research Station, Clam Bay, Washington, and West Point, Seattle, Washington.* *

Species	Growth Rate % fw/day
Iridaea cordata (Turner) Bory	9.5
Iridaea heterocarpa Postels & Ruprecht	8.1
Iridaea cornucopiae Postels & Ruprecht	8.7
Gigartina exasperata Harvey & Bailey	8.3
Gigartina papillata (C. Agardh) J. Agardh	--
Neoagardhiella baileyi (Kützing) Wynne & Taylor	6.0
Plocamium cartilagineum (Linnaeus) Dixon	4.0
Schizymenia pacifica (Kylin) Kylin	1.7
Callophyllis flabellulata Harvey	3.6
Farlowia mollis (Harvey & Bailey) Farlow & Setchell	6.7
Prionitis lanceolata (Harvey) Harvey	1.0
Palmaria palmata f. *mollis* (Setchell & Gardner) Guiry	5.9
Porphyra perforata J. Agardh	8.8

*Growth rates are the maximum rates observed during summer growing conditions over at least one- or two-week intervals. In some instances the rates have been observed over longer time intervals. In large tanks for periods of several months, average growth rates are usually about half the rates shown above; in such cases the growth is usually reported as salt-free dry matter produced per unit of time per lighted surface area.

one side of the tank bottom. The design of these tanks is such that it can be scaled up to larger ponds or raceways of similar design.

Natural seawater was pumped from a depth of 3–5m below mean low low water (MLLW) into a headbox for pressure regulation, then into the culture tanks. During these experiments (1974 to 1976) the salinity ranged from 28 to 32 °/oo. No filtration or other treatment of the seawater was attempted.

The seawater at both cultivation sites ranged from 7°C (late winter) to 14°C (mid- to late summer); daily variations in the tanks was usually ±1°C. Since laboratory experiments with *I. cordata* (44) and *G. exasperata* (45; Merrill and Waaland, in preparation) indicated a temperature increment of 3–5°C above ambient might stimulate growth significantly, such a trial was run using a $2.8m^2$ tank; however, plants grown at elevated temperatures grew no faster than the control plants maintained at ambient temperature.

Because nutrient availability could be growth limiting, an investigation of certain major nutrients in the tank cultures was performed. Measurements were made of available carbon, nitrogen, and phosphorus in the inflowing seawater and in the water leaving the culture tanks. The measurements were made with the algal

Figure 1. *Culture tank (2.8m²) surface area used for seaweed culti-*
vation studies, Puget Sound, Washington. This particular tank is
fitted with a plexiglass side so that the circulation of the sea-
weeds can be observed. The tank is supplied with running seawater;
aeration near the bottom center of the tank causes the water and
plants to circulate. Plants and water move up near the air pipe,
then toward the end walls of the tank where they move down and
finally return toward the air pipe.

density at or near the maximum biomass listed in Table II. Carbon
was determined by the pH and total-alkalinity method. Nitrate,
nitrite, ammonia, and phosphorus were analyzed by standard colori-
metric methods (42). Some of these latter measurements were performed in the Autoanalyzer Laboratory, Oceanography Department,
University of Washington. Carbon determinations showed that typical
seawater contained about 2mM/liter total carbon dioxide; of this
total, most (99%) is in the form of bicarbonate. No significant
changes in its concentration were detectable. Most of the remainder
(1%) is present as dissolved CO_2 of which about 25% was taken up
by the plants. Laboratory experiments (Hansen, personal communication) have shown that CO_2 is limiting photosynthesis in *I. cordata*
at such concentrations, but no attempt has been made to raise the
carbon dioxide content of the seawater in large tank cultures. With
Chondrus crispus, CO_2 supplementation over a 75-day experiment increased the average daily growth rate (assuming exponential growth
occurred) from 1.23%/day to 1.64%/day and resulted in 35% greater
yield compared to the control (38). Such CO_2 additions have not

Table II. *Seasonal biomass per surface area from 1976 to 1977,*
tests at West Point, Seattle, Washington. *

Month	% Summer Optimum Biomass	*Gigartina exasperata* Grams fw/m^2	*Iridaea cordata* Grams fw/m^2
Jan	12	576	480
Feb	25	1200	1000
Mar	50	2400	2000
Apr	70	3360	2800
May	95	4560	3800
Jun	100	4800	4000
Jul	100	4800	4000
Aug	80	3840	3200
Sep	70	3360	2800
Oct	35	1680	1400
Nov	16	768	640
Dec	10	480	400

*This table shows the schedule used for seasonal biomass adjust-
ment per lighted surface area. The summer optimum biomass density
resulting in the maximum production per surface area was determined
experimentally. The schedule above was developed by adjusting the
biomass in proportion to the total average solar radiation for each
month (see Fig. 4). While this schedule could be refined further,
its utility is demonstrated by the results shown in Fig. 5.

been used with mass cultures of *I. cordata* or *G. exasperata*.
       Nitrogen and phosphorus measurements on seawater supplying the
tanks are summarized in Table III; such values are typical for Puget
Sound seawater. Measurements on the water leaving tanks in which
I. cordata or *G. exasperata* were growing showed that significant
amounts of certain nutrients were taken up by the plants (Table III).
When seawater was supplied at less than two tank volumes per day,
nitrogen deficiency symptoms developed in these species; the
symptoms were loss of red and blue phycobiliprotein pigments and a
color shift to yellow-green. If prolonged for several days, such
nutrient shortages stop growth. Resumption of adequate seawater
flow (more than two tank volumes per day) or addition of supplementary nitrogen corrects deficiency symptoms and permits growth
to resume. Supplementary nutrients, above those supplied by the
seawater, did not increase growth; addition of excess nitrogen
resulted in richly pigmented plants.
       Since nitrate nitrogen was typically 95% of the total nitrogen
available in the seawater, nitrate uptake experiments were conducted
with *Gigartina exasperata*; these experiments were conducted during
seasons of active growth and of slower growth. During rapid growth
nearly all the nitrate is removed from the seawater (Fig. 2); during
less rapid growth and when seaweed biomass was lower as in the fall

Table III. *Nitrate and phosphate in inflowing and outflowing Puget Sound seawater; long-term measurements using Iridaea cordata.* *

Date	Seaweed density gfw/m²	gfw/m³	Nitrate (μM/ℓ) inflowing seawater	outflowing seawater	Phosphate (μM/ℓ) inflowing seawater	outflowing seawater
4–1	1640	1426	25.0	21.8	2.18	2.06
4–16	1390	1208	15.5	13.6	1.94	1.78
4–23	2140	1860	21.4	7.2	1.86	1.45
5–7	2940	2556	15.5	2.8	1.53	--
5–14	3490	3034	15.1	2.1	1.53	1.05
5–24	1930	1686	18.2	5.2	1.79	1.31
5–28	2520	2191	14.2	2.8	1.50	1.06
7–12	3110	2704	14.6	2.5	1.75	--
8–28	924	803	15.2	13.8	1.97	1.85
9–19	--	--	14.2	--	1.95	--
10–23	--	--	18.5	--	2.74	--
11–20	--	--	23.9	--	2.40	--

*All samples were taken near midday. These data are from the Manchester, Clam Bay, Kitsap County, Washington, site. Only nitrate values are shown here since nitrate accounted for 85 to 98% of the inorganic nitrogen. The greatest nitrate uptake occurred with high plant density and periods of rapid growth. The inflow samples were made in duplicate; the outflow values are based on single samples. These data show the generally high levels of nutrients present in Puget Sound waters. The flow rate was four tank volumes/day.

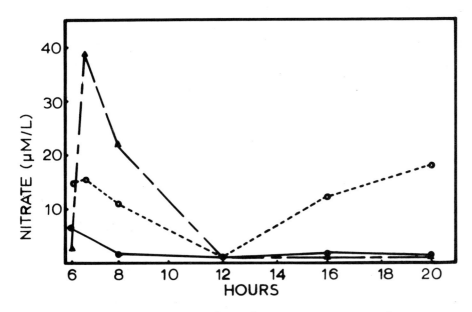

Figure 2. *Nitrate uptake by Gigartina exasperata strain M-11 at a density of 7200 gfw/m^2 (6550 gfw/m^3). Nitrate was measured (a) solid circles, in the incoming water and (b) open circles, in the water leaving a tank supplied with running seawater at four tank volumes per day. In the third experiment (c) solid triangles, nitrate was added at the start of the day so that the concentration was approximately double that of the ambient levels; the running seawater was turned off in this tank after the addition of nitrate. Samples for nitrate analysis were taken at the indicated intervals during daylight hours. This experiment was conducted July 21, 1976, at West Point, Seattle, Washington. It shows that G. exasperata removes substantial amounts of nitrate from sea water in a short time.*

or winter, less nitrate was removed by the plants (Table IV).

Puget Sound waters have an excellent supply of nutrients. In other marine waters, such as the North Atlantic, normal nutrient concentrations are much lower and supplementary fertilizer does result in increased growth (8, 38).

In preliminary experiments conducted in the spring of 1973, cultivation was attempted using cylindrical tanks with no agitation of water or plants. Little or no growth was observed and the plants deteriorated within a few weeks. In late winter of 1974, aerated, slant-bottomed culture tanks were used; the aeration kept plants and water in constant motion. Following inoculation with *I. cordata,* *G. exasperata,* and later other species, continuous sustained growth was observed for several years. It appears that water motion is critical for achieving sustained algal growth in tank culture (26). Water motion can be provided by aeration or paddles (38, 40) and

Table IV. *Nitrate removed from seawater flowing through tanks in which Gigartina exasperata strain M-11 were growing in summer and winter.* *

Season	Plant density gfw/m²	gfw/m³	Nitrate (µM/ℓ) inflowing seawater	outflowing seawater
Summer				
7-13 &				
7-14-76	5750	5000	18.2 ± 1.6	2.9 ± 0.6
n = 14				
Winter				
11-4 &				
11-5-76	1750	1520	24.0 ± 0	17.8 ± 2.5
n = 2				

*Measurements (averages ± standard error) were made on samples of water taken from tanks supplied with running seawater at a rate of 4 tank volumes/day at West Point, Seattle, Washington. In the summer duplicate samples were taken at 4-hr intervals for 24 hrs beginning at 0800 July 13, 1976, a day of bright sun with broken clouds; the difference between inflowing and outflowing nitrate concentration was nearly constant throughout the 24-hr period. The winter samples were taken at 1315 November 4, (cloudy, bright) and 0915 November 5, 1976 (gloomy, fog). Since the plant density must be lower in the winter to maintain positive growth, the amount of nutrient removed per tank surface area is less in winter than in summer.

probably simulates the water currents in natural waters. While slant-bottomed tanks were considered necessary for sustained circu-lation, experimentation revealed that circulation was adequate in flat-bottomed tanks since the moving water sweeps the plants around. Turbulent water motion near the plant surface is believed essential to replenish nutrients (25, 30).

Two attempts were made to grow, in semi-closed culture, *I. cordata* which had colonized an artificial substrate in natural waters; one apparatus which was tried at various angles and positions in the tanks is shown in Fig. 3. Typically, the plants in tanks grew for one or two weeks, then degenerated; control frames in natural waters continued to show good growth for a longer period. No satis-factory explanation for this lack of growth of substrate-affixed

Figure 3. *One type of frame used at Manchester test site, Clam Bay,
Washington, 1975 to hold nylon line that had been colonized by
Iridaea cordata at Minnesota Reef on San Juan Island in 1973. Such
frames were mounted at various angles and depths in the tanks. In
natural waters, a stiff spar buoy was used to keep the frame in a
horizontal position and at a constant depth.*

plants in tank culture has been developed. A growth test of
I. *cordata* which has colonized nylon line fastened to a circulating
substrate slightly denser than seawater is being conducted to see
if growth can be sustained.

Light is a major requirement for seaweed growth. In temperate
latitudes, the average daily solar radiation undergoes a sevenfold
seasonal fluctuation (Fig. 4). Except for shading to reduce intensity and small-scale experiments for special purposes, artificial
lighting is not practical on a large scale. Supplemental lighting
might be used for extension of day-length or some special purpose

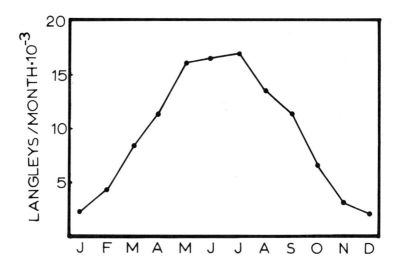

Figure 4. *This graph shows the average daily solar radiation for each month of the year measured at Seattle-Tacoma International Airport, Washington, by the National Weather Service. Data were averaged for 1974 to 1976. The midsummer daily solar radiation is seven times that of the midwinter daily solar radiation in the Puget Sound region.*

with relatively small volume cultures. Transparent-sided culture enclosures such as that in Fig. 1. might be used to provide relatively small volume (several m³) cultures with additional light to maintain stocks of inoculum in a minimum volume of water over winter. Preliminary experiments with supplemental lighting in winter to exten day-length have not resulted in significantly increased growth of *I. cordata* or *G. exasperata*. The average amount of light reaching each plant can be regulated by adjusting the biomass per lighted surface area and by regulating the depth or biomass per culture volume; this is discussed in greater detail later.

In these cultivation experiments, the major competitor algae have been diatoms especially in the spring and "weedy" algae such as *Giffordia, Ulva, Monostroma,* and *Enteromopha* especially in summer and fall. Since these competitors probably all have intrinsically higher growth rates than the desired species, some measures must be taken to prevent them from overgrowing the crop species. The major strategies employed in combating these competitors have been (1) depriving them of light by maintaining the cultivated species at a sufficiently high density per surface area that it absorbs most of the incident light and (2) maintaining the desired species at sufficiently high density per volume that it deprives the weedy species of nutrients in the tank.

With the smooth bladed *Iridaea cordata*, epiphytes only become a problem during slow growth or in the holdfast region which is the oldest, slowest-growing (or nongrowing) part of the plant. Epiphytes colonize the papillate surface of *G. exasperata* more readily than they do *I. cordata,* but if the *G. exasperata* is actively growing, this problem is minimized. Fouling is a more serious problem with most of the more branched species that have been tested, except *Neoagardhiella baileyi* and *Callophyllis flabellulata.* If fouling becomes a severe problem, there are several possible solutions. (1) If the aeration is stopped on a sunny day, the desired species, being denser than the fouling species, will sink, while the rapid photosynthesis of the fouling species produces oxygen bubbles and the fouling algae rise and can be skimmed off. (2) If the fouling algae are epiphytes, the most heavily fouled specimens or parts of specimens can be harvested and the density per surface area and volume adjusted in favor of the desired species. (3) If these treatments fail, a mild chlorine treatment can be used since the fouling algae are usually more susceptible to chlorine; calcium hypochlorite is added at 11.4 gm per kilogram fresh weight of carrageenophyte and allowed to flush out by normal seawater turnover.

RESULTS OF CULTIVATION EXPERIMENTS

Growth was measured by fresh weight determinations of the tank contents at approximately weekly intervals. For comparative purposes, relative growth rates were computed by assuming the long-term growth rate to be exponential (9). For growth comparisons of the carrageenan producers, *I. cordata* and *G. exasperata,* fresh weight measurements were converted to salt-free dry matter remaining after samples were dried, washed in fresh water, then re-dried before weighing. This gives an approximation of carrageenan production rate. For *I. cordata,* salt-free dry matter is 0.15 fresh weight; for *G. exasperata* it is 0.10 fresh weight.

The species tested for growth in semi-closed culture are listed in Table I. The emphasis in this research was on carrageenan producers, but other species were grown to test the broader applicability of this method and to search for species especially well suited to semi-closed culture. Table I also lists the maximum growth rates observed for different species. Most of the maximum rates listed were observed for one- or two-week periods; in many cases these rates have been sustained over many weeks. Long-term growth rate averages for many of the species tested are approximately half the rates listed in Table I. A few species, especially those from the intertidal such as *Iridaea cornucopiae* and *Gigartina papillata,* did not grow well in tank culture. Some species such as *Plocamium cartilagineum* and *Farlowia mollis* grew satisfactorily for many weeks, then became heavily fouled. In addition to *I. cordata* and *G. exasperata,* which grew rapidly in semi-closed culture, a few other species grew well in tank culture: *Palmaria palmata* (=*Rhodymenia palmata*), *Neoagardhiella baileyi* (= *Agardhiella tenera*), and *Callophyllis flabellulata,* all of which had rapid, consistent growth which continued even during the winter months. *C. flabellulata* has no known use. Both *P. palmata,* as dulse (12), and *N. baileyi,* sometimes used as ogo, are edible marketed seaweeds which might be grown on a small scale. Since most of the detailed growth data have been obtained with *I. cordata* and *G. exasperata,* these species are discussed in more detail below.

Because *Iridaea cordata* is widely distributed on the Pacific Coast and has a high content of excellent carrageenan, it has been the subject of numerous recent studies. There have been studies on its growth (44, 45, 47, 48), development (31), reproduction (21, 48), cytology (11), physiology, biochemistry (23, 46), ecology (4, 10, 14, 15, 17, 18), biogeography (21, 22), and taxonomy (1, 2, 21). Because it has somewhat less carrageenan than *I. cordata, Gigartina exasperata* has received less attention, but has been studied as an alternative source. There are recent studies on *G. exasperata's* growth (45, 48), development (31), reproduction, physiology (Merrill and Waaland, in preparation), biochemistry (46), biogeography (21, 22), and taxonomy (21). The majority of the following section deals with experiments with *I. cordata* and *G. exasperata.*

In the Puget Sound region, *I. cordata* typically occurs on rocky reefs and other solid substrates in the intertidal and subtidal with moderate to strong tidal currents. *Gigartina exasperata* occurs in similar habitats, but usually those with weaker tidal currents and greater depths. *I. cordata* in the sheltered, inland waters of the Northwest has a pronounced annual cycle: new blades are initiated from the encrusting holdfast in fall and winter, grow rapidly in spring and summer, mature and reproduce in late summer, and degenerate and disintegrate during the fall. The holdfast remains, overwinters, and produces new blades for the next growing season. This strong seasonal pattern observed in Northern Washington is not so pronounced in other areas (14) but has an important influence on the cultivation of this species in the Northwest. *G. exasperata* differs in that its large blades are present throughout the year. In tank culture, *I. cordata* loses its polarity as evidenced by the curvature of its young blades resulting in several blades radiating outward from the holdfast. In plants raised from spores and maintained in air-agitated culture, blades may be initiated from all sides of the approximately spherical holdfast. Because *G. exasperata* produces new blades from surface and edge proliferations which become detached from their parent blade when they reach a length of 10–15 cm, the loss of polarity is less obvious. There often is no central holdfast to mark the original focus of the radiating blades. In addition, *G. exasperata* blades tear more easily and become detached from their parent blade at an early stage and their original grouping is no longer evident. In culture tanks *I. cordata* frequently matures and produces spores 1 to 2 months earlier than plants in the natural population from which it was originally collected; no such observations of *G. exasperata* have been attempted.

Seasonal Solar Radiation and Optimum Biomass per Unit of Area

Examination of solar radiation data for the Puget Sound region reveals that there is more than seven times as much solar radiation per day in midsummer than in midwinter (Fig. 4); furthermore, the spring and summer growing season (from equinox to equinox) has 3.1

times more solar radiation than does the fall and winter period.
Data from the National Weather Service measured at the Seattle-
Tacoma International Airport and averaged for 1974 to 1976 show an
irradiance of 86,028 Langleys from March 21 to September 21 and
27,516 Langleys from September 21 to March 21. These data, plus
the observation that *I. cordata* died back to negligible biomass per
unit area and *G. exasperata* died back to about 500 gfw/m^2 and ceased
growing, led to the conclusion that the plant biomass per unit area
should be adjusted in proportion to the total solar radiation at
different seasons. The adjustment was figured by determining the
biomass with greatest productivity per unit area during midsummer,
then using lower densities at other times to keep plants growing in
semi-closed culture. This procedure was tested during the fall and
winter of 1976-77 using the biomass per area schedule shown in
Table II. Initial results have been especially encouraging with
G. exasperata; the results have been less conclusive with
I. cordata, although further experiments appear justified. A
further refinement of this density adjustment schedule may produce
even higher overall growth. An advantage of this procedure is that
it allows a prolonged harvest during the fall when failure to
harvest would result in no growth or even a weight loss (Fig. 5).
The top line shows there was a gradual decline in the weight of the
plants in the unharvested control tank which when harvested on day
35 yielded 196 gdw/m^2. The lower line on the graph shows how
periodic harvesting affected overall growth. By periodic density
reduction (harvesting), it was possible to keep the plants growing
actively. The major harvests on day 7 and day 35 and the minor
harvests in between resulted in a net harvest of 280 gdw/m^2 from
this tank, 42% greater than from the unharvested control tank. Thus,
periodic density reduction results in (1) keeping plants growing
actively and (2) increasing total yield. The major accomplishment
of this technique is that it becomes possible to keep substantial
quantities of a particular selected clone of seaweeds through the
winter.

Seasonal adjustment of plant density was based on the idea that
lack of light in winter was limiting growth. Because with high
biomass per area, respiration exceeded photosynthesis and resulted
in a net loss to the plant, a method for short-term measurement of
daily photosynthesis and respiration was sought. It was hoped that
such measurements could be used to develop improved optimum-biomass-
per-area schedules for different seasons. For these measurements,
small seaweed samples are placed in light and dark bottles filled
with enriched sea water ("ES" p. 43 in 41) which has been bubbled
with nitrogen gas to approximately 50% oxygen saturation; the
bottles are ballasted so that they circulate with the plants in the
tank and thus are exposed to the same conditions. Some bottles are
removed after a 4- to 6- hour light period and another set remains
in the tanks for 24 hours. The change in oxygen saturation is
measured using an oxygen electrode probe and gives a measure of
apparent photosynthesis. Table V shows the results of such measurements which indicate that net photosynthesis occurred if the plant

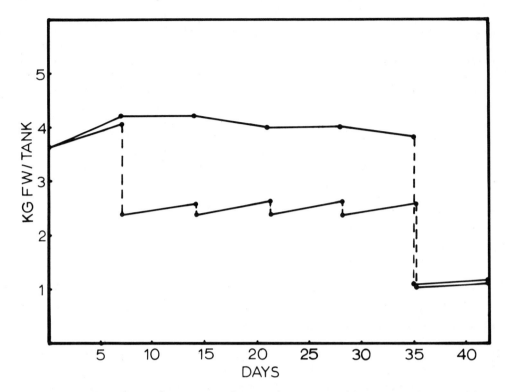

Figure 5. *Results of an experiment at West Point, Seattle, Washington, in which the density of Gigartina exasperata strain M-11 was adjusted weekly (lower line) to the values listed in Table II for October and November. The experiment began on October 13, 1976. The density was adjusted to the October value early in the experiment, the November level near the end. The upper line of the graph represents the control tank, which was not harvested until mid-November; the lower line shows the weight in the experimental tank which was harvested weekly.*

density was sufficiently low during the winter. These measurements also showed that the respiratory losses occurring during long winter nights used up a major portion of the daytime photosynthetic gains.

When plant density is lowered to the winter minimum listed in Table II, plants of the desired species may not use all of the available light and nutrients. This may stimulate the growth of competitor species. To shift the competitive balance in favor of the crop species, the water depth can be lowered to reduce the volume of water and nutrients available per unit of plant biomass.

Methods and Problems of Propagation

A major requirement for a sustained seaweed-aquaculture operation is the availability of a sufficient inoculum each spring at the start of the major growing season in temperate latitudes. This problem has been solved in the case of *Gigartina exasperata*. In this

species, one strain, designated M-11, has been maintained through
two winters. Two major factors make this possible: (1) this species
can be propagated vegetatively by means of cuttings from thalli or
outgrowths from thalli and (2) by adjusting the plant density, growth
can be sustained during the winter when solar radiation is at a minimum. The strain has never been observed to reproduce by means of
spores or gametes so no energy is expended on synthesis of such
reproductive bodies. This strain of *G. exasperata* was selected by
tagging individual plants from natural populations and growing them
in a culture tank under identical conditions. After 4-6 weekly
growth rate determinations it was possible to select fast growing
strains and to propagate them separately. Under the best conditions
the .M-11 strain increases it fresh weight at a rate of 4.25% per
day while wild type *G. exasperata* averages 3.11% per day (49).
Vegetative propagation combined with strain selection can produce
very favorable results and is an excellent propagation method.

Iridaea cordata has proved more difficult for long-term propagation in semi-closed culture. Complete new plants do not develop
from blade cuttings. If cuttings are made, they continue to grow,
mature, sporulate, and senesce as does an intact blade. It is
possible to raise plants from spores as described previously (48),
maintain the sporelings in air agitated cultures, then transfer them
to semi-closed culture tanks for subsequent growth. This method
requires care to retain the juvenile plants in the culture tank,
and it requires controlled-environment culture facilities, making it
more complex than the vegetative propagation method so successful
with *G. exasperata*.

Growing plants on circulating artificial substrates which provide a means of retaining the perennial holdfasts as a source of new
blades is now being tested on a small scale. After the large blades
have been harvested from such substrates, the holdfasts on their
substrates can be returned to a culture tank where they will sprout
new blades.

Another possibility which was successfully tested in 1976 is
to use field-sited artificial substrate cultures as a source of
inoculum at the start of each growing season. Such a method could
provide an inoculum of known characteristics determined from prior
harvests or from colonization of the nets by known strains of seaweeds. In the winter such artificial substrates might be placed in
shallower waters to promote faster growth by allowing better access
to light. Another version of the net-tank hybrid cultivation method
would be to harvest plants from artificial substrates, then give
them a "Neish effect" treatment in tank culture; in this treatment
growth with a limited nitrogen supply promotes carbohydrate synthesis.
In carrageenophytes this results in a 30-40% increase in carrageenan
content in a few weeks (39, 40).

A significant question in evaluating any species for aquaculture
concerns its long-term (annual) product yield. From the short-term
and long-term growth rates (Tables I and V) observed for *I. cordata*
and *G. exasperata*, the former appears to be the better choice since
it has a higher growth rate than *G. exasperata*, a higher carrageenan

Table V. *Photosynthetic rate monitoring experiments: Gigartina exasperata strain M-11.**

Date	Apparent Photosynthesis ($\mu\ell$ O_2/gdw/min)	PAR (Einsteins/m^2)	Dark Respiration ($\mu\ell$ O_2/gdw/min)
1-31-77			
Midday (6-hr)	36.9 ± 1.2	3.3	
Day & Night (24-hr)	10.0 ± 0.6	4.2	-5.0 ± 1.5
3-9-77			
Midday (6-hr)	8.24 ± 3.1	9.7	
Day & Night (24-hr)	14.6 ± 4.7	16.3	-6.0 ± 0.2

*The values are means ± standard error. The experiments commenced in midmorning of the day indicated. The temperature was 7.5°C on 1-31 and 7.0°C on 3-9. On 1-31 the plant density was 785 gfw/m^2 and on 3-9 it was 1625 gfw/m^2. Photosynthetically active radiation (PAR) was recorded using a Lambda 192S quantum sensor. These data show that this technique can be used to determine if net photosynthesis (on a 24-hr basis) is occurring in plants exposed to the conditions of a tank culture adjusted to an appropriate plant density per lighted surface area.

Table VI. *Production of Iridaea cordata, Gigartina exasperata, and some major terrestrial crop plants.*

Species	Annual Growth Rate[4] gsfdw/m^2/yr[5]	ton/ha/yr	kcal/m^2/yr	Mean Daily Growth Rate[1] gsfdw/m^2/day	kcal/m^2/day
Iridaea cordata					
Natural populations					
Pacific Northwest[2]	308	3.08	900	1.7	5.0
California[3]	1133	11.33	3971.	6.3	22.1
Semi-closed Culture					
1974	1948	19.48	9116	10.8	50.6
1976	2154	21.54	10800	11.9	60.0
Gigartina exasperata					
Semi-closed Culture					
1974 – Wild Type	935	9.35	4052	5.2	22.5
1976 – M-11 Strain	1614	16.14	6988	9.0	38.8
Wheat					
Netherlands	1335	13.35	4400	7.4	24.4
Corn					
United States	1281	12.81	4500	7.1	25.0
Rice					
Japan	1524	15.24	5500	8.5	30.6
Sugar Cane					
Hawaii	3297	32.97	12200	18.3	67.8

[1]Based on 180-day growing season for all crops except sugar cane for which a 365-day growing season was used.
[2](4, 10)
[3](14)
[4]Values for terrestrial crops (from Odum, 32); calorific values of seaweeds (from Paine, 33).
[5]gsfdw/m^2/yr = grams salt-free dry weight per year.

content, and a larger fraction of its fresh weight is dry matter (46).
Nevertheless, at the present time, it is more difficult to obtain a
large inoculum of *I. cordata* at the start of a growing season than
of *G. exasperata* which can be propagated vegetatively and overwinters
satisfactorily. A comparison of the annual yields of each species
is given in Table VI. Since *I. cordata* still outgrows the more
manageable *G. exasperata* on an annual basis, an appropriate short-
term strategy would be to begin pilot-scale cultivation of *G. exasperata* while developing improved propagation methods or vegetatively
propagatable strains of *I. cordata*. Table VI also compares the
annual yield of these seaweeds with major terrestrial crop plants and
shows that the productivity of marine crop plants compares favorably
with the productivity of the major terrestrial crops.

Sustained semi-closed cultivation of the carrageenophyte *Gigartina exasperata* appears feasible from a biological standpoint and
would be the best candidate for pilot-scale cultivation especially
if strains with higher growth rates and better yields are developed.
Because of its higher growth rate and carrageenan content, *Iridaea
cordata* merits further research on propagation and overwintering
methods so that sufficient inoculum can be provided at the start of
the spring and summer growing season. For a smaller scale cultivation operation where the product can be marketed directly, the edible
species *Palmaria palmata* (dulse) and *Neoagardhiella baileyi* (ogo)
are attractive prospects for semi-closed cultivation.

REFERENCES

1. Abbott, I.A. 1971. On the species of *Iridaea* (Rhodophyta)
        from the Pacific Coast of North America. Syesis 4:51-72.
2. Abbott, I.A. 1972. Field studies which evaluate criteria used
        in separating species of *Iridaea* (Rhodophyta). *In:* Abbott,
        I.A., and M. Kurogi (eds.) Contribution to the Systematics
        of Benthic Marine Algae of the North Pacific. Japanese
        Society of Phycology, Kobe, pp. 253-265.
3. Allen, J.H., A.C. Neish, D.R. Robson, and P.F. Shacklock. 1971.
        Experiments (1971) on Cultivation of Irish Moss (strain T4)
        in the Sea. Atlantic Regional Lab., Nat. Res. Counc. Can.,
        Halifax, Nova Scotia, Tech. Rept. No. 15, NRCC No. 12254.
        15 pp.
4. Austin, A., R. Adams, A. Jones, and K. Anders. 1973. Develop-
        ment of a Method for Surveying Red Algal Resources in
        Canadian Pacific Waters. Report to the Federal Minister of
        Fisheries and the Provincial Minister of Recreation Conserva-
        tion, University of Victoria, Victoria, British Columbia.
        172 pp.
5. Burlew, J.S. (ed.) 1953. Algal Culture from Laboratory to Pilo
        Plant. Carnegie Inst. Wash. Pub. No. 600. 357 pp.
6. Cheng, T.-H. 1959. Production of kelp -- a major aspect of
        China's exploitation of the sea. Econ. Bot. 23:215-236.
7. Doty, M.S. 1973. Farming the red seaweed, *Eucheuma,* for

carrageenans. Micronesica 9:59–73.
8. Edelstein, T., C.J. Bird, and J. McLachlan. 1976. Studies on
    Gracilaria 2. Growth under greenhouse conditions. Can. J.
    Bot. 54:2275–2290.
9. Evans, G.C. 1972. The Quantitative Analysis of Plant Growth.
    University of California Press, Berkeley. 734 pp.
10. Fralick, J. 1971. The Effect of Harvesting *Iridaea* on a
    Sublittoral Marine Plant Community in Northern Washington.
    M.A. Thesis, Western Washington State College, Bellingham,
    Washington. 60 pp.
11. Fralick, J., and K. Cole. 1973. Cytological observations on
    two species of *Iridaea* (Rhodophyceae, Gigartinales). Syesis
    6:271–272.
12. French, R.A. 1974. *Rhodymenia palmata*: An Appraisal of the
    Dulse Industry. Mimeographed report available from Prof.
    R.A. French, Dept. of Economics, Acadia University, Wolfville,
    Nova Scotia. 49 pp.
13. Goldman, J.C., J.H. Ryther, and L.D. Williams. 1975. Mass
    production of marine algae in outdoor cultures. Nature
    254:594–595.
14. Hansen, J.E., and W.T. Doyle. 1976. Ecology and natural
    history of *Iridaea cordata* (Rhodophyta; Gigartinaceae):
    population structure. J. Phycol. 12:273–278.
15. Hansen, J.E. 1977. Studies on the population dynamics of
    Iridaea cordata (Gigartinaceae, Rhodophyta). *In:* Proceed-
    ings, Eighth International Seaweed Symposium. Inter-Doc.
    Corp., Box 326, Harrison, New York 10528. (in press)
16. Hasegawa, Y. 1976. Progress of *Laminaria* cultivation in Japan.
    J. Fish. Res. Board. Can. 33:1002–1006.
17. Hruby, T. 1975. Seasonal changes in two algal populations from
    the coastal waters of Washington State. J. Ecol. 63:881–890.
18. Hruby, T. 1976. Observations of algal zonation resulting from
    competition. Estuarine Coastal Mar. Sci. 4:231–233.
19. Jones, W.E. 1959. Experiments on some effects of certain
    environmental factors on *Gracilaria verrucosa* (Hudson)
    Papenfuss. J. Mar. Biol. Assoc. U.K. 38:153–167.
20. Jones, W.E., and E.S. Dent. 1970. Culture of marine algae
    using a recirculating sea water system. Helgol. Wiss.
    Meeresunters. 20:70–78.
21. Kim, D.H. 1976. A study of the development of cystocarps and
    tetrasporangial sori in Gigartinaceae (Rhodophyta, Gigartin-
    ales). Nova Hedwigia 27:1–146.
22. Kim, D.H., and R.E. Norris. 1977. Possible pathways of world
    distribution of the Gigartinaceae. *In:* Proceedings,
    Eighth International Seaweed Symposium. Inter-Doc. Corp.,
    Box 326, Harrison, New York 10528. (in press)
23. McCandless, E.L., V.S. Craigie, and J.E. Hansen. 1975.
    Carrageenans of gametangial and tetrasporangial stages of
    Iridaea cordata (Gigartinaceae). Can. J. Bot. 53:2315–2318.
24. Miura, A. 1975. *Porphyra* cultivation in Japan. *In:* Tokida, J.,
    and H. Hirose (eds.) Advance of Phycology in Japan. W. Junk,

136 *J. ROBERT WAALAND*

The Hague, The Netherlands, pp. 273–304.

25. Munk, W.H., and A.G. Riley. 1952. Absorption of nutrients by aquatic plants. J. Mar. Res. 2:215–240.

26. Neish, A.C., and P.F. Shacklock. 1971. Greenhouse Experiments (1971) on the Propagation of Strain T4 of Irish Moss. Atlantic Regional Lab., Nat. Res. Counc. Canada, Halifax, Nova Scotia, Tech. Rep. No. 14, NRCC No. 12253. 25 pp.

27. Neish, A.C., and C.H. Fox. 1971. Greenhouse Experiments on the Vegetative Propagation of *Chondrus crispus* (Irish Moss). Atlantic Regional Lab., Nat. Res. Counc. Canada, Halifax, Nova Scotia, Tech. Rep. No. 12, NRCC No. 12034. 35 pp.

28. Neish, I.C. 1976. The role of mariculture in the Canadian seaweed industry. J. Fish. Res. Board Can. 33:1007–1014.

29. Neushul, M., and A.L. Dahl. 1967. Composition and growth of subtidal parvosilvosa from California kelp forests. Helgol. Wiss. Meeresunters. 15:480–488.

30. Neushul, M. 1972. Functional interpretation of benthic marine algal morphology. *In:* Abbott, I.A., and M. Kurogi (eds.) Contributions to the Systematics of Benthic Marine Algae of the North Pacific. Japanese Society of Phycology, Kobe, pp. 47–73.

31. Norris, R., and D.H. Kim. 1972. Development of thalli in some Gigartinaceae. *In:* Abbott, I.A., and M. Kurogi (eds.) Contributions to the Systematics of Benthic Marine Algae of the North Pacific. Japanese Society of Phycology, Kobe, pp. 265–280.

32. Odum, E.P. 1971. Fundamentals of Ecology. 2nd ed. W. B. Saunders Co., Philadelphia, Pennsylvania. 574 pp.

33. Paine, R.T., and R.L. Vadas. 1969. Calorific values of benthic marine algae and their postulated relationship to invertebrate food preferences. Mar. Biol. 4:79–86.

34. Parker, H.S. 1974. The culture of the red algal genus *Eucheuma* in the Philippines. Aquaculture 3:425–439.

35. Pringsheim, E.G. 1946. Pure Cultures of Algae, Their Preparation and Maintenance. The University Press, Cambridge, England. 119 pp.

36. Provasoli, L., J.J.A. McLaughlin, and M.R. Droop. 1957. The development of artificial media for marine algae. Arch. Mikrobiol. 25:392–428.

37. Saito, Y. 1975. *Undaria. In:* Tokida, J., and H. Hirose (eds. Advance of Phycology in Japan. W. Junk, The Hague, The Netherlands, pp. 304–320.

38. Shacklock, P.F., D. Robson, I. Forsyth, and A.C. Neish. 1973. Further Experiments (1972) on the Vegetative Propagation of *Chondrus crispus* T4. Atlantic Regional Lab., Nat. Res. Counc. Canada, Halifax, Nova Scotia, Tech. Rep. No. 18, NRCC No. 13113. 22 pp.

39. Shacklock, P.F., D. Robson, and F.J. Simpson. 1974. The Propagation of Irish Moss in Tanks. Comparative Growth at Different Locations. Atlantic Regional Lab., Nat. Res. Counc. Canada, Halifax, Nova Scotia, Tech. Rep. No. 19, NRCC No.

14143.   8 pp.

40. Shacklock, P.F., D.R. Robson, and F.J. Simpson.   1975.   Vege-
      tative Propagation of *Chondrus crispus* (Irish Moss) in Tanks.
      Atlantic Regional Lab., Nat. Res. Counc. Canada, Halifax
      Nova Scotia, Tech. Rept. No. 21, NRCC No. 14735.   27 pp.

41. Stein, J.R. (ed.)   1973.   Handbook of Phycological Methods:
      Culture Methods and Growth Measurements.   Cambridge Univers-
      ity Press, London.   448 pp.

42. Strickland, J.D.H., and T.R. Parsons.   1968.   A Practical Hand-
      book of Seawater Analysis.   Fish. Res. Board Can., Ottawa,
      Bull. 167.   311 pp.

43. Venkataraman, G.S.   1969.   The Cultivation of Algae.   Indian
      Council Agricultural Research, New Delhi.   319 pp.

44. Waaland, J.R.   1973.   Experimental studies on the marine algae
      Iridaea and *Gigartina*.   J. Exp. Mar. Biol. Ecol. 11:71-80.

45. Waaland, J.R.   1974.   Aquaculture potential of *Iridaea* and
      Gigartina in Pacific Northwest waters.   *In:*   Seaweed Farming
      in Puget Sound.   Proceedings of a Seminar, Poulsbo, Washing-
      ton, Washington State University, Pullman, pp. 1-17.

46. Waaland, J.R.   1975.   Differences in carrageenan in gametophytes
      and tetrasporophytes of red algae.   Phytochemistry 14:1359-
      1362.

47. Waaland, J.R.   1976.   Growth of the red alga *Iridaea cordata*
      (Turner) Bory in semi-closed culture. J. Exp. Mar. Biol.
      Ecol. 23:45-53.

48. Waaland, J.R.   1977.   Colonization and growth of populations
      of *Iridaea* and *Gigartina* on artificial substrates.   *In:*
      Proceedings, Eighth International Seaweed Symposium.   Inter-
      Doc. Corp.; Box 326, Harrison, New York 10528.   (in press)

49. Waaland, J.R.   (in press).   Growth and strain selection in the
      marine alga *Gigartina exasperata* (Florideophyceae).   *In:*
      Proceedings, Ninth International Seaweed Symposium, Santa
      Barbara, California, August 1977.

- 8 -
Growth of Pacific Northwest Marine Algae
on Artificial Substrates—Potential and Practice[1]

THOMAS F. MUMFORD, JR.

The practice of growing marine macroalgae on artificial substrates has been practiced for centuries. This technique has several advantages over growth on naturally occurring substrates:

1. Shallow waters where natural substrate is not suitable because of sandy or muddy bottoms can be utilized, or a limited substrate can be augmented.

2. Deep water and open ocean areas can be utilized where the benthos is below the photic zone.

3. Desired species or strains of species can be grown by artificially seeding the substrate.

4. More careful management of the algae is possible by moving the substrate to different locations at different times (i.e., removing the substrate from the water for disease and epiphyte control, and fertilization).

5. It is possible to obtain higher yields of biomass per given area than can be achieved from natural populations.

6. Harvesting is often simplified by manipulation of the substrate.

Disadvantages include:

1. High cost of the substrate and the support structures.

2. Placement and maintenance is often labor intensive in present systems, although careful design may reduce this problem.

3. The placement of structures usually precludes other use of the water surface.

---

[1]This work has been funded by the Washington State Department of Natural Resources. Beginning January 1, 1977, partial funding has been from a grant (R/A-12) from the University of Washington Sea Grant Program of the National Oceanic and Atmospheric Administration of the U.S. Department of Commerce.

This paper will briefly review two of the most highly developed systems for growth of macrophytic algae on artificial substrates for *Undaria* (wakame) and *Porphyra* (nori). A more detailed description of the growth of the carrageenophyte, *Iridaea cordata*, on artificial substrates being studied by the Washington State Department of Natural Resources follows.

These three systems, *Undaria*, *Porphyra*, and *Iridaea*, involve seeding net or rope substrate with spores. *Porphyra* and *Iridaea* spore germinate and grow to the mature, harvestable thallus. The spores of *Undaria* produce microscopic gametophytes, which in turn produce the large sporophyte *in situ*. Immature thalli or fragments can be attached to a substrate where they grow to maturity. *Eucheuma* spp. has been succesfully grown by the latter method (7) and is described in detail by Doty in Chapter 11 of this volume. *Macrocystis* has been grown by placing small plants on various substrates where they attach themselves by growth of the holdfast. This method is discussed by Neushul, Wilson, and North in Chapters 9, 10, and 12 of this volume.

CURRENT CULTURE SYSTEMS

Undaria

Undaria of the order Laminariales in the division Phaeophyta is a kelp grown for food in Japan where it is known as "Wakame." Of the three species cultivated, *U. pinnatifida*, *U. undarioides*, and *U. peterseniana*, only the first is used extensively. Recent reviews in English discuss its use and cultivation and are the source of much of the following information (6, 13, 25, 26, 27).

As with most members of the Laminariales, the life history of *Undaria* is characterized by a large sporophyte which produces meiospores in specialized areas called sori. The motile spores swim for some time, then settle on a solid surface where they germinate to produce microscopic gametophytes. Motile spermatozoids produced from the male gametophyte fertilize the non-motile egg *in situ* on the female gametophyte. The resulting zygote develops into the large sporophyte. Understanding the life history of this alga and others is indispensable to the manipulation of the stages in their life histories for aquacultural purposes (26, 27).

The harvest of naturally occurring beds of *Undaria* still provide about 22% of the total harvest or about 25,000 metric tons/year. Suitable areas for growth are characterized by high currents, sandy bottoms with scattered rocks and boulders, and some freshwater influx, presumably to provide more nutrients. Such areas have been fully utilized for years, and methods for their enhancement have been gradually developed, such as weeding out competitors, creating new areas by setting out additional stones or cement blocks in natural beds, or blasting clean surfaces on existing rocks. Later, the technique of sinking sporophylls in suitable areas or tying sporophylls to rocks and sinking them was discovered and production was increased. This led to the specific cultivation of the gametophyte phase. Today, synthetic twine 2-3 mm in diameter is seeded in tanks with meiospores which give rise to the gametophytes. In 1973, 60

million meters of such twine were prepared. Typically the twine is
then wrapped around a larger rope or cut into 5-6 mm lengths which
are then inserted into the strands of the larger rope. These ropes
are then suspended from rafts or floats in long lines or long lines
with vertical sections to produce a hanging curtain. These systems
are anchored in water up to 50 m deep. One hundred fifteen thousand
tons were produced in 1973 with such systems. Production is 5-10 kg
wet weight per meter length, or about 80 tons/hectare (6).

Laminaria and Related Genera

The name "konbu" (kombu) in Japan denotes four species of *Laminaria, Kjellmanniella gyrata, Arthrothamnus bifidus*, and *Cymathere
triplicata*. In China, the term "Haidai" is used for *Laminaria japon-
icus*. These kelp have the same life history as *Undaria,* and many of
the same cultivation techniques are used for their production,
although most of the harvest is from wild populations. In China,
however, great effort has been made in the cultivation of "Haidai,"
and over 100,000 metric tons are produced annually (28). A vigorous
breeding program is underway to create new strains and hybrids with
higher production and iodine content and lower water content. X-ray
irradiation is used to produce mutants, which are then employed in breeding and strain selection in the field. Porous clay pots containing fertilizer are hung from the ropes to increase plant growth. Techniques for
raising the rope-attached plants from the water and immersing them in a
concentrated nutrient solution are being developed (3, 6).

Porphyra

The culture of "nori" (*Porphyra tenera* and closely related species)
is the largest and most sophisticated algal culture system in existence. In 1973, 10 billion sheets of nori were produced, worth about
$330 million. Production involved 50,000 families working a culture
area of 70,000 hectares. Culture techniques are described in a number
of English accounts (6, 13, 18, 20, 22, 25, 27, 29), although almost
all field manuals and literature is in Japanese (20). Current studies
in Japan are concerned with disease control, strain selection, food
chemistry, and net improvement. Harvesting and manufacturing techniques are also being studied with all results applied to production.
The culture of nori began in the 1600s, when brush and twigs
were stuck in estuarine mud flats. Through experience, the best
locations and times for a good set on the brush was discovered,
although the source of the spores was not known. More elaborate and
efficient structures such as bamboo slats and fiber nets replaced
brush as a substrate. But the success of the sets still depended
upon placing nets in the right place at the right time. In 1949,
Drew (8) discovered a microscopic, shell-inhabiting phase in the
life history of *Porphyra*, known as *Conchocelis*. Further research
quickly revealed that spores from the leafy thallus grow into a minute
alga previously known as *Conchocelis*. The *Conchocelis* stage is the
previously mysterious source of the spores (conchospores) that attach
to nets and produce the leafy thallus again. Japanese scientists

within years revolutionized the industry by controlling the production
of conchospores used to set the nets. Conditions for optimal conchocelis growth and spore formation and release were delimited (18, 24).
Today, nets are artificially seeded in the fall either by immersing the nets in a large tank containing a spore suspension, or by
hanging bags containing conchocelis-bearing shells from the nets to
be seeded. To spread production over a longer period of time in the
winter, some seeded nets, after a short growing period, are stored
at -20° C until needed. Traditionally, nets were suspended from stakes
driven into the bottom. This allows the nets to be out of water for
a portion of the tide cycle, which helps control disease and epiphytes
Suitable shallow water sites, however, are decreasing because of
pollution and land-fills. In order to meet increasing demands for
nori, methods for floating culture were developed and put into production around 1964. Nets are strung tight at the surface by
a float system (6, 21). In 1973, this method accounted for about
40% of the total number of nets placed.

The technology of net structures and anchoring systems has been
worked out to produce a system that is cost effective. Mechanical
harvesters have largely replaced hand picking, and the manufacture
of nori sheets is also mechanized.

The nori industry is a mixture of traditional methods, modern
technology, and science. Yields are now achievable at 35-105 kg wet
weight/18 m x 1.2 m net or 16.2-48.6 metric ton/hectare. Current pro
duction exceeds demand, but efforts are being made to increase produc
tion per unit area to grow a high quality product which will reduce
production costs (13, 27).

IRIDAEA CULTURE IN PUGET SOUND

Because of its high carrageenan content, the marine alga *Iridaea
cordata* (Turner) Bory[2] (Rhodophyta, Gigartinales) has been studied
repeatedly to clarify not only its basic biology and ecology, but its
potential as a cultivated species.

I. cordata is distributed from Baja, California, to the Aleutian
Islands, Alaska, on the west coast of North America, and occurs in
the low intertidal and shallow subtidal regions. Its ecology in
natural populations has been studied in California (11, 12), Washington (9, 14, 15, 16), and British Columbia (5). Details of its life
history in laboratory culture (17), thallus development (23), and
cytology (10) have been examined. Waaland has determined optimal
growth conditions (31, 32) and how it may be grown on artificial
substrates (35) and in semi-enclosed tank culture (34).

I. cordata thalli contain 52-66% carrageenan (dry weight). The
tetrasporophyte contains only lambda-carrageenan and the gametophyte
thalli kappa-carrageenan (19, 33). Because of their commercial value

---

[2]Abbott (1, 2) recognizes two forms: *I. cordata* var. *cordata*
and *I. cordata* var. *splendens*. Most plants involved with this study
seem to be var. *cordata*. Kim (17) has proposed a new combination:
Gigartina cordata and *Gigartina cordata* var. *splendens*.

Figure 1. *Thallus of Iridaea cordata, showing holdfast, bladelets arising from holdfast, and one large blade at top. Polypropylene netting, December 21, 1976.*

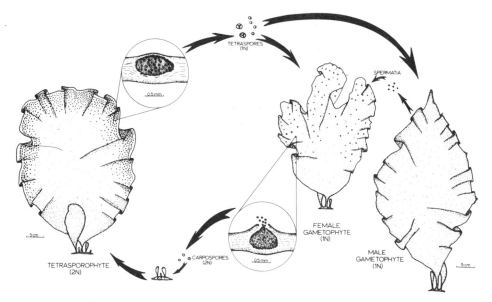

Figure 2. *Life history of Iridaea cordata.*

Iridaea populations in nature have been examined for harvesting
potential (5, 9) and were commercially harvested for a short time in
1970-1971 in the San Juan Islands, Washington, by the Lummi Indian
Tribe. These efforts failed because of low density of the beds and
high cost of hand-harvesting by scuba divers. The Washington State
Department of Natural Resources initiated a program at this time
under the direction of Dave Jamison and Ralph Beswick to develop a
commercially feasible aquaculture system for *I. cordata*. This report
is the first published account of this program.

An *I. cordata* thallus consists of a holdfast, which is a pad
of tissue of varying width and 1-2 mm thick which fastens the plant
to the substrate, and up to 25 bladelets less than 2 cm in length
which arise from the holdfast. In spring, one or two of these
bladelets begins to grow rapidly and by mid-summer reaches a length
of up to 1 m, although most blades are 30-50 cm long (Fig. 1). These
large blades become fertile, release their spores, and then decay and
fragment in the fall, leaving the holdfast and bladelets to over-
winter. Thalli (holdfasts and blades) are one of three types: tetrasporophytes, female gametophytes, or male gametophytes. How these
thalli are related in a life history is illustrated in Fig. 2. Non-
motile spermatia are produced on the male thalli and fertilize the tri-
chogyne on the female thallus. After a series of cytological events,
diploid carpospores are produced and released from the female thallus.
Carpospores germinate to give tetrasporophyte thalli, where tetraspore
are produced after meiosis and released. These spores germinate to
give the gametophyte thalli. Holdfasts and immature blades of gametophytes and tetrasporophytes are indistinguishable, except by chemical
analysis of the carrageenans, and occur together in natural populations, although the tetrasporic phase appears to be dominant (12).

When large blades are harvested during the summer, regrowth of the
cut blade or growth of a bladelet takes place. Only some parts of
this life history have been demonstrated conclusively (17). The
rest is hypothesized from other, more well-known, red algal life
histories.

What follows is a description of the successful seeding of nets
with tetraspores and carpospores of *I. cordata* and the yields and
growth rates of the alga on nets in various locations in Puget Sound
and the San Juan Islands, using different netting materials and mesh
size with two different harvesting times.

Materials and Methods

Three types of netting were used in experiments begun on September 10, 1974: white type 66 nylon seine netting made of No. 60 cord
with a 3" (7.6 cm) stretched mesh, black plastic "Vexar" (Dupont, Inc.)
with a 1" (2.54 cm) mesh, and orange 1/4" (0.64 cm) polypropylene
netting with a 6" (15.2 cm) stretched mesh.

The "Vexar" netting was suspensed by a thin stainless steel rod
run through both ends. The nylon and polypropylene nettings were
attached peripherally to a frame made of 1/2" (1.3 cm) diameter white
PVC pipe and fittings. The nets were anchored by cement blocks
(30.5 cm each side, weighing 68 kg) in which pieces of chain were
embedded. The finished structure is shown in Fig. 3.

The structures were placed at Minnesota Reef, San Juan Islands,
(Fig. 4) on September 10-11, 1974, over natural beds of *I. cordata*.
Because of the initial success of these nets, more nets were placed
in the fall of 1975 at Sucia, Barnes, and Sinclair Islands (Fig. 4).
These were single 1 m^2 PVC pipe frames with 3" nylon netting. Additional nets were placed at Minnesota Reef, 1 each of expanded and
collapsed nylon and polypropylene netting in 1 m^2 frames. The collapsed netting was used to put 2-3 times the amount of netting on a
single frame for seeding by collapsing the mesh in one direction.
At Reef Point, one frame of expanded and collapsed polypropylene and
one expanded nylon netting were placed using concrete anchors as
above. The above work was done by Mr. Clifford Kemp.

The nets at Barnes Island and Sinclair Island were removed on
March 25-26, 1976, and placed at Minnesota Reef. The net at Sucia
Island could not be located. On April 7-8, 1976, the nets at Minnesota Reef were cut into four pieces each and stretched on 0.5 m x
0.5 m plastic pipe frames (0.25 m^2 nets). Ten such frames were left
at Minnesota Reef and are shown *in situ* in Fig. 5. Seventeen frames
were packed in plastic bags and placed on ice for transport to Manchester (Fig. 4) where they were suspended under a raft until they
were placed in shallow subtidal waters on April 15. The Manchester
site is located at -2.5' (-0.76 m) below mean low water (MLW) in an
area behind a *Nereocystis* bed. *Iridaea* and *Gigartina exasperata*
occur in the immediate area. At low tide considerable silt is stirred
up by wave action, but it is washed away at other times. On April 29-
30, 1976, two of these nets were transported to Olympia on ice and
placed at Squaxin Reef (Fig. 4). This site is a wide reef located

Figure 3. *Three types of netting used in experiments at Minnesota Reef, San Juan Islands, Washington. Three-inch stretched nylon netting (nearest to camera), and 6" mesh polypropylene netting (middle) are placed on 4' x 4' frames made of 1/2" PVC plastic pipe. Plastic "Vexar" netting is shown at the back. Anchors are 1-ft. cubes of concrete. These nets were placed over a natural bed of Iridaea cordata and the spores from these plants attached and grew on the nylon and polypropylene netting. (Clifford Kemp shown)*

in the middle of a narrow channel with high currents. The reef top is hardpan and covered with *I. cordata* and other red algae. These nets were placed at -2.5' (0.76 m) below MLW.

The weight of the *Iridaea* on the nets was determined by first emptying the water from the frames, pulling off the kelp and ulvoid contaminants, if necessary, and then weighing the frame, net, and algae with a spring scale. The tare weight of 0.8 kg was used for the $0.25 m^2$ 6" mesh nets and 1.1 kg for the $0.25 m^2$ nylon nets.

Growth rate is calculated by the formula:

$$R \text{ (g salt-free dry matter day}^{-1} m^{-2}) = \frac{(W_2-W_1) \ (0.6)}{\Delta t}$$

Where: W_1= total weight in grams of $0.25 m^2$ nets and frames at beginning of period.
W_2= total weight in grams of $0.25 m^2$ nets and frames at end of period.
Δt= time period on days.
[Salt-free dry weight is 15% of wet weight of *Iridaea* (34)]

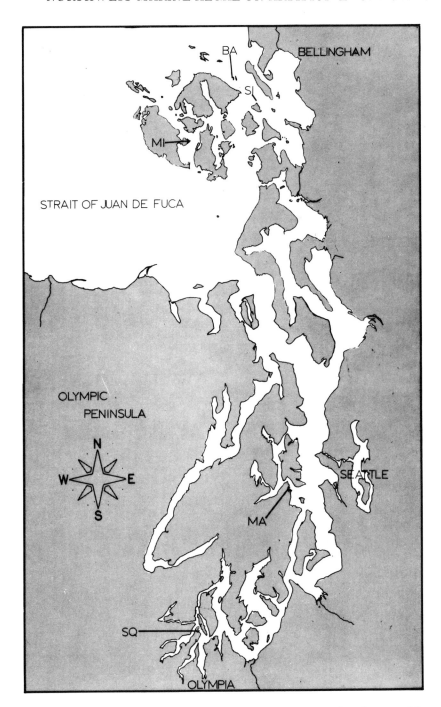

Figure 4. *Map of Puget Sound and the San Juan Islands, Washington, showing locations where nets were placed for the growth of Iridaea cordata. BA - Barnes I.; SI - Sinclair I.; MI - Minnesota Reef, San Juan I.; MA - Manchester; SQ - Squaxin Reef.*

Results

   Nets at Minnesota Reef, September 1974. By April 1975 it had
become evident that "Vexar" plastic netting was not suitable. It
was covered with the kelp *Pleurophycus gardneri* and the filamentous
green alga *Cladophora* sp. The "Vexar" net examined in February 1976
was covered with various kelp species (*P. gardneri*, *Laminaria saccharina*, and *Alaria marginata*) and was removed at that time. No *Iridaea*
was visible after two years in place.
   A good set of *Iridaea* was noted on the nylon nets in April 1975
and a moderate set on the polypropylene nets. These nets were monitored throughout the summer of 1975 and harvested on August 7. Each
net was divided into four 1-ft^2 quadrats. Two were selected at
random and harvested. The results are shown on Table I.

Table I. *Yield of Iridaea cordata from 4-ft^2 nets, August 7, 1975,
placed at Minnesota Reef, San Juan Island, Washington, September 1974.*

Net Type	Yield* g/ft^2	g/m^2
3" nylon (Net 2)	554.2	5970
" "	382.6	4120
6" poly (Net 3)	405.5	4365
" "	485.6	5227

   *Ave. yield (mean ± S.E.) 457.0 ± 39.2 g/ft^2, 4921 ± 423 g/m^2
   Data from Clifford Kemp

   These nets were observed through October 6, 1975, when the harvested areas had grown to almost the same length as those not harvested. Two harvests per year were seen to be possible and a projected
yield of 6-8 kg dry matter m^{-1} year^{-1} was estimated, although the
second harvest was not made. Work to this point was performed by
Clifford Kemp.
   Yields and Growth Rates from Nets, Summer, 1976. The netting
that was collapsed had become badly fouled with silt and no algae
were seen on it in February 1976. The "Vexar" netting was covered
with kelp but had no *Iridaea* on it.
   Growth of *I. cordata* on the expanded nylon and polypropylene
netting was excellent. Fig. 6 illustrates the polypropylene netting
and *Iridaea* as it appeared on May 12, 1976. Plants had already
reached a length of 35 cm. The harvest yields from selected nets in
1976 are given the Table II. These include only nets with 100% cover
of *Iridaea*, thus eliminating any comparisons with those with partial
cover. Highest yield was from the June harvest, somewhat less from
the July harvest, and much less from the October harvest.
   The standing drop of *Iridaea* on two similar nets (3" nylon,
Minnesota Reef) and the regrowth after harvesting is shown in Fig. 7.
Net 2A was harvested in mid-June before a maximum standing crop was
reached and regrowth was rapid during the rest of June. The second

Figure 6. *Frames (0.25m²) and 6" polypropylene netting with Iridaea cordata. May 12, 1976, (Nets 3A, 3B, see Table II) Minnesota Reef, San Juan Island, Washington.*

Figure 5. *Anchors, frames (0.25m²) and netting with Iridaea cordata in place at Minnesota Reef, San Juan Island, Washington, May 12, 1976, -2.0 ft below MLLW.*

Table II. *Yields of Iridaea cordata from 0.25m^2 nets, placed at Minnesota Reef, off San Juan Island, and Manchester and Squaxin Reefs, in Puget Sound, Washington.*

Net No.	Net Type	Location	Harvest Date			Yield (g/0.25m^2)
			12 Jun '76	26 Jul '76	18 Oct '76	
2A	3" nylon	Minn.[1]	646.3	--	274.1	920.4
2B	"	"	--	630.0	261.0	891.0
2C	"	Man.[2]	--	617.1	116.9	734.0
2D	"	"	350.8	--	61.6	412.4
7A	"	Minn.	--	276.9	166.6	443.5
13A	"	"	--	174.4	253.6	427.5
15A	"	Man.	--	278.1	160.9	439.0
15B	"	Squax.[3]	--	118.9	--	118.9
15C	"	Man.	--	245.2	230.7	475.9
3A	6" poly	Minn.	485.5	--	212.7	698.2
3B	"	"	--	617.1	189.7	866.8
3C	"	Man.	--	270.6	126.2	396.8
3D	"	"	290.0	--	207.1	498.0
Total harvest (g/0.25m^2)			1773.5	3228.3	2260.6	7322.4
Ave. yield			443.4	358.7	188.4	563.3
(g/0.25m^2+S.E.)			±78.9	±67.9	±18.6	±66.1

[1]Minnesota Reef, San Juan Island, Washington

[2]Manchester Reef, Puget Sound, Washington

[3]Squaxin Reef, Squaxin Passage, Puget Sound, Washington

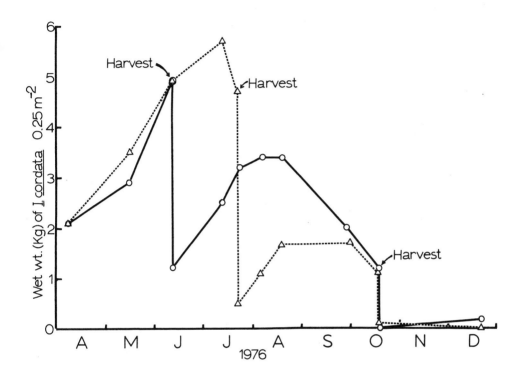

Figure 7. *Wet weight of Iridaea cordata per 0.25m² netting during summer, 1976. The 2 similar nets were located at Minnesota Reef, San Juan Island. Netting material was 3" mesh nylon. (open circles, Net 2A; open triangles, Net 2B).*

biomass peak was reached in mid-August. The second harvest was de-layed until mid-October when much of the standing crop had been lost through sloughing of blades. Net 2B was not harvested until late July when some loss had occurred. Regrowth was not as rapid as on Net 2A, although the biomass peaked in mid-August, the same time as that on the net harvested earlier.

If harvesting had been made at the times of peak standing crop, i.e., in mid-July for Net 2B and in mid-August for the final harvest for both nets, the theoretical yield of 1278 g dry matter/0.25m² (7.1 kg x 18%) for net 2A and 1242 g dry matter/0.25m² for net 2B (6.9 kg x 18%) would have been possible. This is 39% more than what was harvested.

Comparison of the yields of the two harvesting strategies (June-October vs. July-October) indicates that there is no significant difference between them (Table III).

Table III. *Yields of Iridaea cordata from nets placed at Minnesota Reef, San Juan Island, Washington, harvested at different dates in 197*

Harvest Times	g Air-Dried Matter/0.25m^2 $\pm$ S.E.
12 June and 18 Oct.	580.3 $\pm$ 142.0 (N=4)*
28 July and 18 Oct.	722.2 $\pm$ 113.8 (N=4)*

*Yields not significantly different, Student's t-test (t=-0.78)
[1]Nets harvested 12 Jun and 18 Oct: 2A, 2D, 3A, 3D; nets harvested 28 Jul and 18 Oct.: 2B, 2C, 3B, 3C.

In Table IV, yields from one harvest in August 1975 are compared with the combined harvest of the same nets in 1976. The much lower yield in 1976 may be explained by the cool, rainy August of 1976, although no solar irradiance or other environmental factors have been quantitatively compared. There may be a very high yearly variance in growth rates and standing crops of *Iridaea*, and its correlation to physical parameters should be examined.

Table IV. *Yields of Iridaea cordata on nets harvested in 1975 and 1976 grown on Minnesota Reef, San Juan Island, Washington.*

Year	Yield (g/m^2 $\pm$ S.E.)
1975[1]	4920.5 $\pm$ 422.8 (N=4)
1976[2]	3383.5 $\pm$ 203.5 (N=4)

[1]Nets 2 and 3 before cut up; Minnesota Reef, San Juan Island, Washington, harvested 7 Aug 75.
[2]Nets 2A, 2B, 3A, 3B; Minnesota Reef, total of two harvests; June-July and October, 1976.

The 1976 yields of nets seeded in 1974 is not significantly different from nets seeded in 1975 (Table V).
The yield from different mesh size nets is of great importance in relating the optimal yield to cost. Three-inch netting gives twice the linear area of a 6" mesh net but weighs twice as much. As netting is bought by the pound, not area, 6" mesh gives twice the area for the same cost as a 3" mesh net. Although not strictly comparable because of different materials, yields from the 3" mesh nets were not significantly different from 6" mesh (Table VI). This would suggest that crowding and stunting of growth on the 3" mesh net offsets the advantage of increased linear area available.
Three sites were chosen for comparison of growth rates: Minnesota Reef, Manchester, and Squaxin Reef (Fig. 4). Nets placed at Squaxin Reef were damaged when they became unfastened from the anchor and the *Iridaea* never covered 100% of the nets. Yields at this site

Table V. *Yields of Iridaea cordata from nets seeded in 1974 and 1975 at Minnesota Reef, San Juan Island, Washington, and harvested in 1976.*

Year Seeded	Yield (g/0.25m^2 $\pm$ S.E.)
1974[1]	677.2 $\pm$ 76.2 (N=8)
1975[2]	446.4 $\pm$ 10.4 (N=4)

Yields not significantly different, Mann-Whitney "U" test (U=24)
[1]Nets 2A, 2B, 2C, 2D, 3A, 3B, 3C, 3D, see Table II
[2]Nets 7A, 13A, 15A, 15C, see Table II

Table VI. *Yields of Iridaea cordata from 3" mesh nylon nets and 6" polypropylene nets at Minnesota Reef, San Juan Island, Washington.*

Net Types	Yield (g/0.25m^2 $\pm$ S.E.)
3" nylon[1]	540.5 $\pm$ 86.2 (N=9)
6" polypropylene	599.8 $\pm$ 104.7 (N=4)

Yields not significantly different, Student's t-test (t=-0.50)
[1]Nets 2A, 2B, 2C, 2D, 7A, 13A, 15A, 15B, 15C, see Table II
[2]Nets 3A, 3B, 3C, 3D, see Table II

are not comparable directly, but appear to be about one-half of that at the other sites. Yields from Minnesota Reef and Manchester are not significantly different (Table VII).

Table VII. *Yields of Iridaea cordata harvested from nets at Minnesota Reef, San Juan Island, and Manchester, Puget Sound, Washington.*

Location	Yield (g/0.25m^2 $\pm$ S.E.)
Minnesota Reef[1]	645.3 $\pm$ 78.8 (N=7)
Manchester[2]	510.3 $\pm$ 77.8 (N=4)

Yields not significantly different, Student's t-test (t= -1.08)
[1]Nets 2A, 2B, 3A, 3B, 15C, 7A, 13A, see Table II
[2]Nets 2C, 2D, 3C, 3D, see Table II

The growth rates of *Iridaea* for 2- or 4-week periods from April 8 to December 21, 1976, are shown in Table VIII. Variance between nets was high, particularly during the July 11-26 period when many of the nets were exposed during an extremely low tide and the plants were sunburned. The highest growth rate at Minnesota Reef was 38.7 g

Table VIII. Growth rates of *Iridaea cordata* on nets at Minnesota Reef (San Juan Island) and Manchester and Squaxin Reef (Puget Sound, Washington) in the summer of 1976.

Location — Growth Rate (mean g salt-free dry matter day^{-1}m^{-2} ± S.E.)[1]

Minnesota Reef

8 Apr–12 May	12 May–12 Jun	12 Jun–11 Jul	11 Jul–26 Jul	26 Jul–10 Aug	10 Aug–24 Aug	24 Aug–1 Oct	1 Oct–18 Oct	18 Oct–21 Dec
1.83 ±5.42 (N=7)	23.90 ±4.16 (N=7)	19.70 ±2.76 (N=6)	-1.33 ±10.10 (N=6)	22.30 ±3.13 (N=7)	11.90 ±3.74 (N=7)	-3.70 ±3.99 (N=7)	-16.7 ±3.0 (N=7)	-0.27 ±0.49 (N=7)

Manchester

9 Apr–14 May	14 May–11 Jun	11 Jun–9 Jul	9 Jul–28 Jul	28 Jul–11 Aug	11 Aug–30 Sep	30 Sep–18 Oct
-0.44 ±1.67 (N=4)	11.30 ±3.23 (N=4)	20.40 ±4.20 (N=5)	-3.20 ±7.01 (N=5)	9.40 ±4.96 (N=5)	-4.30 ±3.77 (N=5)	-11.40 ±2.57 (N=5)

Squaxin Reef

8 Apr–17 May	17 May–10 Jun	10 Jun–13 Jul	13 Jul–29 Jul
-6.0 (N=1)	10.0 (N=1)	10.9 (N=1)	-13.1 (N=1)

[1] Calculated from the differences of the total wet weight of nets, frames, and algae on the dates indicated. Salt-free dry matter calculated as 15% of wet weight of algae.

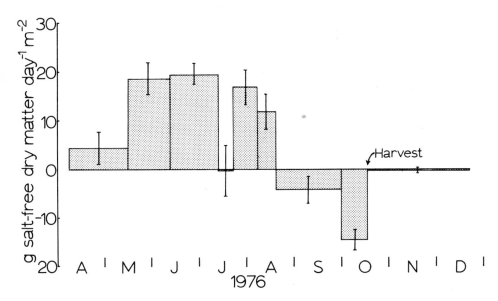

Figure 8. *Average growth rates for summer 1976 of all nets with Iridaea cordata at Minnesota Reef, San Juan Island, Manchester, and Squaxin Reef, Puget Sound, Washington.*

salt-free dry matter m^{-1} day^{-1} (Net 2A Table II, May 12–June 12) and at Manchester 43.9 g salt-free dry matter m^{-1} day^{-1} (Net 2C Table II, June 11–July 9).

The average growth rates of *Iridaea* in 1976 are shown in Fig. 8. In 1976, growth on the nets was seasonal, with the maximum rates occurring from mid-May to mid-August. By the end of August, negative growth rates as a result of senescence and decay were rapid. The growth curve was artificially terminated by harvesting in October.

Discussion

The standing crop and yields of *Iridaea cordata* from various localities on the west coast of North America is shown in Table IX. Density on netting is about 10 times that in natural populations in Washington and British Columbia and about 1.5 times that at Año Nuevo Island, California.

Only about one-half of the nets placed in 1975 were seeded with enough *I. cordata* to give 100% cover the following year. There is no appreciable increase in cover from additional seeding the second and third years. Fouling by ulvoids and kelp on clear areas prevents further seeding. Therefore, studies are underway to devise a practical way for seeding nets artificially in tanks on a large scale. Efforts so far have been rewarded with plant mortality when nets with germlings are placed in the field. By ensuring a dense, even set of plants on the nets, fouling should be avoided. Particular strains or different species or genera of algae may be used. It should be possible to seed for one phase of the life history to get a particular

Table IX. *Standing crops and yields of Iridaea cordata from the west coast of North America.*

Date	Location	Yield (g dry matter/m² + S.E.)[1]		Reference	Note
7 Aug 1975	Minn. Reef	4920.4 + 422.8	(N=4)	This paper	Standing crop
June, Oct 1976	Minn. Reef	3383.5 + 203.5	(N=4)	This paper	Two harvests
June, Oct 1976	Puget Sound[2]	2253.0 + 264.6	(N=13)	This paper	Two harvests
12 June 1976	Puget Sound	1773.5 + 315.8	(N=4)	This paper	Standing crop
26 July 1976	Puget Sound	1434.8 + 271.8	(N=9)	This paper	Standing crop
2 July 1976	Barnes Island	196.6 + 61.0	(N=20)	This paper	Standing crop
10 July 1970	Barnes Island	294.3		(9)	Standing crop
1 Sep 1976	Barnes Island	425.6 + 50.9	(N=10)	This paper	Standing crop
1972-73	N. Strait of Georgia	284.4 + 9.0	(N=?)	(5)	Peak density
1972-73	N. Strait of Georgia	172.8 + 9.0	(N=?)	(5)	Ave. density
1972-73	Año Nuevo I., Calif.	1320.0 + 244.8	(N=9-16?)	(12)	
1975	Tanks, Washington	1947.9		(34)	

[1]The energy content of *Iridaea cordata* is approximately 2.92 kcal/g dry wt.
[2]Nets at Minnesota Reef, San Juan Island, Washington, Manchester, and Squaxin Reef, Puget Sound, Washington.

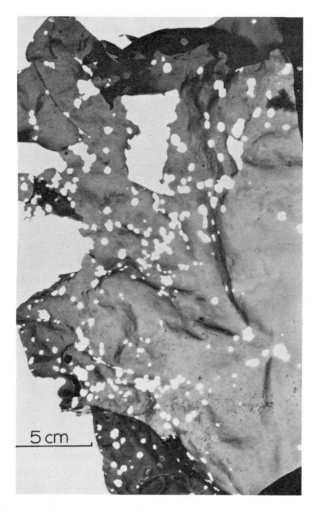

Figure 9. *Blade of Iridaea cordata showing damage from the herbivorous snails, Lacuna variegata, and Lirularia lirulata. Besides decreasing the photosynthetic area, damage may weaken the blade so that it is lost entirely. (Barnes Island, Washington)*

type of carrageenan.

The harvesting work done in 1976 indicated that two harvests are possible per growing season, and harvests should be timed so that the first occurs early in the growing season and the second occurs when the standing crop peaks in August or September. Further experimentation is needed to determine if several closely spaced harvests will yield more biomass or extend the growing season. The costs of a harvesting run must be considered in multiharvesting.

Techniques for fertilizing and the effects of fertilizing on growth and carrageenan content have not been studied. Ammonium is the preferred source of nitrogen (11) and a high ammonium level in the fall has been noted by Hansen (11) to delay blade loss by several weeks.

Effects of herbivores have already been noticed. Thalli in natural beds at Barnes Island were extensively damaged by the snails *Lacuna variegata* and *Lirularia lirulata* (Fig. 9). These snails also colonized nets placed at Barnes Island in September 1976 and so effectively grazed the netting that virtually no macroscopic algal growth was seen from September 1976 to February 1977. The effects of sea urchins on algal standing crops is well known but no problems have been encountered yet.

No major diseases have been noted. Minor infestations of the endophytic green alga *Endophyton regenerans* and the encrusting bryozoan *Membranipora* were present but seemed to have little effect on growth at their present levels. Many organisms now unnoticed or at low levels may present major problems when large monocultures are available. Disease problems have been recently reviewed (4, 30).

The major obstacles to any aquaculture program may be social and legal rather than biological, engineering, or economic. This subject is presented in Chapter 21 of this volume by Benner, but a few observations are given here. The present permit system, while set up with the laudable goals of protecting the environment, creates time-delays that make it almost impossible to perform research within a biological time frame. Any aquaculture program must carefully consider conflicts with commercial, recreational, and esthetic values placed on marine waters. Investment and interest will come only after clear legal guidelines are established regarding aquaculture's place in our society.

SUMMARY AND CONCLUSIONS

The use of artificial substrates for the culture of marine algae has a long history. The edible seaweeds *Porphyra*, *Laminaria*, and *Undaria* are now grown on nets in a large and technologically advanced industry. More recently, algae used for the supply of phycocolloids are being grown on nets, *Eucheuma* in the Philippines and *Iridaea* and *Macrocystis* on the west coast of North America. Techniques being developed will open to utilization wide areas of the oceans for the production of marine algae for energy, sources of hydrocarbons, and chemical, food, and pharmaceutical products.

ACKNOWLEDGMENTS

I would like to thank C. Boswell, Nanci Brose, Don Melvin, Doug Magoon, and Fred Winningham for their help with the various aspects of the field work. For use of the facilities of the University of Washington Friday Harbor Laboratories, my thanks to Dr. Dennis Willow and Dr. Dick Strathmann. Carl Lange of Friday Harbor has permitted the use of his tidelands for the Minnesota Reef site. My thanks to him for his interest and cups of hot coffee when they were most neede Dr. J. Hansen and Dr. J.R. Waaland have provided many stimulating discussions and suggestions. And my special thanks to Ralph Beswick for his years of support, insight, and guidance, without whom this project would not have been successful.

REFERENCES

1. Abbott, I.A. 1971. On the species of *Iridaea* (Rhodophyta) from the Pacific coast of North America. Syesis 4:51-72.
2. Abbott, I.A. 1972. Field studies which evaluate criteria used in separating species of *Iridaea* (Rhodophyta). *In:* Abbott, I.A., and Kurogi, M. (eds.) Contributions to the Systematics of Benthic Marine Algae of the North Pacific. Japanese Society of Phycology, Kobe, pp. 253-264.
3. Academica Sinica. 1975. The breeding of new varieties of Haidai (*Laminaria japonica* Aresch.) with high production and high iodine content. *In:* Proceedings, Third International Ocean Development Conf., Aug 5-8, 1975, Tokyo. Oceanographical Society of Japan, Box 5050, Tokyo Intl. 100-31, Japan.
4. Andrews, J.H. 1976. The pathology of marine algae. Biol. Rev. 51:211-253.
5. Austin, A., and R. Adams. 1975. Red algal resource studies in Canadian Pacific waters. Carrageenophyte Inventory and Experimental/Cultivation Phase 1974/75. Vol. I, Text; Vol. II, Appendices. Report to Fed. Minister of Fisheries, and Provincial Minister of Recreation and Conservation, British Columbia.
6. Bardach, J.E., J.H. Ryther, and W.O. McLarney. 1972. Aquaculture: The Farming and Husbandry of Freshwater and Marine Organisms. Wiley-Interscience, New York, pp. 790-836.
7. Deveau, L.E., and J.R. Castle. 1976. The Industrial Development of Farmed Marine Algae: The Case History of *Eucheuma* in the Philippines and U.S.A. FAO Tech. Conf. Aquaculture, Kyoto: FIR:AQ/Conf/76 E.56.
8. Drew, K. 1949. Conchocelis-phase in the life history of *Porphyra umbilicalis* (L.) Kütz. Nature 164:748.
9. Fralick, J.E. 1971. The Effect of Harvesting on a Sublittoral Marine Plant Community in Northern Washington. Masters Thesis, Western Washington State College, Bellingham. 60 pp.
10. Fralick, J.E., and K. Cole. 1973. Cytological observations on two species of *Iridaea* (Rhodophyceae, Gigartinales). Syesis 6:271-272.
11. Hansen, J.E. 1976. Population biology of *Iridaea cordata* (Rhodophyta, Gigartinaceae). Ph.D. Thesis, University of California, Santa Cruz. 341 pp.
12. Hansen, J.E., and W.T. Doyle. 1976. Ecology and natural history of *Iridaea cordata* (Gigartinaceae): Population structure. J. Phycol. 12:273-278.
13. Hasegawa, Y., and Y. Kuwatani. 1974. Present status of major marine cultivation and propagation in Hokkaido and some problems of research activities. *In:* Shaw, W.N. (ed.) Proceedings, First United States--Japan Meeting on Aqua-

culture, Tokyo, 18-19 Oct., 1971. NOAA Tech. Rep. NMFS Circ-388, pp. 3-6.

14. Hruby, T. 1974. A Study of Several Factors Influencing the Growth and Distribution of *Iridaea cordata* (Turner) Bory in Coastal Waters of Washington State. M.S. Thesis, University of Washington, Seattle.

15. Hruby, T. 1975. Seasonal changes in two algal populations from the coastal waters of Washington State. J. Ecol. 63:881-890.

16. Hruby, T. 1976. Observations of algal zonation resulting from competition. Estuarine Coastal Mar. Sci. 4:231-233.

17. Kim, D.H. 1976. A study of the development of cystocarps and tetrasporangial sori in Gigartinaceae (Rhodophyta, Gigartinales). Nova Hedwigia 27:1-146.

18. Kurogi, M. 1961. Species of cultivated *Porphyras* and their life histories. Bull. Tohoku Reg. Fish. Res. Lab. 18:1-115. (in Japanese with English summary)

19. McCandless, E.L., J.S. Craige, and J.E. Hansen. 1975. Carrageenans of gametangial and tetrasporangial stages of *Iridaea cordata* (Gigartinaceae). Can. J. Bot. 53:2315-2318.

20. Miura, A. 1975. *Porphyra* cultivation in Japan. *In:* Tokida, J., and H. Hirose (eds.) Advance of Phycology in Japan. Junk, The Hague, The Netherlands, pp. 273-304.

21. Nationwide Federation of Nori Cultivation Cooperatives (Zen-Nori). 1970. Manual of Nori Cultivation with Apparatus of Floating System. Zen-Nori, Tokyo, Japan. 110 pp. (in Japanese)

22. Neish, I.C. 1976. Culture of Algae and Seaweeds. FAO Tech. Conf. Aquaculture, Kyoto, FIR:AQ/Conf/76/R.I.

23. Norris, R.E., and D.H. Kim. 1972. Development of thalli in some Gigartinaceae. *In:* Abbott, I.A., and M. Kurogi (eds.) Contributions to the Systematics of Benthic Marine Algae of the North Pacific. Japanese Society of Phycology, Kobe, pp. 265-275.

24. Ogata, E. 1975. Physiology of *Porphyra*. *In*: Tokida, J., and H. Hirose (eds.) Advance of Phycology in Japan. Junk, The Hague, The Netherlands, pp. 151-160.

25. Okazaki, A. 1971. Seaweeds and Their Uses in Japan. Tokai University Press, Tokyo. 165 pp.

26. Saito, Y. 1975. *Undaria*. *In*: Tokida, J., and H. Hirose (eds.) Advance of Phycology in Japan. Junk, The Hague, pp. 304-320

27. Saito, Y. 1976. Seaweed Aquaculture in the Northwest Pacific. FAO Tech. Conf. Aquaculture, Kyoto, FIR:AQ/Conf/76/R.14.

28. Silverthorne, W., and P.E. Sorensen. 1971. Marine algae as an economic resource. *In:* Preprints, Seventh Annual Conference. Marine Technological Society, 1730 "M" St. N.W., Washington, D.C., 20036, pp. 523-533.

29. Suto, S. 1974. Mariculture of seaweeds in Japan and its problems. *In:* Shaw, W.N. (ed.) Proceedings, First United States--Japan Meeting on Aquaculture, Tokyo, 18-19 Oct., 1971. NOAA Tech. Rep. NMFS CIRC-388, pp. 7-17.

30. Suto, S., Y. Saito, K. Akiyama, and O. Umebayashi. 1972. Text-

book of Diseases and Their Symptoms in *Porphyra*. Tokai
Reg. Fish Res. Lab., Contrib. No. 18. 37 pp. (in Japanese)

31. Waaland, J.R. 1973. Experimental studies on the marine algae
Iridaea and *Gigartina*. J. Exp. Mar. Biol. Ecol. 11:71-80.

32. Waaland, J.R. 1974. Aquaculture potential of *Iridaea* and
Gigartina in Pacific Northwest waters. *In:* Seaweed
Farming in Puget Sound. Proceedings of a seminar, Oct.,
1974, Poulsbo, Washington. Washington State University,
Pullman, pp. 1-17.

33. Waaland, J.R. 1975. Differences in carrageenan in gametophytes
and tetrasporophytes of red algae. Phytochemistry 14:1359-
1362.

34. Waaland, J.R. 1976. Growth of the red alga *Iridaea cordata*
(Turner) Bory in semi-closed culture. J. Exp. Mar. Biol.
Ecol. 23:45-53.

35. Waaland, J.R. 1977. Colonization and growth of populations of
Iridaea and *Gigartina* on artificial substrates. *In:* Proceedings, Eighth International Seaweed Symposium, Inter-
Document Corp., Box 326, Harrison, New York 10528. (in
press)

-9-
The Domestication of the Giant Kelp, *Macrocystis,* as a Marine Plant Biomass Producer

MICHAEL NEUSHUL

Mankind depends for food on 15 major crop plants, these being the cereals, legumes, and the tree, sugar, and root crops. Almost all of these were first domesticated over 10,000 years ago by prehistoric agriculturalists. In contrast, the domestication of marine crop plants is being undertaken by our own generation. There are now perhaps four domesticated marine plants, definable as such since the amount harvested from cultivated sources exceeds that taken from wild populations. These are the red algae *Porphyra* and *Eucheuma* and the brown algae *Laminaria* and *Undaria*. *Macrocystis,* the largest known marine plant, is still a wild plant, although sizeable amounts of it are harvested and processed for the production of alginates (21). As indicated elsewhere in this volume, the possibility of using *Macrocystis* as a biomass producer is being seriously considered.

With the end of the fossil-fuel era in sight, it is time to design a new world food and energy system geared to renewable energy (3, 31). Solar energy input, prior to the fossil-fuel era, was in balance with agricultural output. Today it is not in balance, since fossil-fuel subsidies are used to increase agricultural yield to the extent that in some areas the population exceeds that which can be supported by solar-based agricultural yields. Thus, it is certainly worthwhile to explore the potential of marine plants as collectors of energy and concentrators of nutrients, since these might contribute significantly to a solar-based world food and energy production system.

Freshwater algae are able to fix and store considerable amounts of energy. It has been suggested that algal culture in sewage ponds

---
[1]This work has been supported by the U.S. Department of Commerce (USDC, NOAA 04-6-158-44110) Sea Grant Program and the National Science Foundation (NSF OCE-76-24360).

might yield as much as 30,000 kilowatt hours per acre per year (10,
23). The preliminary phases of an effort to use macroscopic marine
algae as collectors of energy in the open sea are now under way (12,
13). It has been assumed that it will be economically feasible to
construct large floating structures to which kelp can attach and grow
in the open ocean. Some major constraints to both open-sea and tank
culture of marine algae have been recently discussed by Jackson
(personal communication), who emphasizes that other constraints are, as
yet, not fully recognized. However, since floating macroscopic algae
occur naturally along our coasts and even in the open sea, there
might be room for optimism about such a scheme. Also, there have
been very encouraging recent advances in the domestication of benthic
marine algae that may lead to the establishment of effective plant
mariculture in many places (1, 9, 14, 15, 24). It is the purpose of
this chapter to examine the reasons for such optimism and to consider
specifically the current status of our knowledge of the life history
of *Macrocystis*, because control of the life history is essential if
this plant is to be domesticated.

FLOATING BROWN ALGAE IN THE OPEN SEA

Brown algae floating in the open sea are mentioned in the writing
of the early navigators of both the Atlantic and Pacific. The largest
accumulation of open-ocean macro-vegetation is in the Sargasso Sea in
the Atlantic. This mass of vegetation extends over some 2 million
square nautical miles, where from 1.9 to 0.6 tons of *Sargassum* occur
per square mile, with the plants occurring in windrows and patches (25)

A warm, nutrient-poor area like the Sargasso Sea is an unlikely
place to find large amounts of floating macroscopic algae, since
nutrient limitation and high-temperature damage are the most obvious
problems encountered by algae growing there. It has been suggested
that the blue-green alga, *Trichodesmium*, growing on *Sargassum* can fix
nitrogen (4). It is also likely that animals living in the floating
clumps of algae contribute to the recycling of nutrients. Studies
of drift seaweeds in open water, like those in the Sargasso Sea,
would be extremely useful for one attempting to produce artifically,
a cultivated Sargasso Sea.

Although drift seaweeds occur in the Pacific, there is no accumulation of drift plants like that in the Sargasso Sea. The Pacific
coasts of North and South America support large submarine forests
of kelp, with *Macrocystis* being a dominant plant. However, only
in the northern hemisphere does one find the bull kelp, *Nereocystis*,
and the elk kelp, *Pelagophycus*. Both of these plants produce large
gas-filled floats.

Setchell (29) pointed out that the drifting Pacific Coast kelps
occur in sufficient quantities to have been noted by the Spanish.
On their voyages from the Philippine Islands, they included the sighting of these drift algae in their sailing instructions. The plant
called "Porra" by the Spaniards and a "Sea Leek" by the English,
as pictured in their texts, is recognizable as the Californian elk
kelp, *Pelagophycus porra*. When galleons from Manila reached a
longitude of about 96° W, the sighting of a floating *Pelagophycus*

would signal an immediate turn southward towards Panama without wait-
ing to sight land.

The bull kelp *Nereocystis* also drifts for long distances and
was recorded 8 or 9 leagues from land along the entire Pacific Coast.
A most remarkable series of records are those associated with the
occasional stranding of drift *Nereocystis* in Japan. Since this plant
occurs in the Aleutians but not in Japan, its stranding there suggests
that it is capable of drifting for many months and many hundreds of
miles. Floating kelp also become entangled with attached plants in
Californian kelp beds, where they are known to live unattached as
"floaters" for long periods of time.

The natural occurrence of floating drift kelps in both the
Pacific and Atlantic suggests that open-sea cultivation of macroscopic
algae is at least a possibility. Recent advances by the Japanese and
Chinese in seaweed cultivation provide techniques through which both
open-sea and improved coastal cultivation might be accomplished.

CHINESE AND JAPANESE CULTIVATION OF BROWN ALGAE

The Chinese have used the brown alga, *Laminaria*, as a source of
iodine for 1,500 years. As much as 45,000 dry metric tons were
imported yearly from Japan and Korea, because *Laminaria* did not then
occur naturally on the Chinese coast. As a measure of their recent
successes in cultivation the Chinese now export *Laminaria* to Japan.
Their cultivation strategy is one that has been developed over the
past two decades (5, 30). Sporelings are produced in the laboratory
and held at low temperatures in tanks during the summer, when seawater temperatures are high. When conditions are suitable, they are
placed in the sea. Low nutrient levels in the sea are countered by
periodically dipping plants in nutrient-enriched seawater and return-
ing them to the plantations, or by dispersing fertilizer in the sea
around them. *Laminaria* in China is grown on rope lines, supported
by floats or rafts, in a manner similar to that used in Japan. The
amounts now harvested are said to be in excess of the amount of *Macrocystis* presently harvested in California (Fig. 1).

The Japanese cultivate *Laminaria* and also another kelp, *Undaria*.
For the former they use a new "forcing" technique (11), where plants
suitable for harvest are obtained in one year rather than two, as was
the case previously. Saito (26) has developed a series of culture
techniques for *Undaria* and has been able to produce hybrids between
some of the forms. The successes of the Japanese and Chinese are
based on their ability to control the life histories of their plants
and their appreciation of basic functional aspects of marine plants
generally.

FUNCTIONAL ASPECTS OF MARINE VEGETATION

The uninitiated are almost always amazed at the striking differences between benthic marine algae and land plants. The former have
no flowers, no seeds, no fruits, no roots, no stems *per se*, and only
leaf-like structures called blades. The major structural element in
a community of marine algae is the water surrounding the plants,

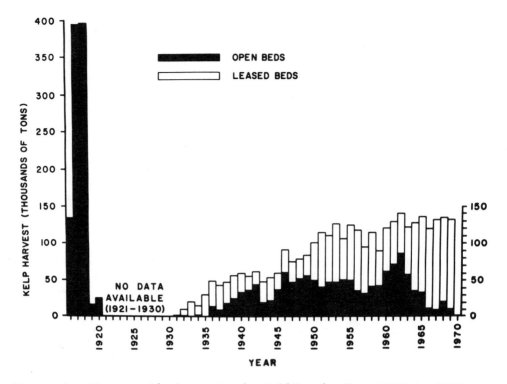

Figure 1. *Macrocystis harvests in California from 1900 to 1970. California kelp beds are either individually leased by specific harvesters or open to all. Amounts taken prior to 1920 were large, presumably because entire plants were cut rather than only the floating canopy as is done at present. Harvested amounts continue to increase and in 1977 are at ca 200,000 tons/year.*

rather than the lignin and cellulose within them as with land plants. Plants submerged in water are constrained with regard to nutrient uptake. The water also acts as a selective filter to change both the quantity and quality of the light reaching the plants. In view of these and other differences between land and marine plants, it is not surprising that in marine plants one encounters "combinations" of organs. For example, the blade of a kelp is obviously a leaf-like structure, but because it also takes up nutrients from the surrounding water, it has a root-like function. Clearly, to understand how marine vegetation functions, it is essential to set aside the compartmentalized thought patterns of the land-plant botanist. The reproductive processes of marine algae are also very different from those of land plants, as the absence of flowers, fruits, and seeds suggests. Instead of one basic life history, as with the flowering land plants, one encounters an often bewildering array of life histories, involving the alternation of similar and/or dissimilar phases, all with their own unique limitations and requirements. Many life histories, like those of the kelps, involve microscopic as well as macroscopic

phases. In the life history of *Macrocystis* one finds the largest
known marine plant, on one hand, alternating with one of the smallest
unicellular gametophytic phases, on the other. An understanding of
the morphological and reproductive features of marine algae in
functional terms is the key to their domestication.

THE LIFE HISTORY OF MACROCYSTIS

Macrocystis is the only Pacific coast kelp now commercially
harvested. If it is desirable to insure a stable crop from the
natural kelp beds, to expand them, or to grow *Macrocystis* elsewhere,
it is essential to have detailed knowledge of its life history and
to be able to control it. The macroscopic and microscopic stages
occur in habitats characterized by different water-motion "climates."
The relationship between the life history phases and their "water-
motion-defined" habitats is shown in Fig. 2. The sexual life history
can also be diagrammatically represented in a more traditional way,
as in Fig. 3.

The complete sequence of events from one generation to the next
can occur in as little as 8-12 months. However, both the gametophytes
and sporophytes as individuals can live for years. Estimates of
sporophyte age range from 16 to 30 years, and for gametophytes in
culture one can assume that 3-4 years, if not longer, is possible.
Gametophytes, studied both in dish-culture and in the sea, appear to
require a minimum of 13 days to form gametes. Fertilization itself
occurs in a matter of minutes. Methods for the induction of spore
release, spore collection, and the isolation and long-term cultivation of gametophytic "seed stock" are described by Sanbonsuga
and Neushul (27, 28), who have also developed methods for the induction
of gametogenesis. This manipulation of the microscopic phases has
been effectively combined with new greenhouse-rearing techniques
where larger post-embryonic plants can be grown to reproductive maturity. As will become evident, these techniques now make it possible
to start a genetic improvement program for *Macrocystis*.

W. J. North, in discussing the life history of *Macrocystis*, has
spoken metaphorically of "environmental turnstiles" when describing
factors that contribute to the high mortality of young sporophytes
in the sea. One can also postulate that similar factors operate in
culture dishes in the laboratory, but here conditions can more easily
be altered experimentally so as to identify such "turnstiles." One
might then refer to them as "control points" in the life history,
where two or more alternative developmental pathways are inducible.
The points where the *Macrocystis* life history can be controlled are
indicated in the legend to Fig. 3. These are points where a "control
trigger" can change subsequent patterns of resource allocation. Such
a trigger might be a change in light intensity (Lüning and Neushul,
manuscript), a change in light quality, or a change in photoperiod
shown to be the case in receptacle formation in the brown alga *Fucus*
by Bird and McLachlan (2). By interpreting the patterns of growth,
development, and reproduction in *Macrocystis* as resource allocation
"strategies" which either individually or collectively represent
adaptations for survival over a range of environmental conditions,

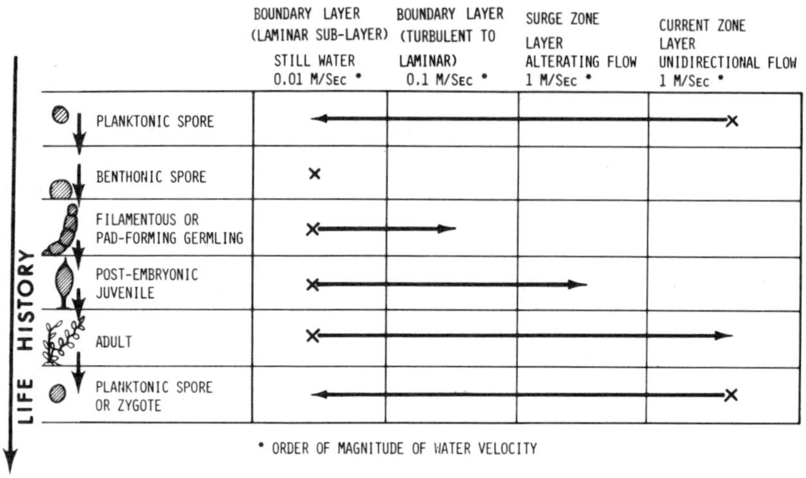

	BOUNDARY LAYER (LAMINAR SUB-LAYER) STILL WATER 0.01 M/Sec *	BOUNDARY LAYER (TURBULENT TO LAMINAR) 0.1 M/Sec *	SURGE ZONE LAYER ALTERATING FLOW 1 M/Sec *	CURRENT ZONE LAYER UNIDIRECTIONAL FLOW 1 M/Sec *
PLANKTONIC SPORE				X
BENTHONIC SPORE	X			
FILAMENTOUS OR PAD-FORMING GERMLING	X			
POST-EMBRYONIC JUVENILE	X			
ADULT	X			
PLANKTONIC SPORE OR ZYGOTE				X

* ORDER OF MAGNITUDE OF WATER VELOCITY

Figure 2. *The relationship between the microscopic and macroscopic stages in the life history of Macrocystis and the water motion rates in the habitats of these stages. After Neushul (20).*

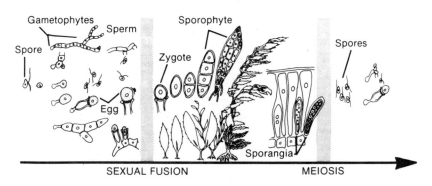

Figure 3. *The life history of Macrocystis, diagrammed in a linear way from left to right. The haploid spore goes through a dumbbell stage to form either multicellular male or female gametophytes or unicellular female plants or few-cellular male plants. After sexual fusion the diploid zygote develops first into an embryonic sporophyt then post-embryonic stages, to ultimately form a large sporophyte, which when fertile produces sporangia on basal sporophylls. The life history can be controlled at the spore-to-gametophyte phase, gametophyte-to-gamete phase, and to some extent in the young sporophyte phase.*

it should be possible to relate a given developmental pathway to a
given trigger, and so to build gradually an appreciation of the
adaptive significance of each phase in the life history and its
developmental potential.

The Haploid Microscopic Phases

Spores, gametophytes, and gametes of *Macrocystis* can be collectively referred to as the haploid microscopic phases of the life
history (Fig. 3). Spores are formed in sporangia that occur on the
surfaces of special blades called sporophylls. The sporangia cover
areas of the blade called sori. The formation of the sporangial sori
of the floating kelps has been recently studied by D. L. Walker (unpublished manuscript), who describes sorus formation as a development of meristoderm (Fig. 4) into the two cell types (sporangia and
paraphyses) found in the sorus. Microscopic and ultrastructural
features of the sporangia of *Macrocystis* have been studied by Chi
(7) and Chi and Neushul (6). These contain specialized attachment
vesicles, an eyespot, two plastids, mitochondria, and vacuoles of
several sorts (Figs. 4, 5, 6). The spores are motile, each bearing
two laterally inserted flagella. Spore release occurs through a
thickened, and presumably weakened, region at the terminal end of
the sporangium. The settling and germination of the spores of *Macrocystis* can occur almost immediately or even after a period of as long
as 12 hours, during which time the spore remains in the free-swimming
condition. When the spores adhere, the flagella are lost. Germination begins within 3 days. A characteristic "dumbbell" stage is
formed (Fig. 3). After this stage the germinating spore can follow
alternative developmental pathways, like those described for the
Laminaria spore by Lüning (personal communication). Either a filamentous gametophyte can be formed, as is the usual case in dish culture, or if conditions are optimal a single-celled gametophyte can
be formed by the female. The male under similar conditions forms a
few-celled plant. In the female the single-celled gametophyte becomes an oogonium and an egg is released. Since many workers have
described the multicellular gametophytes that are commonly formed
in dish culture, it was surprising to find that unicellular gametophytes were formed in tank culture (16). In preliminary experiments
carried out at Santa Barbara, *Macrocystis* spores on slides, placed
in the sea at a depth of ca 5m, formed unicellular gametophytes, and
sporophytes were produced in 13 days. It is likely that unicellular
gametophytes are the "natural" form of this life-history phase in
the sea. If so, this might answer the long standing question as to
why kelp gametophytes have not been found in nature. If multicellular gametophytes occur in the sea, they are either ephemeral or
assume forms that cannot be easily distinguished from other filamentous or crustose brown algae (32).

There has been a lack of agreement among those who have studied
kelp gametophytes as to what effects various light, temperature,
and nutrient levels have on them. Recent work in this laboratory
(Lüning and Neushul, manuscript) has focused on the response of 9
California kelps. Included among these are *Macrocystis pyrifera*

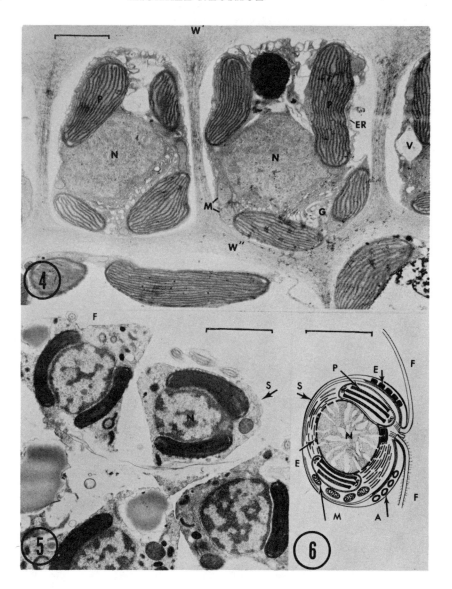

Figure 4. *Macrocystis spores are formed in sporangial cells derived from the isodiametric meristoderm cells of the sporophyll. Shown here are the nucleus (N), plastids (P), mitochondria (M), endoplasmic reticulum (ER), golgi (G), vacuoles (V), and outer (W') and inner (W'') cell walls. A 10μm scale is given. From Chi (7).*

Figures 5, 6. *The spores of Macrocystis within sporangia (left) and a diagram of a single spore (right) show the same organelles within the spores (S) as in the "parental" meristoderm cells along with specialized flagella, and an eyespot (E); 10μm scales are given. From Chi (7).*

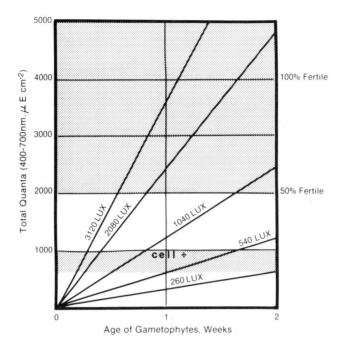

Figure 7. *Cell division and egg production of laminarian gametophytes, including those of Macrocystis, under continuous cool-white fluorescent light over a period of two weeks in laboratory dish culture. Five light levels in Lux are shown. These correspond to 0.5, 1.0, 2.0, 4.0, and 6.0 nE/cm^{-2}/s^{-1}. The cumulative amounts of light over the two-week period are indicated on the left; the percentage of plants producing eggs is indicated on the right.*

and *M. integrifolia*. It was found that, with some exceptions, the gametophytes behaved in similar ways. Vegetative gametophyte growth is light-saturated at a very low level, about 2nE/cm^{-2}/s^{-1}. At the other extreme plants are killed by 4 minutes of direct sunlight. In contrast with the light requirements for vegetative growth, the induction of gametophyte fertility requires much higher light levels. Egg production in *Macrocystis* exhibits an almost linear dependence on quantum irradiance up to 6nE/cm^{-2}/s^{-1}. Blue light, not red or green, induces fertility. Gametophyte responses to light (at five irradiance levels) over a two-week period are shown diagrammatically in Fig. 7.

The Diploid Sporophytic Phases

It scarcely needs to be said that the microscopic-to-macroscopic transition which occurs after an egg is fertilized is a spectacular process, since under optimum conditions the plant grows from the single-celled zygote to ca 20m in length within 8 to 12 months (17). The development of the zygote and embryonic sporophyte follows a well-defined pathway, with filamentous, monostromatic, and anatomically complex phases following in rapid succession. Along with this blade

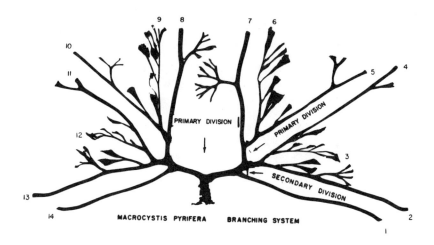

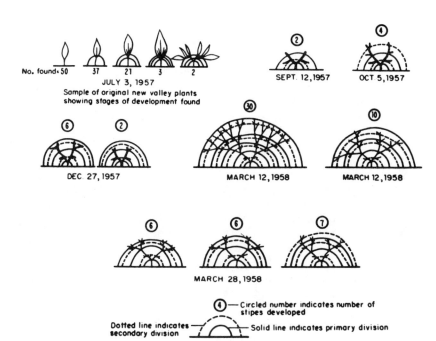

Figure 8. *The symmetrical basal branching system of a 14-frond Macrosystis, showing primary and secondary divisions (above). A diagrammatic representation of basal-system-development in a population of similarly sized plants growing in the sea (below). The plants all presumably started growing at the same time. Samples from the population were taken on five dates over a 9-month period.*

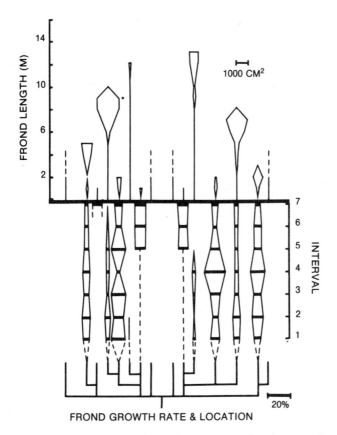

Figure 9. *A diagram representing the symmetrical growth pattern of a 15-frond Macrocystis plant and the dimensions (length and area) of its fronds when harvested. The growth of fronds from specific positions on the basal branching system (shown below) was measured over six 5- to 7-day intervals between Aug. 7 and Sept. 23, 1975. Length increase is given as % of original size. A 20% scale is given. Area measurements were also made and are shown in the upper "baloon" diagrams, where area relative to position along the length of the frond is shown as width of the "baloon" at a given point. A scale bar representing an area of 1,000cm² is shown. There is symmetry in area and, to a lesser degree, in growth responses with time, on either side of the primary branching system. From Coon and Wheeler (8).*

development, one can see the holdfast change from a rhizoidal one, through a discoid phase, to one characterized by many large root-like haptera.

The growth of a natural population of similarly sized *Macrocystis* plants in the sea was studied in 1957 when a submarine landslide exposed rocks at the head of a submarine canyon near La Jolla, California (Fig. 8). The population was sampled in early July with a total of 113 plants taken. These ranged from single- to six-bladed stages. After about 9 months, some of the plants in the population had reached

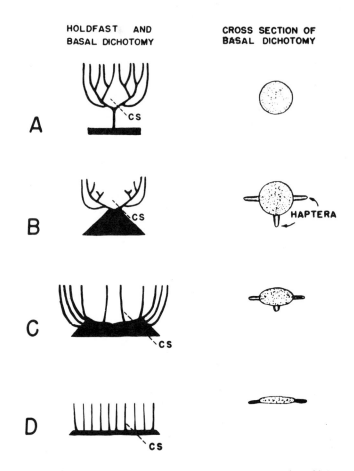

Figure 10. *The Pacific coast forms of Macrocystis differ in the configuration of the basal branching system and holdfast. Type A is the more southerly and type D the most northerly. Types A–C are presently considered to be M. pyrifera, and type D, M. integrifolia.*

60 ft in length, had basal branching systems with as many as five primary dichotomies, and had as many as 30 fronds.

The growth of an individual *Macrocystis* plant, measured at 3- to 5-day intervals, has been studied recently by Coon and Wheeler (8) They have shown that the symmetry seen in the single-bladed juvenile is also found in larger plants bearing many fronds. Fronds that are in comparable positions on opposite sides of the primary division are comparable in their stage of development. They are also similar in their growth rates, when these are considered as a percentage increase in initial size (Fig. 9). The change in a given frond from fast-, to slow-, to fast-growing, suggests considerable and, as yet, poorly understood complexity in the internal resource allocation that occurs as the plant enlarges.

The basal branching systems of the species of *Macrocystis* have also been the subject of study from a taxonomic point of view (18).

Figure 11. *Surge-tank culture of Macrocystis accomplished in a sea-water supplied greenhouse. The plants being held up on the right (ca 2m long) have been bred from dish-cultured gametophytes and grown in the tank shown. A waterbroom (WB) is moved back and forth over the tank on a track (T) by a belt running over pulleys (PY). It sprays water from a head (H) keeping the plants in moving water.*

On the west coast of North America, plants occur with very different holdfast and basal-branching systems. The more southerly plants have long, extended basal branching systems and flattened holdfasts. As one progresses north, a more peaked holdfast is formed, and ultimately loaf-formed ones occur. All of these forms are presently considered to be *M. pyrifera*. To the northern end of the range one finds the strap-like basal branching system and prostrate holdfast of *M. inte-grifolia* (Fig. 10).

Presently there is disagreement as to whether these very different plants are ecotypes or species. These variations may represent endpoints of different environmentally triggered developmental path-

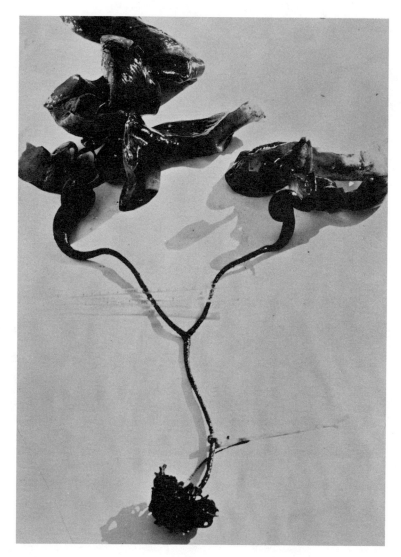

Figure 12. *An intergeneric hybrid between Macrocystis and Pelago-phycus, with two unusual coiled pneumatocysts (P) above a primary division.*

ways or may be genetically defined. Answers to these questions would seem to be attainable through controlled dish and tank culture of genetically defined strains and the outplanting of such plants in a variety of environmental conditions.

Experimental Manipulation

As indicated above, it is possible to obtain spores and to grow gametophytes in the laboratory, both in dishes and in large aquaria. A special green house has been constructed at Santa Barbara, where

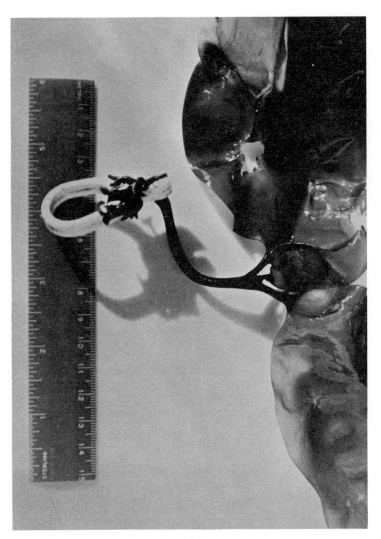

Figure 13. *An intergeneric hybrid between Macrocystis and Nereocystis with no pneumatocysts and an unusual tripartite division of the primary blade.*

water motion is provided in large tanks. Here it is possible to grow *Macrocystis* sporophytes to sizes that would be out of the question in dish culture (Fig. 11). By combining the ability to raise sporophytes through embryonic and post-embryonic stages to large sizes in these tanks, with techniques for the manipulation of gametophytes so as to make controlled crosses in dish culture, it is possible to make various crosses and then raise the progeny to verify the cross. This hybridization work (27, 28) has taxonomic and other implications. It is also of special interest in that where intergeneric hybrids are produced, one can follow the development of a plant representing a "combination" of the different resource allocation strategies of the

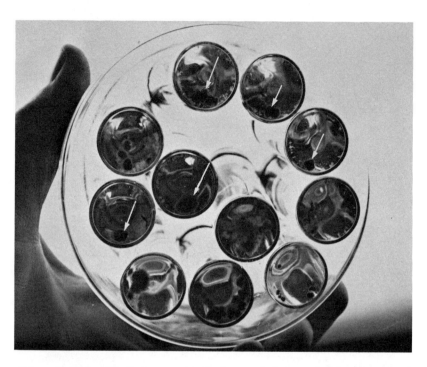

Figure 14. *Multicellular gametophytes of Macrocystis (arrows) grown from individual unicellular germlings are held for long-term storage in vials (seen from below) from which the same specific crosses can be repeatedly produced.*

two parents. To date, only two types of intergeneric hybrid have been made. These are a cross between *Macrocystis* and *Pelagophycus* (M x P) (Fig. 12) and between *Macrocystis* and *Nereocystis* (M x N) (Fig. 13). It is obvious that in each case these hybrids are very different from either parental type. There is certainly a basic symmetry in the hybrids, although in the case of the M x N cross, a tripartite branching occurs that suggests some disturbance of blade and branching symmetry. Also in the examples of this hybrid grown to date (4 specimens), there is considerable deformation and wrinkling of the blades after the initial divisions, giving the plants a Hedo-phyllum-like appearance. The M x P cross is more consistent and has regular symmetry, but here also the hybrid is very different from either parent, both in being much smaller and in having a distinctive morphology.

   While the implications of intergeneric hybridization are not yet fully appreciated, the ability to produce these plants suggests that *Macrocystis* itself, when the known varieties are crossed and the progeny grown, will prove to be amenable to genetic manipulation. It is particularly interesting that under appropriate culture conditions the gametophytes of all the kelps studied to date can be held for long periods in culture vials, where vegetative growth occurs, therein providing a means of stockpiling specific genetic strains as a gene bank for later use (Fig. 14).

SUMMARY AND CONCLUSIONS

Although there are many basic questions to be answered about the taxonomy, physiology, and morphology of *Macrocystis*, the plant is certainly amenable to further study using available techniques. It is likely that progress will be rapid. Any culture scheme that might be visualized for *Macrocystis* should be based on a rational selection of germ plasm. The ability to store single-cell isolates (gametophytes) will allow the storage of specific strains for long periods followed by the production of large numbers of genetically identical progeny at will. This is analogous to holding stores of seeds until needed.

Many problems remain for the would-be marine algal farmer. While rapid progress is being made with those phases of the life history that can be grown in laboratory dishes or tanks, it is not yet possible to "seed" *Macrocystis* on natural substrates in the sea, although W. J. North (22) is making considerable progress. We have introduced an experimental technique involving the use of fouling plates that may also be useful. On these plates we have been able to measure natural rates of recruitment and loss of several benthic algae under natural conditions in the sea (19). The reasons for the magnitude of the loss rates observed are still not known, but it is likely that experimental manipulation of fouling-plate populations of *Macrocystis* and other plants will allow identification and control the "environmental turn-stiles" postulated by Dr. North.

In spite of the many problems still remaining, the very rapid progress of the last few years would seem to justify an optimistic outlook for the future of marine plant aquaculture.

REFERENCES

1. Abbott, I.A., and M. Kurogi (eds.) 1972. Contributions to the Systematics of Benthic Marine Algae of the North Pacific. Japanese Phycological Society, Department of Botany, Kobe University, Kobe, Japan. 279 pp.
2. Bird, N.L., and J. McLachlan. 1976. Control of formation of receptacles in *Fucus distichus* L. subsp. *distichus* (Phaeophyceae, Fucales). Phycologia 15:79-84.
3. Calvin, M. 1976. Photosynthesis as a resource for energy and materials. Photochem. Photobiol. 23:424-444.
4. Carpenter, E.J. 1972. Nitrogen fixation by a blue-green epiphyte on pelagic *Sargassum*. Science 178:1207-1208.
5. Cheng, T.H. 1969. Production of kelp--a major aspect of China's exploitation of the sea. Econ. Bot. 23:215-236.
6. Chi, E.Y., and M. Neushul. 1972. Electron microscopic studies of sporogenesis in *Macrocystis*. *In:* Nisizawa, K. (ed.)

Proceedings, Seventh International Seaweed Symposium. University of Tokyo Press, Tokyo, Japan, pp. 181–187.

7. Chi, E.Y. 1973. The Development of Reproductive Cells of Brown Algae. Ph.D. Dissertation, University of California, Santa Barbara.

8. Coon, D., and W.M. Wheeler. 1976. Ecological studies in kelp forests: short term variations in the growth of *Macrocystis*. J. Phycol. (suppl.) 12:14. (abstr.)

9. Doty, M.S. 1973. *Eucheuma* Farming for Carrageenans. Sea Grant Advisory Report, University of Hawaii. Sea Grant AR 73-02, U.S. Department of Commerce, Washington, D.C.

10. Golueke, C.G., W.J. Oswald, and H.B. Gottaas. 1957. Anaerobic digestion of algae. Appl. Microbiol. 5:47–55.

11. Hasegawa, Y. 1976. Progress of *Laminaria* cultivation in Japan. J. Fish. Res. Board Can. 33:1002–1006.

12. Jackson, G.A., and W.J. North. 1973. Concerning the Selection of Seaweeds Suitable for Mass Cultivation in a Number of Large Open-ocean Solar Energy Facilities ("Marine Farms") in order to Provide a Source of Organic Matter for Conversion to Food, Synthetic Fuels, and Electrical Energy. U.S. Naval Weapons Center, China Lake, California. Final Report, Contract No. N60530-73-MV176. 135 pp.

13. Leese, T.M. 1976. The conversion of ocean farm kelp to methane and other products. *In:* Symposium Papers, Clean Fuels, Jan. 27–30, Orlando, Florida. Institute of Gas Technology, Chicago, Illinois, pp. 253–266.

14. Mathieson, A.C. 1975. Seaweed aquaculture. Mar. Fish. Rev. 37:2–14.

15. Neish, I.C. 1976. Role of mariculture in the Canadian seaweed industry. J. Fish. Res. Board Can. 33:1007–1014.

16. Neushul, M. 1959. Studies on the Growth and Reproduction of the Giant Kelp, *Macrocystis*. Ph.D. Thesis, University of California, Los Angeles.

17. Neushul, M., and F.T. Haxo. 1963. Studies on the giant kelp, *Macrocystis*. I. Growth of young plants. Am. J. Bot. 50:349–353.

18. Neushul, M. 1971. The species of *Macrocystis*. *In:* North, W.J. (ed.) The Biology of the Giant Kelp Beds (*Macrocystis*) in California. Nova Hedwigia 32:223–228.

19. Neushul, M., M.S. Foster, D.A. Coon, J.W. Woessner, and B.W.W. Harger. 1976. An *in situ* study of recruitment, growth and survival of subtidal marine algae: techniques and preliminary results. J. Phycol. 12:397–408.

20. Neushul, M. 1977. The ocean as a culture dish: experimental studies of benthic algae. *In:* Proceedings, Eighth International Seaweed Symposium, Inter-Doc. Corp., Box 326, Harrison, New York 10528. ·(in press)

21. North, W.J. (ed.) 1971. The Biology of Giant Kelp Beds (*Macrocystis*) in California. Nova Hedwigia 32: suppl. 600 pp.

22. North, W.J. 1976. Aquacultural techniques for creating and

restoring beds of giant kelp, *Macrocystis* spp. J. Fish.
Res. Board Can. 33:1015-1023.

23. Oswald, W.J. 1976. Gas production from micro algae. *In:*
Symposium Papers, Clean Fuels, Jan. 27-30, Orlando, Florida.
Institute of Gas Technology, Chicago, Illinois, pp. 311-324.

24. Parker, H.S. 1974. The culture of the red algal genus *Eucheuma*
in the Philippines. Aquaculture 3:425-439.

25. Parr, A.E. 1939. Quantitative observations on the pelagic
Sargassum vegetation of the western north Atlantic. Bull.
Bingham Oceanogr. Coll. 6:1-94.

26. Saito, Y. 1972. On the effects of environmental factors on
morphological characteristics of *Undaria pinnatifida* and
the breeding of hybrids in the genus *Undaria*. *In:* Abbott,
I.A., and Kurogi (eds.) Contributions to the Systematics
of Benthic Marine Algae of the North Pacific. Japanese
Phycological Society, Department of Botany, Kobe University,
Kobe, Japan, pp. 117-130.

27. Sanbonsuga, Y., and M. Neushul. 1977. Hybridization and Genetics of Brown Algae. Phycological Handbook, Phycological Society of America. (in press)

28. Sanbonsuga, Y., and M. Neushul. 1978. The hybridization of
Macrocystis with other float bearing kelps. J. Phycol.
(in press)

29. Setchell, W.A. 1908. *Nereocystis* and *Pelagophycus*. Bot. Gaz.
45:125-134.

30. Tseng, C.K. 1975. The breeding of new varieties of Haidai
(*Laminaria japonica* Aresch.) with high production and high
iodine content. *In:* Proceedings, Third International Ocean
Development Conference, Tokyo, Japan. Oceanographical Society of Japan, Box 5050, Tokyo, Intl. 100-31, Japan.

31. Ward, R.F. 1976. Federal fuels from biomass energy program.
In: Symposium Papers, Clean Fuels, Jan. 27-30, Orlando,
Florida. Institute of Gas Technology, Chicago, Illinois,
pp. 427-436.

32. Yabu, H. 1964. Early development of several species of
Laminariales in Hokkaido. Mem. Fac. Fish. Hokkaido Univ.
12:1-72.

- 10 -
Kelp Restoration in Southern California[1]

*KENNETH C. WILSON, PETER L. HAAKER,
AND DOYLE A. HANAN*

Macrocystis, or giant kelp which belongs to the division
Phaeophyta, order Laminariales, is among the largest and most elabor-
ately organized of all algae. It is related to other kelps such as
Laminaria, Nereocystis, Pelagophycus, Postelsia, and *Eisenia* (7).
Abbott (1) recognizes two species of *Macrocystis* in California, *M.
pyrifera* and *M. integrifolia,* and suggests that a variation of *M.
pyrifera*, resembling *M. angustifolia* of the southern hemisphere,
may be found in southern California.

    Macrocystis plants consist of two parts: the fronds and hold-
fast. The fronds are branch-like structures composed of a central
stipe and subalternately arranged blades. The number of blades on
a mature frond varies between 150 and 250, the higher totals occurring on plants in deeper water. Each blade is comprised of a distal
leaf-like lamina and proximal gas-filled float, called a pneumatocyst.
The pneumatocysts buoy the fronds upright in the water. Blades of
mature fronds, at or near the surface, form the primary region of
photosynthesis for adult plants. Photosynthetic products are trans-
located from this primary region down the stipes to portions of the
plant exposed to lower light intensities (26).

    The number of fronds relates to season, age, and condition of
individuals and ranges from two on a juvenile plant to over 400 on
larger ones. Fifty to 150 healthy fronds were typical of adult kelp
plants in the San Diego area. Growth rates of individual fronds
reach 61 cm per day (26). Growth rates of *Macrocystis* fronds in

[1]Kelp restoration work of the California Department of Fish and
Game was conducted by the Sportfish-Kelp Habitat Project (D-J F27D)
and supported by Dingell-Johnson Federal Aid to Sportfish Restoration funds.

Monterey Bay have been observed between 2.8 cm and 32.5 cm per day to average 14 cm per day (18).

The holdfast, a root-like structure, varies in appearance and is composed of many slender finger-like haptera. Unlike a true root which serves to absorb nutrients, the holdfast functions primarily in anchoring the plant to the substrate.

Macrocystis reproduces either asexually or sexually by dimorphic alternation of generations. The large plant viewed along the coast is the diploid sporophyte stage which produces free-swimming spores from specialized sporophyll blades near the base of the adult plant. These spores finally settle on suitable substrate and grow into haploid, dioecious gametophytes producing eggs and sperm, which combine to develop into a new sporophyte generation (26).

Giant kelp grows at depths between 2 and 40.3 m (26) but most commonly between 6 and 24 m (30) through central and southern California. Plants are generally anchored to rock substrate in open coastal areas, but may occur on sandy or muddy bottoms in protected areas.

Plants form aggregations known as kelp beds or kelp forests. These kelp forests are complex marine communities providing food and habitat for a myriad of organisms including many fishes and invertebrates of both sport and commercial importance. Quast (30) states, "In regions with similar rocky substrates of low and moderate relief, the areas with kelp give estimates of the standing crop of fishes that is two to three times as great as that of similar habitats barren of kelp."

The *Macrocystis* community and its associated subtidal plant distribution patterns are quite complex and approach the dimensions and degree of complexity of terrestrial plant communities (9, 19). Many species of algae, although associated with kelp beds, are not found in profusion beneath a moderate to dense surface canopy of *Macrocystis* where the light intensity may be 1/100 of surface values (6). The shading effect of a kelp canopy tends to reduce the standing crop of benthic algae below and appears to encourage the growth of low-light tolerant algae (6). North (6) states, "The standing crop of benthic algae in an open area may be seven times that existing below a *Macrocystis* canopy, but if *Macrocystis* were included, the standing crop of plant material was three to four times as great within the bed as outside."

HISTORY OF KELP BED FLUCTUATIONS IN SOUTHERN CALIFORNIA

Although historical information relating to size and condition of kelp beds is scarce, available information suggests that, aside from natural fluctuations, kelp beds along the southern California coast (Fig. 1) were relatively stable prior to 1940 (Table I). Regression of kelp forests along portions of the southern California coastline was first reported in 1945 when beds off Pt. Loma, in San Diego County, and Palos Verdes Peninsula, in Los Angeles County, began to deteriorate in areas nearest major sewage outfalls. This deterioration progressively affected beds at increasing distances

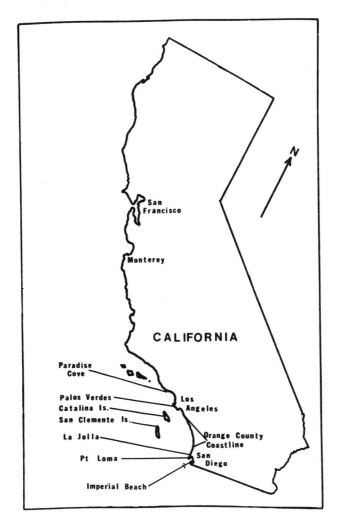

Figure 1. *Map of California showing major Macrocystis restoration areas.*

from the discharges until all that remained of formerly extensive beds was a few isolated stands of kelp at Pt. Loma, La Jolla, and Palos Verdes Peninsula. Deterioration of kelp accelerated when an influx of warm oceanic water persisted off California from 1957 through 1959 (6). Kelp beds along the Pt. Loma coast decreased from a maximum recorded area of 1541 ha in 1911 to approximately 3 ha in 1961. Beds in the La Jolla area decreased from 673 ha in 1934 to only a few small patches between 1960 and 1965. Kelp beds along Palos Verdes Peninsula decreased from 748 ha in 1928 to only two known adult *Macrocystis* plants in 1967 (23)   (Table I).

Initial investigations concerning this decline of kelp beds began at the Institute of Marine Resources (IMR) in 1956 and continued through 1963 (24). These studies, funded by the California

Table I. Variations in kelp coverage of selected beds in southern California given in hectares.

Palos Verdes Peninsula[1]		La Jolla[2]		Pt. Loma[3]		Orange County, Newport Harbor to San Mateo Pt.[4]	
Year	Area	Year	Area	Year	Area	Year	Area
1911	627	1911	490	1857	1541	1911	420
1928	748	1934	673	1911	1541	1955	637
1945	422	1941	648	1934	945	1958	207
1947	271	1949	150	1941	684	1961	62
1953	114	1955	199	1942	904	1967	98
1957	34	1956	122	1949	547		
1958	13	1958	80	1955	306		
1967	No measureable surface canopy.	1969	52	1956	202		
1968		1970	52	1958	158		
1969		1971	78	1959	62		
1970		1972	130	1960	8		
1971		1973	155	1961	3		
1972		1974	155	1963	26		
1973				1964	311		
1974	1.5			1965	414		
1975	9.5 ⎱ 5			1966	518		
1976	26.1			1967	907		
1977	34.4			1968	777		
				1969	829		
				1970	803		
				1971	544		
				1972	570		
				1973	285		

1 From California State Water Quality Control Board (6)
2 From California Institute of Technology (3)
3 From California Institute of Technology (3)
4 From North (20)
5 California Fish and Game Department Estimates

Department of Fish and Game, California State Water Quality Control
Board, and the National Science Foundation, were collectively known
as the IMR Kelp Program. After 1963 some of the IMR work continued
as the Kelp Habitat Improvement Project at the California Institute
of Technology with grants and cooperation from Kelco Company of
San Diego, Fish and Game Commissions of San Diego, Los Angeles, and
Orange Counties, and other contributors (26).

The IMR program and the Kelp Habitat Improvement Project yielded
considerable information about the kelp bed environment. These
studies suggest that kelp beds undergo natural fluctuations in area
as well as in density of plants. Large variations in area may occur
due to natural conditions, but should return to an average value.
These fluctuations may relate to several factors including the quantity and quality of light, water temperature, age of plants, grazing
damage, storms, surge, current, and sedimentary drift (6).

Man-caused changes in the marine environment can also be responsible for loss or deterioration of kelp beds. These changes include
discharge of both domestic and industrial wastes, thermal effluents
from power plants, and siltation resulting from construction and
dredging in the marine environment (6). Miller's studies (18) in
Monterey Bay indicated that repeated, heavy kelp harvests can result
in diminished hapteral growth and subsequent loss of plants through
holdfast deterioration.

These and other yet unknown factors may act independently or collectively to influence the size, distribution, and condition of kelp
beds. North (6) felt that declines in the Pt. Loma and Palos Verdes
Peninsula kelp beds were most directly related to benthic grazing
and increased turbidity resulting from marine outfalls. Although
this study discussed in detail the effects of grazing by sea urchins
and other macro-invertebrates, it did not investigate the possible
effects of micro-consumers on microscopic stages of *Macrocystis*.
This component of the kelp environment needs to be thoroughly
investigated.

KELP RESTORATION

Three kelp restoration programs were in progress in southern
California through 1976. They were affiliated with the California
Institute of Technology, Kelco Company, and the California Department of Fish and Game. Dr. Wheeler North directed the California
Institute of Technology program which was known as the Kelp Habitat
Improvement Project. This project operated continuously from 1964
through 1976 to reestablish kelp bed communities along the southern
California coastline. The project is currently investigating other
aspects of kelp biology. Kelco, a kelp harvesting concern, has been
conducting kelp restoration operations in San Diego County since 1963
to increase their kelp harvest. Kelco's project is currently under
the direction of Ronald H. McPeak. The California Department of Fish
and Game has been actively engaged in kelp restoration work in Los
Angeles County since 1970. Beginning in 1971 the Department intensified its kelp restoration activities with the establishment of the

Sportfish-Kelp Habitat Project. The goal of this project is to enhance sportfish populations through kelp restoration on Palos Verdes
Peninsula.

    Kelp restoration techniques have evolved considerably since 1963
when the first kelp restoration operations were initiated off Pt. Loma.
Kelp restoration operations may be divided into four major aspects:
1) transplantation, 2) predator control, 3) competitor control, and
4) monitoring. A fifth technique, nutrient level modification, is
currently being investigated by the California Institute of Technology. The goal is to determine the nutritional needs of kelp and
whether man-made changes in nutrient levels can prevent deterioration
of kelp beds.

Transplantation

    The technique of transplantation requires the collection, prep-
aration, transportation, and placement of plants into selected locations. Plants may be obtained from existing kelp beds and from laboratory cultures.

    North (25, 29) described laboratory culture operations for
Macrocystis used by the Kelp Habitat Improvement Project at Kerckhoff
Laboratories. Spores, collected from sporophyll blades, are allowed
to attach to plastic films and develop normally through the gametophyte stage into embryonic sporophytes. These embryonic plants are
scraped from the plastic film and dispersed into suitable habitats.

    North's studies suggest that approximately one embryonic sporo-
phyte per 100,000 will reattach to develop into a juvenile plant.
To compensate for this high mortality, several billion kelp embryos,
cultured over a period of 10 to 20 days, are dispersed in the restoration site. Environmental conditions at the time of dispersal and
the age of embryos may affect the success of operations, and results
may not be known for 4 to 18 months. The advantage of this technique
is that large numbers of kelp embryos, genetically adapted to specific environmental conditions, may be released into selected locations
without handling large amounts of plant material.

    Transplantation of adult, sub-adult, and juvenile *Macrocystis*
plants from existing healthy beds has been employed extensively since
the late 1950s by kelp restoration workers throughout southern California. The method requires handling large amounts of plant material
and difficulties are incurred in anchoring plants. Advantages of
using adult plants are great, because up to 200 plants may be moved
into a transplant site per day, representing a wet weight of as much
as 2700 kg of plant material. This substantial biomass affords protection to transplants and their offspring against damage from kelp-
grazing fishes and invertebrates and provides instant reproductive
potential.

    The first step in the California Department of Fish and Game's
transplantation operations was to locate a suitable kelp bed of sufficient size that periodic removal of plants would not cause damage.
Collection and preparation of kelp plants by divers usually began in
water less than 6 m in depth, where plants were selected on the basis

Figure 2. *Basal region of a Macrocystis transplant with rubber ring used for anchoring on the holdfast.*

of condition and size. Plants generally ranged in length from 2 to 6 m, had at least 4 to 6 fronds showing little damage from fish graz- ing, had new hapteral growth, and, when possible, developed sporo- phyll blades. Prior to removal from the substrate, old or deteriora- ting fronds were removed from the plant to facilitate handling and reduce drag. The holdfast was pried from the substrate with a diver's knife and trimmed if necessary to a manageable size. However, when trimming the holdfast, efforts were made to avoid damaging new hap- tera thus prolonging reattachment of the plant.

After two to ten plants, depending upon size, were collected, they were towed to the transfer boat and handed to surface support personnel. If surface personnel were not available to haul plants aboard, metal or wooden rods with circlets of inner tube attached at intervals were suspended at water level alongside the collecting boat. The divers could then hang holdfasts of one to several plants through each of the rubber loops until sufficient numbers had been collected to haul aboard after the dive.

Care of the kelp during transportation is important to assure optimum chances of a successful transplant. Plants can be laid on the deck of a boat but should not be piled more than a few plants deep. Keeping the kelp wet is critical; plants may be immersed, doused regularly, or sprinkled continuously with sea water. Covering the plants with burlap will protect them from drying and burning by the sun. Kelp can also be placed in burlap sacks, a method especially

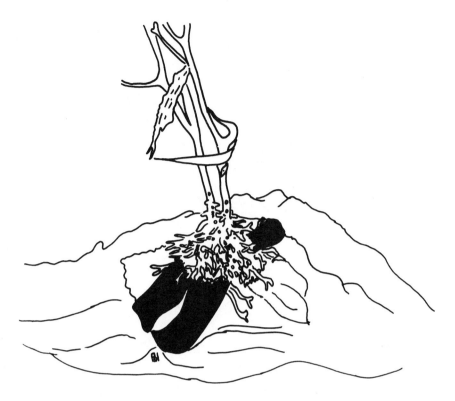

Figure 3. *Basal region of a Macrocystis transplant attached with a rubber ring.*

useful when handling large numbers of small plants.

Important in the transplantation of kelp is the reattachment of the plant in a suitable habitat. In areas where sea urchin grazing is not a problem, plants can be attached directly to the substrate with rubber bands or rubber rings cut from inner tubes using a method developed by Kelco (15) (Figs. 2 and 3). Plants have been success- fully secured to anchored floats using a variation of the method described by North (22) (Fig. 4). The floats suspend the plants above the bottom so that urchins cannot forage upon them.

McPeak (14) developed a method for laboratory culture of *Macro- cystis* plants on rings cut from 1.27 cm plastic pipe. Rings could be taken below by divers in substantial numbers and attached to suitable substrates. They were originally secured directly to rocky substrate using epoxy cement, but this technique was unsuccessful because of damage to kelp plants by starfish which were attracted to the epoxy. A newer method, in which rings were forced tightly over stumps of *Eisenia* and *Pterygophora*, proved more successful (15). Plants were also successfully secured by Number 32 rubber bands directly to stumps or to nails hammered into the substrate (17).

Transplantation programs do not attempt to replace whole kelp beds in any area, rather, kelp transplants are intended as "seed

Figure 4. *Basal region of a Macrocystis transplant secured to a float on an anchor chain with a rubber ring.*

stock" to form the nucleus of a kelp bed. It is important to transplant into areas where the substrate, oceanic conditions, and water quality are suitable for growth and reproduction. Periodic visits must be made to the transplant site to control predators and check the tie-downs of the plants for chafing.

Grazer Control

Many species of invertebrates and several fishes utilize tissues of *Macrocystis* for food. Leighton (11) listed 11 benthic invertebrate species observed to feed upon kelp among which were red and purple sea urchins, *Strongylocentrotus franciscanus* and *S. purpuratus;*

pink, green, and red abalone, *Haliotis corrugata, H. fulgens,* and
H. rufescens; and the common kelp crab, *Pugettia producta.* His
studies suggest that most benthic grazers prefer *Macrocystis* to
other algae.

Of the benthic grazers observed by Leighton (11), sea urchins
were the most detrimental to kelp. Urchins coexisting with vegetation usually cluster in rock crevices or among cobbles and feed
upon drift kelp. But these animals can aggregate in dense concentrations, moving as fronts through algal beds, feeding upon *Macrocystis*
plants and associated algae. Urchin grazing often severs the primary stipes of the plants allowing them to drift away. The area
through which the fronts have passed is usually denuded of algae.

To reestablish the kelp beds and natural species diversity,
high urchin densities in restoration sites had to be reduced. Urchin
control activities may be divided into three phases related to restoration activities: 1) clearing urchins from the restoration site and
adjacent areas prior to installation of transplants; 2) controlling
any remaining or invading urchins which might damage transplants and
their offspring; and 3) monitoring and controlling urchin densities
in the established reproducing kelp bed.

Two principal techniques have been employed by kelp restoration
workers to control dense concentrations of sea urchins. The first is
to destroy urchins physically by crushing them with hammers. Although
positive, urchin crushing is time consuming in relation to the area
cleared. For this reason manual urchin controls are employed only
during reconnaissance operations when more efficient techniques are
not immediately available.

The second technique for sea urchin control is chemical destruction using calcium oxide (quicklime). The use of quicklime to destroy
sea urchins was originally conceived by Leighton (10, 12) and was
based upon Loosanoff and Engles' (13) report of its use to control
starfish in oyster beds. The Kelp Habitat Improvement Project and
the Kelco Company adapted this method for sea urchins. The technique
proved to be quite effective in reducing dense concentrations and is
now used exclusively where extensive populations preclude or endanger
kelp. During 1976, Kelco Company began using a device, developed by
George Beason and John Pickens, for dispersing a slurry of quicklime
and sea water by a diver-controlled hose. Divers selectively direct
the slurry onto target species bypassing the major portion of the
water column and using 40 percent less lime per unit area than in
the earlier method.

The Sportfish-Kelp Habitat Project is also using quicklime on
Palos Verdes Peninsula to control dense populations of sea urchins.
A dispersal device (Fig. 5) fitted to the stern of the project boat
HALFMOON consists of water pump, hopper, quicklime control valve,
sea water mixing bucket, aspirator, and hose. The quicklime is
mixed with the sea water to form a slurry which is sucked out of the
mixing bucket through a 6.25 cm diameter hose. Suction is created
by an aspirator through which sea water is pumped. The slurry is
then carried to the bottom through a 30 m lenth of 6.25 cm flexible
hose which is buoyed every 5 meters to keep it off the bottom.

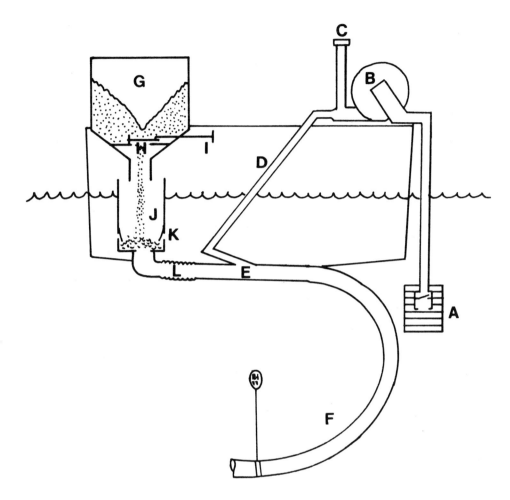

Figure 5. *Diver-operated quickliming device for sea urchin control.
Centrifugal water pump (B), driven by a 2 hp gasoline engine, draws
seawater through a strainer basket and non-return valve assembly (A).
Standpipe (C) is used to prime pump. Seawater is pumped through 3.8
cm I.D. hose (D) to aspirator (E). Quicklime, stored in hopper (G),
drops through control valve (N) which is regulated by handle (I),
into mixing bucket (J). Seawater enters bucket through flapper
valves (K) and mixes with quicklime. Aspirator draws quicklime-
seawater slurry through hose (L) and pumps it through 7.6 cm hose
(F). Diver holds F and spreads quicklime on the ocean bottom.*

Studies have been conducted by Kelco and the California Department of Fish and Game to establish the relative effectiveness of the
two methods for dispersal of quicklime and their effects on kelp,
sea urchins, and other species. This information has been compiled
in a 1976 unpublished report by C. M. Parsons and J. M. Duffy on
chemical sea urchin control activities for the California Department

of Fish and Game.   These studies suggest that the new technique is
more effective in killing sea urchins than surface dispersal (80 –
90% kill ratio vs 50-60% ratio).   Few negative effects were observed
on conspicuous non-target organisms.   These investigations are now
being altered to include smaller and cryptic animals.

Commercial harvesters are removing large numbers of red urchins
from the Palos Verdes Peninsula.   The effects of this harvest are not
yet known but should have a positive influence on kelp restoration.

With the return of the kelp beds, there has been an increase in
fish populations such as the sheephead, *Pimelometopon pulchrum*, a
natural predator of juvenile red and purple urchins.   The effects of
natural predators other than sea otters on urchin populations are not
known, but studies (31) indicate that high sheephead densities may
exert some control on urchin populations.

Fishes such as the opaleye, *Girella nigricans*, and the halfmoon,
Medialuna californiensis, feed upon *Macrocystis* (8, 23, 30) often
causing considerable damage to kelp especially when plant numbers
are low and production of plant tissue does not exceed grazing demands.   North (23) suggests that herbivorous fish prefer *Macrocystis*
to other species of kelp.   Attempts were made to control herbivorous
fish by spearfishing and traps; however, these techniques proved inefficient and their use was discontinued.

California Institute of Technology and Kelco used gill nets to
control grazing fishes in Abalone Cove from 1971 through 1973. Although
the nets succeeded in removing substantial numbers of kelp grazing
fishes, loss of desirable fish species was high and the nets were a
hazard to sport divers; thus, gill netting was discontinued (2, 16).

Fish exclosures were constructed around transplanted kelp (23,
28) to exclude kelp grazing fishes.   These exclosures required considerable maintenance and are no longer used.

Competing Vegetation

Low growing algae, including *Pterygophora, Egregia, Cystoseira,
Sargassum,* and *Laminaria,* can become so dense as to inhibit growth
of juvenile *Macrocystis*.   Less desirable kelps were removed using a
hand held scythe or a fish filet knife to prepare additional substrate for *Macrocystis*.   Some species of competing vegetation provided substrate for settling and growth of kelp plants.   Before this
competitive vegetation was removed, juvenile *Macrocystis* plants were
collected and secured to more suitable substrate.   Where urchins pose
a threat to *Macrocystis*, concentrations of other algae slow the movement of urchins and may thus protect adjacent stands of kelp from
urchin foraging.

Monitoring

Monitoring of kelp beds is an integral part of kelp restoration
work.   Periodic inspection and maintenance of restoration sites is
critical, particularly in early stages, to ensure survival of transplants and their offspring.   Once a stand of kelp is established,
monitoring efforts must continue to detect fluctuations in kelp

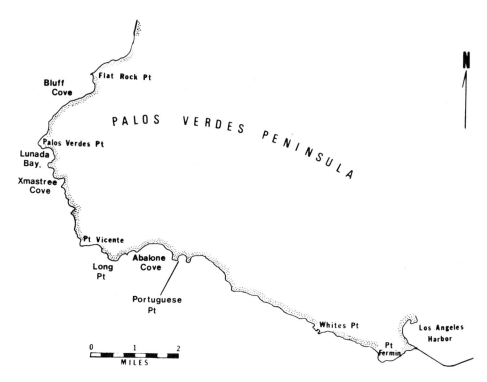

Figure 6. *Kelp restoration sites on the Palos Verdes Peninsula, California.*

bed size and condition and to determine possible causes of change.
    Monitoring operations include aerial, boat, and diving obser-
vations. Aerial observations show fluctuations in surface canopy,
while boat and diving operations provide information on condition,
density, growth rate, grazing damage, and recruitment.
    Aerial infrared photography is used for documentation of lo-
cation, condition, and size of kelp beds. Photographs were taken
using a hand-held 35 mm reflex camera with color infrared slide film
and No. 12 Wratten or No. 25A filter to enhance surface canopies.
Photographs were taken through a viewing port at altitudes ranging
from approximately 1.5 to 2.7 km. To minimize reflected glare, surveys were conducted before or after apparent noon. A 10-20% over-
lap was included between adjacent shots to assure full coverage and
minimize barrel distortion at film edges. Slides of the kelp beds
were projected onto charts of the coast and the surface areas of canopies were calculated using a polar planimeter or a measured grid-
network. Photographic surveys of kelp beds on Palos Verdes Peninsula
are conducted by the California Department of Fish and Game on a
quarterly basis.

RESULTS OF RESTORATION EFFORTS

    Kelp restoration in Los Angeles County has been restricted to

Palos Verdes Peninsula. The California Institute of Technology began its work in 1967 in Abalone Cove (23) (Fig. 6). The California Department of Fish and Game joined in cooperative kelp restoration efforts beginning in 1971 with its Sportfish-Kelp Habitat Project. Unfavorable environmental conditions hindered kelp restoration operations prior to 1973. In 1973 both cooperating institutions intensified kelp restoration activities with the introduction of several thousand adult and juvenile kelp plants, dispersal of mass cultures, and gill netting to reduce large numbers of kelp grazing fishes. A small bed of approximately 60 naturally attached adult *Macrocystis* plants was in the vicinity of the restoration site in February 1974. This small stand of kelp grew to 2.8 ha by March 1975, to 9.06 ha by January 1976, and 13.83 ha by January 1977 (Table II).

In 1974 the California Institute of Technology began work at Whites Pt., approximately 6 km to the east of Abalone Cove (5). Kelp restoration operations at Whites Pt. involved installation of transplants from Abalone Cove, juveniles cultured at Kerckhoff Laboratories, and dispersal of mass cultures of kelp. A small *Macrocystis* canopy calculated to be 0.12 ha was first noted at Whites Pt. in August 1976; by January 1977 this canopy had grown to 0.69 ha.

Encouraged by rapid growth of kelp in Abalone Cove, the Sportfish-Kelp Habitat Project established six additional kelp restoration sites along Palos Verdes Peninsula between 1974 and 1976. These were located at Pt. Vicente, Christmas Tree Cove, Lunada Bay, Marineland, Portuguese Pt., and Pt. Fermin (Fig. 6). A total of 1,307 plants, representing approximately 3700 kg of biomass, was transplanted into the Pt. Vicente site in 1974 and 1975 (Table III). A surface canopy, measuring 0.08 ha, appeared at this site in July 1975, 10 months after work began. This small canopy grew to 0.86 ha by January 1976 and to 2.87 ha by January 1977. Similar growth was noted at Marineland where recruitment from three adult plants grew into a 1.09 ha kelp bed by January 1977. Kelp beds in new sites at Lunada Bay and Portuguese Pt. are also progressing well.

Macrocystis was first found in Bluff Cove on the northwest side of Palos Verdes Peninsula in May 1975. This bed grew to 0.24 ha in July 1975, to 3.92 ha in January 1976, and to 15.8 ha in January 1977.

Kelp restoration work in San Diego County was primarily conducted by Kelco and the Kelp Habitat Improvement Project.

Large-scale kelp restoration operations began at Pt. Loma in 1963. Efforts were directed toward chemical treatment of large numbers of sea urchins which were preventing natural reestablishment of kelp beds throughout the area. By 1968, these efforts were successful in restoring kelp beds to their pre-1940 levels (Table I). However, winter storms of 1973 heavily damaged the Pt. Loma beds and substantially reduced the surface canopy. McPeak (15) notes that damaged plants have begun to be replaced by natural regrowth and that kelp canopies in 1975 were similar to the pre-storm canopies of 1972. Sea urchins still threaten kelp beds along portions of central and south Pt. Loma; consequently, Kelco has a continuing sea urchin control program to maintain and expand these beds.

Table II. *Area occupied by Macrocystis beds around Palos Verdes Peninsula, California, 1974–1977, given in hectares.*

Year	1974		1975			1976				1977
Month	6	10	3	7	10	1	4	8	11	1
Pt. Fermin to Whites Pt.								0.12	0.89	0.69
Portuguese Pt.								0.03	0.04	0.08
Abalone Cove	0.57	1.50	2.80	4.17	5.71	9.06	10.36	9.51	11.61	13.83
Marineland (Long Pt.)						0.24	0.28	0.28	1.34	1.09
Pt. Vicente				0.08	0.76	0.86	1.11	1.40	3.47	2.87
Christmas Tree Cove			–	–	–	–	–	–	0.04	–
Lunada Bay							–	0.01	0.04	0.02
Flat Rock Pt. to Palos Verdes Pt.				0.24	3.03	3.92	3.92	3.23	8.78	15.80
Totals	0.57	1.50	2.80	4.49	9.50	14.08	15.67	14.58	26.21	34.42

Table III. Source, number, and biomass of Macrocystis transplanted to restoration sites on Palos Verdes Peninsula by the California Department of Fish and Game, 1971-1976.

Year	From	To	# Plants	Biomass kg
1971	Paradise Cove	Abalone Cove	53	723
1972	Paradise Cove	Abalone Cove	104	1,418
1972	Catalina Island	Abalone Cove	874	11,918
1973	Catalina Island	Abalone Cove	1,830	24,545
1974	Catalina Island	Abalone Cove	1,415	2,700
1974	Catalina Island	Pt. Vicente	1,086	4,000
1974	Catalina Island	Christmas Tree Cove	195	2,200
1975	Catalina Island	Abalone Cove	110	272
1975	Catalina Island	Pt. Vicente	221	545
1975	Abalone Cove & Pt. Vicente	Christmas Tree Cove	100	272
1975	Abalone Cove & Pt. Vicente	Lunada Bay	185	1,682
1975	Abalone Cove	Portuguese Pt.	30	364
1976	Abalone Cove	Christmas Tree Cove	30	341
1976	Abalone Cove	Lunada Bay	100	455
1976	Abalone Cove	Portuguese Pt.	283	777
1976	Abalone Cove	Pt. Fermin	75	582

Totals by Year

Year	# Plants	Biomass kg
1971	53	723
1972	978	13,336
1973	1,830	24,545
1974	2,696	8,182
1975	646	3,136
1976	488	2,255

Restoration work began off La Jolla in 1964 (21). Efforts included both chemical and manual destruction of sea urchins, transplantation of adult and juvenile kelp plants, control of competing
vegetation, and dispersal of kelp embryos (29). A total of 28,000
juvenile and adult *Macrocystis* plants was transplanted into the
La Jolla area by Kelco and the California Institute of Technology
between 1973 and 1975 (15). An additional 7,870 plants were transplanted into this area in 1976. (This information has been compiled
by Ron McPeak, Kelco Company, 8355 Aero Drive, San Diego, California
92123.) Kelp canopies off La Jolla increased substantially in 1974
and 1975, and restoration efforts continue to further increase and
stabilize existing beds.

Kelp restoration and enhancement operations were initiated along
other portions of the San Diego County coastline. Kelco and California Institute of Technology began kelp restoration work off Imperial Beach in 1963. These operations proved unsuccessful because of
heavy loss of transplants and their offspring to fish and sea urchin
grazing. Restoration operations were subsequently discontinued so
efforts could be concentrated in the Pt. Loma and La Jolla beds. In
1974 North stated that *Macrocystis* no longer existed off Imperial
Beach (3). Kelco has also been conducting a large-scale sea urchin
control program since 1964 to enhance and maintain kelp beds at San
Clemente Island.

The Kelp Habitat Improvement Project began work in 1967 on kelp
beds along the Orange County coastline to hasten recovery of *Macrocystis* in barren areas (20). Transplantation of adult kelp plants
in 1967 and 1968 (22) was not successful because of losses of plants
from fish grazing. Dispersal of kelp embryos produced at Kerckhoff
Marine Laboratory began in Orange County in 1970 and continued
through 1976 (4, 29). Kelp beds in the Orange County area are still
only a fraction of their former abundance and it appears that much
more work will have to be done to restore them (27).

SUMMARY AND CONCLUSIONS

Kelp restoration operations continue to play an important role
in restoring kelp communities along the southern California coastline. However, even with current kelp restoration techniques, only
a few of the many biological and physical factors that affect kelp
survival, growth, and reproduction can be controlled. These techniques have evolved to a point where workers can exert a powerful
influence on competing vegetation and sea urchin grazing. Nevertheless, efforts have often been unsuccessful because of an inability
to fully comprehend and control adverse environmental conditions.
Ultimately, the success of any kelp restoration effort will be decided by the identification and protection or the enhancement and
maintenance of a healthy environment.

ACKNOWLEDGMENTS

Many people have assisted with this work, especially Ken
Hashagen, D-J Coordinator, who provided invaluable assistance in

all phases of project work. Wheeler North and Ron McPeak provided
up-to-date information on their work. Thanks are also due to Herb
Frey who edited the manuscript and offered valuable suggestions and
to Laura Cartner and Charel Cueva who typed the manuscript.

REFERENCES

1. Abbot, I. A., and G. J. Hollenberg. 1976. Marine Algae of
       California. Stanford University Press, Stanford, Califor-
       nia. 827 pp.
2. California Institute of Technology. 1972. Kelp restoration in
       Los Angeles County. *In:* Kelp Habitat Improvement Project,
       Annual Report 1971-1972. W. M. Keck Engineering Labora-
       tories, California Institute of Technology, pp. 57-70.
3. California Institute of Technology. 1974. Operations in San
       Diego County. *In:* Kelp Habitat Improvement Project,
       Annual Report 1973-1974. W. M. Keck Engineering Labora-
       tories, California Institute of Technology, pp. 13-28.
4. California Institute of Technology. 1974. Operations in
       Orange County. *In:* Kelp Habitat Improvement Project, Annual
       Report 1973-1974. W. M. Keck Engineering Laboratories,
       California Institute of Technology, pp. 29-40.
5. California Institute of Technology. 1977. Kelp restoration
       activities in Los Angeles County. *In:* Kelp Habitat Im-
       provement Project, Annual Report 1974-1975. W. M. Keck
       Engineering Laboratories, California Institute of Tech-
       nology. (in press)
6. California State Water Quality Control Board. 1964. An In-
       vestigation of the Effects of Discharged Wastes on Kelp.
       California Water Quality Control Board, Sacramento, Pub.
       26. 124 pp.
7. Dawson, E. Y. 1966. Marine Botany: An Introduction. Holt,
       Reinhart and Winston, Inc., New York. 371 pp.
8. Feder, H. M., C. H. Turner, and C. Limbaugh. 1974. Observa-
       tions on Fishes Associated with Kelp Beds in Southern
       California. California Dept. Fish and Game, Fish Bull.
       160. 138 pp.
9. Foster, M. S. 1975. Regulation of algal community develop-
       ment in a *Macrocystis pyrifera* forest. Mar. Biol. 32:331-
       342.
10. Institute of Marine Resources. 1963. Kelp Habitat Improvement
       Project, Final Report. University of California, IMR Ref.,
       63-13. 123 pp.
11. Leighton, D. L. 1971. Grazing activities of benthic inverte-
       brates. *In:* North, W. J. (ed.) Biology of Giant Kelp
       Beds *(Macrocystis)* in California. Nova Hedwigia 32:421-453.
12. Leighton, D. L., L. G. Jones, and W. J. North. 1966. Ecologi-
       cal relationships between giant kelp and sea urchins in
       southern California. *In:* Young, E. G., and J. L. McLachlan
       (eds.) Proceedings, Fifth International Seaweed Symposium.

Pergamon Press, London, England, pp. 141–153.

13. Loosanoff, V. L., and J. B. Engle. 1942. Use of Lime in Controlling Starfish. U. S. Fish and Wildl. Serv., Res. Rep. 2. 29 pp.

14. McPeak, R. H. 1972. Use of PVC rings as substrate for culturing *Macrocystis* transplants. *In:* Kelp Habitat Improvement Project, Annual Report 1972–1973. W. M. Keck Engineering Laboratories, California Institute of Technology, pp. 100–107.

15. McPeak, R. H. 1977. Observations and transplantation studies at Pt. Loma and La Jolla. *In:* Kelp Habitat Improvement Project, Annual Report 1974–75. W. M. Keck Engineering Laboratories, California Institute of Technology. (in press)

16. McPeak, R. H., T. G. Stephan, and W. J. North. 1972. Details of grazer control operations at Abalone Cove. *In:* Kelp Habitat Improvement Project, Annual Report 1971–1972. W. M. Keck Engineering Laboratories, California Institute of Technology, pp. 71–99.

17. McPeak, R. H., H. Fastenau, and D. Bishop. 1973. Observations and transplantation studies at Pt. Loma and La Jolla. *In:* Kelp Habitat Improvement Project, Annual Report 1972–1973. W. M. Keck Engineering Laboratories, California Institute of Technology, pp. 57–72.

18. Miller, D. J., and J. J. Geibel. 1973. Summary of Blue Rockfish and Lingcod Life Histories; a Reef Ecology Study; and Giant Kelp, *Macrocystis pyrifera*, experiments in Monterey Bay, California. Calif. Dept. Fish and Game, Fish Bull. 158. 137 pp.

19. Neushul, M. 1971. The kelp community of seaweeds. *In:* North, W. J. (ed.) Biology of Giant Kelp Beds (*Macrocystis*) in California. Nova Hedwigia 32:265–267.

20. North, W. J. 1967a. Kelp restoration in Orange County. *In:* Kelp Habitat Improvement Project, Annual Report 1966–1967. W. M. Keck Engineering Laboratories, California Institute of Technology, pp. 24–32.

21. North, W. J. 1967b. Field studies in San Diego County. *In:* Kelp Habitat Improvement Project, Annual Report 1966–1967. W. M. Keck Engineering Laboratories, California Institute of Technology, pp. 5–23.

22. North, W. J. 1968a. Kelp restoration in Orange County. *In:* Kelp Habitat Improvement Project, Annual Report 1967–1968. W. M. Keck Engineering Laboratories, California Institute of Technology, pp. 28–33.

23. North, W. J. 1968b. Kelp restoration in Los Angeles County. *In:* Kelp Habitat Improvement Project, Annual Report 1967–1968. W. M. Keck Engineering Laboratories, California Institute of Technology, pp. 34–38.

24. North, W. J. 1968c. Foreward. *In:* North, W. J., and C. L. Hubbs (eds.) Utilization of Kelp Bed Resources in Southern California. Calif. Dept. Fish and Game, Fish Bull. 139, pp. 7–12. (out of print)

25. North, W. J. 1971a. Mass-cultured *Macrocystis* as a means of increasing kelp stands in nature. *In:* Nisizawa, K. (ed.) Proceedings, Seventh International Seaweed Symposium. University of Tokyo Press, Tokyo, Japan, pp. 394–400.

26. North, W. J. 1971b. Introduction and background. *In:* North, W. J. The Biology of Giant Kelp Beds (*Macrocystis*) in California. Nova Hedwigia 32:1–97.

27. North, W. J. 1973a. Operations in Orange County. *In:* Kelp Habitat Improvement Project, Annual Report 1972–1973. W. M. Keck Engineering Laboratories, California Institute of Technology, pp. 28–40.

28. North, W. J. 1973b. Kelp restoration activities in Los Angeles County. *In:* Kelp Habitat Improvement Project, Annual Report 1972–1973. W. M. Keck Engineering Laboratories, California Institute of Technology, pp. 41–56.

29. North, W. J. 1976. Aquacultural techniques for creating and restoring beds of giant kelp, *Macrocystis* spp. J. Fish. Res. Bd. Can. 33:1015–1023.

30. Quast, J. C. 1968. Some physical aspects of the inshore environment, particularly as it affects kelp bed fishes. *In:* North, W. J., and C. L. Hubbs (eds.) Utilization of Kelp Bed Resources in Southern California. Calif. Dept. Fish and Game, Fish Bull. 139, pp. 25–34. (out of print)

31. Tegner, M. J., and P. K. Dayton. 1977. Sea urchin recruitment patterns and implications of commercial fishing. Science 196:324–326.

- 11 -

Eucheuma—Current Marine Agronomy[1]

MAXWELL S. DOTY

Eucheuma has aroused great commercial interest because
the carrageenan produced by some species of this red alga is exclusively kappa-carrageenan and that synthesized by other species is
solely iota-carrageenan (7). Farm production of *Eucheuma* for its
kappa-carrageenan is a labor-intensive family[2] operation (2, 6, 10).
Although production is very successful (Fig. 1), four major problem
areas have been identified and are receiving attention. The nature
of these problems and the work toward their solution are the subjects
of this paper. The problem areas, although specifically related to
tropical seaweed farming, have been selected for discussion because
they must be understood as parts of any programmatic seaweed production development wherever it may be undertaken. The four major
concerns will be treated consecutively.

THE DANGER OF PRODUCTION MONOPOLIES

All commercial production of *Eucheuma* is now located in rapidly developing nations. Such countries frequently change their poli-
tics, their policies, and their regulations. Thus, industry runs
considerable risk in relying entirely on the farm production from
any one such developing country. In fact, the shift in the national
policies of the former major source of wild *Eucheuma* led to the
development of marine farming for this genus.
The solution to being forced to deal with a single national
monopoly is to encourage industry's support of farm development in

[1]This research was supported in part by U.S. Sea Grant Contract
No. 04-6-158-44114.
[2]In the sense of U.S. Public Law 94-161.

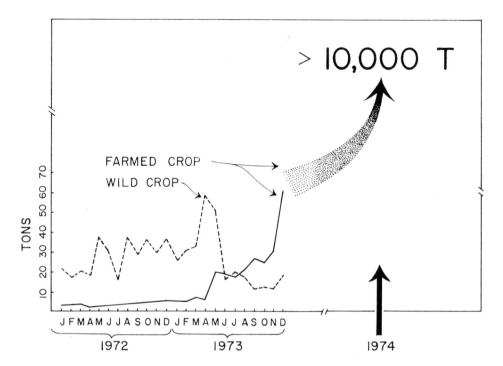

Figure 1.   *The record of success in the farming of a species of Eucheuma, a marine alga producing kappa-carrageenan.   Eucheuma is one of the many kinds of seaweed that might be cultivated—each producing a product of economic value.*

several different countries.   At present, support is being given to *Eucheuma* farm development in the Society Islands, the Line and Gilbert Islands, Micronesia, both Samoas, and Guam (Fig. 2).   The hope is that as the needs for carrageenans continue to grow, major increases in raw product will be derived from the buildup of farm-ing in several of these islands as well as in the Philippines so that significant amounts of the world's production will come from each.   Such competition is hedging, in an economic sense, and is expected to benefit commerce in the United States by stabilizing the price of the raw seaweed material and improving the reliability of the supply.

THE TRADITIONAL ADVANTAGE OF MERCHANTS OVER FARMERS

Family farming of *Eucheuma* returns to the producers who first sell their products in Southeast Asia between 10 to 60% of the f.o.b. value obtained when it is exported.   The return appears to relate to the export price variation during recent years.   The range between $250–$700 U.S. dollars per ton (30% water) is largely due to the different buying tactics of the ultimate processors. There is no organization among the farmers.   A lag in producing

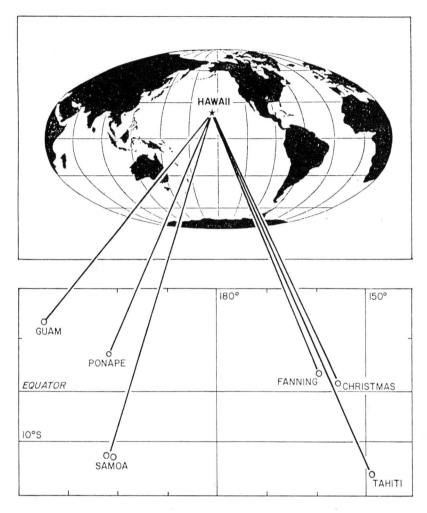

Figure 2. *Areas into which Eucheuma farming is being introduced. Industries in the United States can look forward to greater reliability of crops and stability of prices from producers in different countries or regions which are politically and economically independent.*

a crop after an increase in demand occurs causes a radical "boom or bust" cycle. The unreliability of the price of *Eucheuma* has discouraged farmers who cannot entrust their family's livelihood to the vagaries of a wildly fluctuating market.

Achieving economic stability for seaweed farmers is a current research theme of the University of Hawaii Marine Agronomy Program. This program includes the developing of cultural practices, the obtaining of agreements that will enable a coastal community to decide who can farm the cultivatable reef flats in their area, and the establishing of advance contract sales with processors. These steps

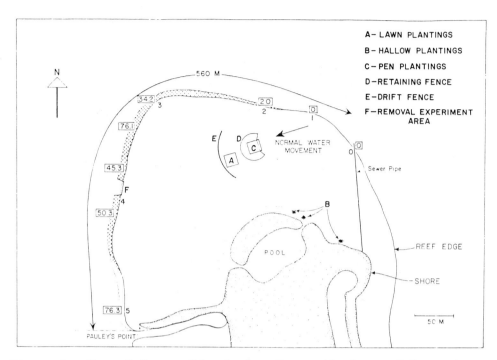

Figure 3. *Map of the reef bordering the north shore of Coconut Island, Oahu, Hawaii, where experimentation toward rational utilization of the subtidal environment is underway.*

should lead to cooperative farm production and community management to reduce the often exhorbitant profiteering by middle men. From 1.5 to 3 kg of algae, producible per man hour by farming (6), is economically rewarding for it interferes little with the normal subsistence efforts of the farmers. This is especially true in countries where the family income of the native is approximately $50 per year. Cooperation coupled with better management will both increase and stabilize prices, as well as improve satisfaction with seaweed farming as a way of life.

THE NEED TO IMPROVE FARMING METHODS

Various cultivation techniques now used successfully include tying thalli to the intersections of nets, tying thalli at intervals along lines or poles, and merely putting sprigs of *Eucheuma* under convenient rocks to grow (2, 6, 13). These methods, respectively, result in either orderly or barely recognizable farms. Despite the use of the very successful TAMBALANG strain of kappa-carrageenan producing *Eucheuma* (6), such labor-intensive methods would be non-economic in developed countries.

Therefore, finding techniques for more capital-intensive farming of *Eucheuma* has become a key part of the University of Hawaii Marine Agronomy Program. Many methods have been tried and abandoned in the Philippines and on the reef bordering the north shore of Coconut Island in Hawaii (Fig. 3). The three methods of culture now selected for further testing can be called *Lawn planting*, *Hollow planting*, and *Pen planting*.

Lawn Planting

In *lawn planting* (A in Fig. 3), a roll of polyvinylchloride-covered wire is hooked to the hard horizontal sand bottom just below the extreme low water level by inverted J-shaped 3/8" or 1/4" diameter reinforcing-iron pegs. The seedstocks are pushed under the wire and, thus, held rigidly in place. Growth rates are excellent, usually exceeding 5% per day. As the crop increases, a progressively higher proportion of the thalli break off and drift away until an equilibrium is achieved.

Initially, harvesting of lawn plantings was attempted by mechanical methods. Considerable time was spent designing a harvester that would operate in waters that were often less than 30 cm deep, that would be stable, and that required but one man to operate and adjust efficiently for tide heights, bottom unevennesses, cross currents, contrary cross winds, and other exigencies. However, an easier harvesting method was soon devised by installing a drift fence (E in Fig. 3) downstream of the planting. The mass of thalli which were cut or broken from the beds accumulated against it and could easily be loaded into a boat by hand or with a pitchfork. An injection pump, such as the one devised by Dr. Maurice Dube for *Iridaea* harvesting, could also pick up the harvest.

Hollow Planting

Another successful method of culture involves growing the algae loose in hollows (B in Fig. 3) on the otherwise horizontal reefs. This technique followed three observations: (a) a negative regression between light intensity and the standing crop on reef flats (5), (b) the habit of the iota-carrageenan producing *Eucheuma denticulatum* ("spinosum" of commerce) to collect in reef hollows where it grows into sizeable colonies, and (c) the similar habit of commercial *Furcellaria* as observed growing in a similar habitat in Danish Kattegat waters (10). The subtidal to lower eulittoral zone populations of *Phyllophora* (3) and other seaweeds (8, 15), some of which are commercially harvested, also occupy similar habitats. In the hollows producing commercial crops of *Eucheuma*, no reproduction by spores has ever been reported. In fact, sporulation has not been observed in the TAMBALANG strain of *Eucheuma* anywhere during its four years of cultivation.

The farming of hollows requires little labor, other than protection from poaching, and only a boat for gathering at harvest time. It takes about 1-1.2 man-days to remove 1000 kg from a hollow and to spread it to dry on the shore, usually only 50-100 feet away. Since growths are dense, mechanical harvesting with equipment such as an unusually shallow draft barge would be easy.

Problems do arise when culture in hollows is attempted. At the research site in Hawaii the very dense crops lose vegetative propagules with every small disturbance and cause proliferation of new colonies where they are not wanted. However, in nature few hollows suitable for *Eucheuma* production are available and the problem of overproliferation does not arise.

Pen Planting

In the Sulu Archipelago bamboo-slatted fences in various sizes, some surrounding areas up to five meters square, have been employed as a promising way of creating artificial hollows. In Hawaii polyvinylchloride-covered wire fencing is wired to 3/8" diameter reinforcing-iron stakes driven into the bottom (C in Fig. 3). The resulting pens form long rectangles which are divided to distribute the seaweed evenly to facilitate mechanical harvesting.

The lack of economically feasible control of fouling organisms, largely *Enteromorpha* and barnacles which clog the fences, has been a block to effective *pen planting*. The pressure from currents increases as the fencing clogs with fouling algae and shell. The currents then push the fences like sails and destroy the pens. The crop may escape, but if the pens hold, the water circulation may become so limited that the seaweeds grow slowly. Fouling by barnacles may be dangerous to personnel, for their scute edges are razor-sharp. However, excellent growth in pens of over 5% per day has been obtained. Control of fouling by economical means is now being attempted.

Growth obtained by all three farming methods has been very satisfactory for *Eucheuma*. All produce plants which are amenable to economical mechanical harvesting. These methods have been so successful with the TAMBALANG strain of *Eucheuma* that a series of unforeseen complications has arisen. They are described below as the fourth major problem area.

THE GOVERNMENTAL BUREAUCRACY AND THE MANAGEMENT OF INTRODUCED SPECIES

The current concern for environmental quality and the lack of crucial information needed for the rational utilization of the remaining natural resources has generated public support for research on algae regarding their commercial production for human food, industrial chemicals, energy sources, and water purification. Introducing marine plant farming into consuming countries such as the U.S. could be of significant economic importance. It would help by stabilizing the supply, shortening the distance between farm and factory, and aiding water purification as waste water discharge into the sea increases. In Hawaii the potential significance is not only economic but social. Labor, made available by plantation closures, is becoming plentiful and its employment to achieve rational social advances is of increasing importance.

Even an advanced country such as the USA is faced with complex series of problems when considering the adoption of a new technology. In Hawaii all subtidal lands are zoned for conservation which leads

to a commendable ratio of conserved to unprotected or non-conserved
state land.   The effort to obtain clearance to bring a new seaweed
crop to Hawaii is currently involving eleven federal, state, and
local agencies.   On Guam starting a mariculture project requires
clearance from at least seven federal agencies.   Bureaucratic complexity is a major barrier to progress and could serve to deprive
Hawaii's unemployed coastal residents of income.   In France the introduction of seaweed crops to reduce imports is being blocked by a
similar bureaucracy.

It is, of course, true that new crop species should not be introduced without careful consideration of possible damaging effects.
Species being considered for introduction must first be studied in
their native habitats.   Their roles in the old and the new ecological structure should be understood.   For example, it was discovered
that the species developed for *Eucheuma* culture in the Philippines
failed to become naturalized when introduced into Hawaii.   Colonies
died out when a farm was abandoned as corn disappears from a fallow
corn field.   Sometimes, unexpected enhancements of the biota have
been observed.   In the case of *Macrocystis* plantings (11, 12) and
in the *Eucheuma* farms, fish populations increase.

Careful biological investigation of the life cycles of the potentially valuable exotic species which might be introduced to a new
area may be most instructive.   An example of the considerations that
must be reviewed can be taken from the proposal to introduce *Macrocystis* into Hawaiian waters.   Such an introduction appeared advan-
tageous because of the plant's saleability, its aesthetic values,
its capability to protect coasts from erosion, and its known ability
to shelter large fish populations.   The obligate gametophytic stage
of *Macrocystis* doesn't succeed when the temperature rises above 18°C.
In Hawaii water temperatures as low as 19°C at the depths in which
Macrocystis will grow are rare.   Warm water strains of the saleable
sporophyte stage of *Macrocystis* will survive in waters 5-7 degrees
higher.   If this stage could survive, it might thrive in Hawaii and
create a desirable open-ocean mariculture.   A necessary complication,
although probably not a real barrier in view of North's simple technique for overcoming it (11, 12), would be that introduced by attempting to maintain the farms through gametophyte-seeding.   This involves
production of haploid spores under artificially refrigerated conditions in order to produce the relatively short-lived microscopic
gametophytic stage which dies or fails to function above 18°C, but
which is required to generate the macroscopic sporophyte that would
constitute the commercial crop.

In Hawaii efforts are underway to study the control of farmed
seaweeds, both before they are introduced and after, to prevent damage
to the environment (A-F Fig. 3).   Several methods for the eradication
of weedy invasion of introduced species beyond the culture plots have
been considered.   The discovery of a selective toxicant has proved a
dream thus far.   Control by removing oxygen cheaply to near zero under a plastic cover can be achieved simply by introducing sugar.
However, that is disastrous to the long-lived sessile animals, such
as corals.   The mobile and short-lived animal or botanical populations

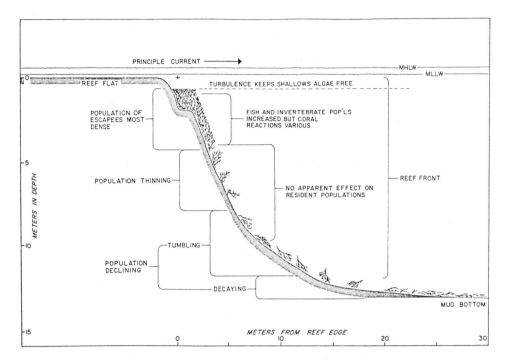

Figure 4. *The nature of the reef-edge populations of the TAMBALANG strain of Eucheuma on Cocount Island, Oahu, Hawaii. Procedures are being developed for removal of the escaped algal population on the reef front.*

can be expected either to move out or recover quickly. Another procedure using an opaque covering placed over unwanted photosynthetic algae starves them out, but it kills the hermotypic corals which depend on their endozooic algae which are also dependent on photosynthesis. In Oregon or any other area lacking significant coral, it could be less difficult to remove unwanted seaweed by shading.

In Hawaii experimental cultivation on a reef flat has provided an opportunity for studying the control of an exotic seaweed population (F in Fig. 3). Beyond its edge the reef front drops precipitously and irregularly to approximately 13 meters (Fig. 4). There it consists primarily of dead coelenterate coral and sediment. The currents are very slow in the deeper water and the bottom is covered with a fine sediment. This sediment originates mostly on land and it is thought to have been the major cause of death of nearly all the coral on this particular reef edge (1).

In the experiments with the introduced TAMBALANG strain of *Eucheuma*, branches that broke off were often lost and drifted over the downstream edge of the reef. The repeated introduction of propagules on the reef front resulted in what is now a splendid stand between, and often covering, the dead coral. A population of fish species feeding primarily or secondarily on *Eucheuma* has appeared. The fish as well as the invertebrate population have both increased

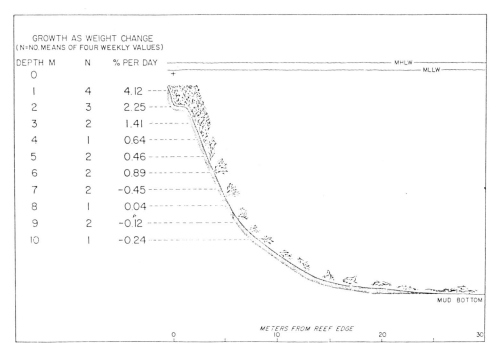

Figure 5. *Growth rates of the TAMBALANG strain of Eucheuma on the reef front on Coconut Island, Oahu, Hawaii, as a function of depth. The column headed "N" shows the number of measurements made over a four-week period which were averaged to yield the growth rates in the "% per day" column.*

beyond those in the non-vegetated and dead coral stands nearby.

Concern has arisen that *Eucheuma* may prevent the return of the corals as water quality restoration efforts become effective. In fact, some small corals can be seen already. Generally there is little competition between the seaweed and the corals. However, both can hardly occupy the same space at the same time. In the tropics, coral-rich and seaweed-rich areas are usually more or less distinct from one another--perhaps this is a response to differences in water quality. Such a relationship would suggest that water over the present reef does not now encourage coral regrowth, but it obviously encourages seaweed growth as well as high fish and invertebrate populations. Nevertheless, a unique opportunity has been presented to develop and test methods of controlling, eliminating, or reducing seaweed populations where they are not wanted.

Research on the most promising methods of controlling, eliminating, or reducing *Eucheuma* populations to insignificance in Hawaii has been aimed at mechanical techniques and the physiology of the alga itself. Harvesting, with no market for the crop, is costly. However, when brought ashore and spread thinly, *Eucheuma* dries and decomposes in the sun and rain with no objectionable odor or appearance. Therefore, mechanical removal and disposal presents no

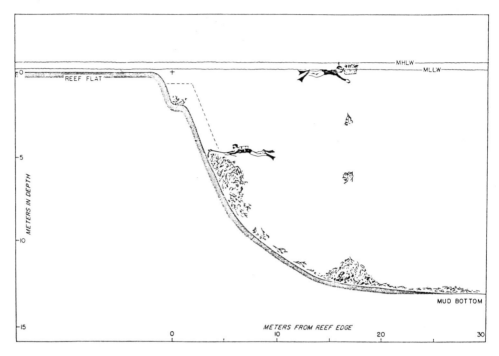

Figure 6. *The least expensive way of reducing the reef front popu-lation of the TAMBALANG strain of Eucheuma from the reef front of Coconut Island, Oahu, Hawaii. In the dark, sediment-laden depths it slowly dies.*

problem except that of cost.

An ecological approach to control of the escaped *Eucheuma* popu-lation in Hawaii has revealed the following. Below 6 meters the thalli lose weight (Fig. 5). They decay in a few weeks at 10–12 meters beneath the surface. The increasing herbivore population is holding the shallower fast-growing populations at a near stand-still through the removal of the growing branch tips. Fish not formerly considered herbivores are eating it. They browse the young growth back to the coarse basal branches. These old branches in-crease to 4 cm in diameter and, while not as persistent as some brown algae in the dark (4, 9), they will retain almost their normal capacity for photosynthesis after storage in the dark for as long as 4 weeks. There is a tendency for the thallus fragments that settle below the 4-meter level to slide down the steep reef front without attaching. Furthermore, when revegetation from the reef flat is prevented, the population of *Eucheuma* that has taken hold begins to dwindle.

The control treatment now being used experimentally is shown in Fig. 6. The thalli are simply dislodged and raked below the 10-meter depth. This is a labor-intensive practice, but is proving the least expensive in the long run. It requires 2–8 man hours to clear a 10-meter-wide transect. An area once cleared by divers requires monitoring to remove laterally borne or overlooked vegetative pieces. Such work has taken a total of 2 hours during one 3-month period for a 10-meter length of reef margin. Reintroduction by fragments from the experimental plantings on the reef flat was greatly reduced by a fence at the reef margin. It is likely that the unwanted reef front population can be reduced to insignificance or even eliminated. Monitoring to prevent its interfering with coral regeneration can be terminated in less than a year after the reef front has been cleaned and the reseeding potential has been eliminated.

SUMMARY AND CONCLUSIONS

In resume, the physical environment and the biological characteristics of algae may not be the severe barriers to economic seaweed production that the bureaucratic and socio-political ones are. The only places where marine agronomy has succeeded have been those where by tradition it is separated from animal aquaculture. It is today the sixth largest dollar earner of all mariculture activities in the world and this does not include wild crop production (14). *Porphyra* production alone is the largest of all maricultural practices in Japan today. There, it is worth over half a billion dollars yearly. Worldwide, marine plant cultivation in the early 1970s appears to have been the key to an annual production of 2.2 billion dollars of product sales. It would seem that seaweed research should be uncoupled from year-to-year project funding, from traditional thinking, and from its all too frequent submersion in other programs. Universities should be encouraged to organize marine agronomy as an independent discipline with ongoing funding.

To fulfill its natural role, a marine agronomy effort must be programmatically planned to include the gamut of activities normally recognized in advanced terrestrial agriculture. These can be listed as: a) *taxonomy and plant exploration*--the marine botanical organisms are very poorly known; b) *natural product chemistry*--early efforts should concentrate on the most abundant and widely distributed marine plants because wide distribution suggests broad ecological tolerances that improve the prospects for successful culture; c) *development of production methodology*--wild crop harvesting, as in the present case of *Macrocystis*, or farming, as in the case of *Eucheuma*, should be explored; d) *an extension program for the encouragement of the industry*--knowledge about the production and marketing of the crop should be distributed and publicized; e) *education and training*--finally, marine agronomy must assume the responsibility for providing more trained personnel to play major roles in the development of the industry and to assure reasonable conservation of the environment in which the algae are cultured. A major effort should be mounted to ensure a continuation and expansion of this renewable resource.

REFERENCES

1. Banner, H. A. 1974. Kaneohe Bay, Hawaii: Urban pollution and
       a coral reef ecosystem. *In:* Cameron, A. M., *et al.* (ed.)
       Proceedings, Second International Coral Reef Symposium.
       Great Barrier Reef Committee, Dr. P. Mather, Queensland
       Museum, Fortitude Valley, Qld. 4067, Australia. 2:685-702.
2. Braud, J. P., R. Perez, and G. Lacherade. 1974. Etude des
       Possibilities d'Adaptation de l'Algue Rouge *Eucheuma spinosum*
       aux Cotes du Territoire Francais des Afars et des Issas.
       Science et Peche, Bull. Inst. Peches Marit. No. 238. 20 pp.
3. Chapman, V. J. 1970. Seaweeds and Their Uses. Second Ed.
       Methuen & Co., Ltd., London, England. xii + 304 pp.
4. Dahl, A. L. 1971. Development, form and environment in the
       brown alga *Zonaria farlowii* (Dictyotales). Bot. Mar. 14:76-
       112.
5. Doty, M. S. 1971. Antecedent event influence on benthic marine
       algal standing crops in Hawaii. J. Exp. Mar. Biol. Ecol.
       6:161-166.
6. Doty, M. S., and V. B. Alvarez. 1975. Status, problems, ad-
       vances and economics of *Eucheuma* farms. Mar. Tech. Soc. J.
       9(4):30-35.
7. Doty, M. S., and G. A. Santos. 1977. Carrageenans from tetra-
       sporic and cystocarpic *Eucheuma* species. Aquatic Botany
       (in press)
8. Harger, B. 1972. Studies on the benthic algal flora seaward
       from the reef flat at Waikiki, Oahu, Hawaii. Masters Thesis,
       University of Hawaii, Honolulu. viii + 185 pp.
9. Mathieson, A. C. 1965. Contributions to the life history and
       ecology of the marine brown alga *Phaeostrophion irregulare*
       S. et G. on the Pacific Coast of North America. Ph.D. Dis-
       sertation, University of British Columbia, Vancouver.
       113 pp.
10. Naylor, J. 1976. Production, trade and utilization of seaweeds
       and seaweed products. Food and Agriculture Organization of
       the United Nations, Rome, Italy. FAO Fisheries Tech. Pap.
       No. 159. V + 73 pp.
11. North, W. J. 1971. The biology of giant kelp beds (*Macrocystis*)
       in California. Nova Hedwigia 32:xiii + 600 pp.
12. North, W. J., *et al.* 1976. Ocean Food and Energy Farm Project:
       ...Biological Studies of *M. pyrifera* Growth in Upwelled
       Oceanic Water...Final Report. California Institute of Tech-
       nology, Pasadena. VI + 259 pp.
13. Ohmi, H., and I. Shinmura. 1976. Growth of *Eucheuma amakus-
       aensis* in the field culture. Bull. Jap. Soc. Physol. 24(3):
       98-102.
14. Pillay, T. V. R. 1976. The State of Aquaculture, 1975. FAO
       Technology Conference on Aquaculture, Kyoto, Japan, 1976.
       13 pp.
15. Smith, L. K. 1968. The role of benthic marine plants in the
       littoral phosphate cycle. Ph.D. Dissertation, Stanford
       University, California. viii + 323 pp.

- 12 -

Algal Nutrition in the Sea—

Management Possibilities[1]

WHEELER J. NORTH

Considerable information is available concerning algal
nutrition under laboratory culture conditions (4, 7, 8, 15). There
are also several compendia that summarize the many analyses of elemental composition of marine algae (e.g., 3, 18). In contrast, relatively little is known about management of algal nutrition under
natural conditions. Specialists can usually identify unhealthy
plants in nature, but only rarely are able to prove that the cause
of a problem does indeed arise from a nutritional deficiency. The
Japanese and Chinese have undertaken fertilizing experiments and
operations among their algal crops (2, 17). Judging from descriptions
which have been published in English, fertilizing activities in the
Orient have largely been trial and error operations. They do not
seem to be based on an extensive body of knowledge that permits
assessment of the condition of the water or the needs of the crop in
a given estuarine or coastal environment.

Occasionally plant nutrients, typically nitrogen and phosphorus concentrations, are monitored in aquaculturing operations. It is
widely and probably correctly assumed that nitrogen is an important
limiting nutrient in the marine environment (11). Inorganic N often
falls to low, even undetectable, concentrations in oceanic surface
waters. Levels of both N and P tend to rise as depth increases because nutrient regeneration processes below the euphotic zone are no
longer coupled to nutrient accumulation by growing plants. The

[1] In part, this work is a result of research sponsored by NOAA
Office of Sea Grants, Department of Commerce, under Grant No. GH-92.
The U.S. Government is authorized to produce and distribute reprints
for governmental purposes notwithstanding any copyright notation
that may appear.

ratios of maximum N and P concentrations at various depths to the
surface concentrations cover a broad range of values--from about 10
to more than 400 in one survey (Table I).

Table I. *Concentration ratios for N, P, and Ni in ocean waters*[1].

Station No.	Shallowest Depth Samples (m)	Element Ratio and (Depth)		
		P	N	Ni
3	13	1.88 (232)	1.84 (441)	2.26 (3730)
343	59	14.20 (791)	444.00 (1033)	3.18 (3717)
204	57	107.00 (1039)	434.00 (1039)	3.27 (1839)

[1]Ratios of the maximum elemental concentrations below the top 60 m
of the water column to the concentration in the top 60 m for P (as
PO_4), N (as NO_3), and Ni. Depths in meters where maximal concentrations occurred are shown in parentheses. The ratios indicate to
what degree a particular element in the upper layer of the ocean
could be enriched by upwelling from the depth of maximum concentrations. Data from Sclater, Boyle, and Edmond, 1976 (12).

It is useful to determine which element or elements might become
limiting in situations where N is plentiful. With such information,
it might be possible to enhance algal growth under certain conditions if fertilizing operations with elements other than, or in addition to, nitrogen were conducted.
Knowledge of nutritional trace minerals other than N and P in
seawater is quite limited because of assay difficulties. Only recently have reliable data become available. The vertical distributions of many of the micronutrients appear to be quite different
from the distributions of N and P. The ratios of maximum concentrations at various depths to surface concentrations for six micronutrients commonly range from 0.5 to 2 but rarely exceeds 10, (Table
II, also see ratios for Ni in Table I), contrasting with the range
of 10 to more than 400 for N and P shown in Table I. Vertical variation appears restricted for plant nutrients such as Mn, I, and Mo.
Vertical variation is somewhat greater for Cu and Zn (though still
small compared to P and N) and erratic for Co. Concentrations of
trace metals are frequently distributed vertically in random fashion.
The well-defined patterns of steady increase with depth that characterize vertical distributions of P and N are poorly defined or alto-
gether lacking for most of the trace metals. Probably, physical and

Table II. *Concentration ratios for ten elements in ocean water*[1].

Date, or Location, or Station No.	Element Ratio and (Depth)										Reference
	Cd	Cu	Fe	Mn	Mo	Ni	V	Zn	Co	I	
6905	1.42 (150)	1.83 (250)	1.38 (1000)	1.47 (500)	0.55 (250)	1.21 (150)	1.37 (750)	1.36 (250)			Riley and Taylor (9) [northeast Atlantic Ocean]
6917	1.27 (2500)	4.10 (2500)	1.16 (2000)	1.06 (150)	1.14 (2000)	1.06 (150)	2.22 (1500)	0.50 (500)			"
6930	1.22 (2000)	1.06 (1000)	0.94 (2500)	1.00 (750)	1.30 (500)	1.48 (750)	1.18 (150)	1.17 (250)			"
6947	1.89 (500)	2.57 (1500)	1.05 (2500)	1.10 (150)	1.40 (150)	1.38 (1000)	1.29 (750)	1.05 (500)			"
6982	1.11 (450)	3.40 (150)	1.41 (150)	1.19 (150)	1.01 (450)	1.00 (450)	1.75 (150)	1.65 (150)			"
6998	1.56 (2500)	6.40 (1500)	1.09 (2500)	1.31 (1500)	1.43 (1500)	1.35 (500)	2.11 (2500)	2.09 (750)			"
Geosecs Test Sta. NE Pacific		5.60 (3500)	4.16 (1000)			2.28 (650)		2.31 (2500)	1.80 (1000)		Spencer *et al.* (14) [northeast Pacific Ocean]
Geosecs Test Sta. II-Sargasso		4.65 (3232)	4.31 (190)					3.92 (3232)	0.32 (390)		Brewer and Spencer, (1)
62H 13-1		0.80 (900)		0.69 (900)				0.92 (900)			Slowey and Hood (13) [Gulf of Mexico]
63A 1-7		7.70 (500)		1.54 (500)				5.90 (500)			"

Table II – cont'd

Date, or Location, or Station No.	Cd	Cu	Fe	Mn	Mo	Ni	V	Zn	Co	I	Reference
				Ratio and (Depth)				Element			
63A 1-3		1.36 (700)		1.26 (350)				2.56 (1000)			Slowey and Hood (13) [Gulf of Mexico]
64A 5-4		5.15 (1000)		1.10 (300)				2.90 (1000)			"
63A 1-4		1.50 (1000)		0.66 (470)				3.06 (1000)			"
65A 11-25		3.00 (750)		1.07 (750)				2.90 (750)			"
10/30/68									3.18 (400)		Robertson (10) [northcentral Pacific Ocean]
11/2/68									17.20 (5350)		"
11/6/68									0.20 (100)		"
11/8/68									0.59 (15)		"
6/25/67									0.43 (300)		[northeast Pacific Ocean-- off Newport, Oregon]
6/26/67									2.86 (500)		"

Table II – cont'd

Date, or Location, or Station No.	Cd	Cu	Fe	Element Ratio and (Depth) Mn	Mo	Ni	V	Zn	Co	I	Reference
Central Atlantic										1.29 (100 to 200)	Wong and Brewer (19)

[1]Ratios of the maximum elemental concentrations below the top ten m of the water column to the concentration in the top ten m from various oceanic stations as reported in the literature. Depths (in m) of the maximum concentrations are shown in parentheses.

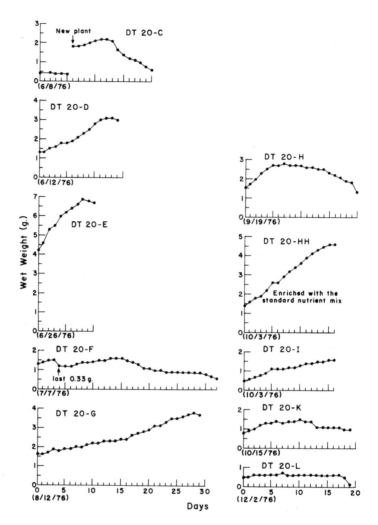

Figure 1. *Growth records from juvenile Macrocystis sporophytes held in unenriched deep seawater (except for Experiment DT-20HH) pumped up from depths of 300 m. Experiment DT-20HH consisted of deep water enriched with a standard nutrient mix (see Table III). All experiments excepting DT-20HH grew slowly or not at all.*

chemical factors control total amounts of trace metals in seawater, and biological processes play a minor role. It appears that up-welling processes, which renew supplies of N and P to the euphotic zone, may be much less effective in restoring other micronutrients.

Little or no variation in the vertical distribution of elements such as Mn, I, and Mo might indicate that supplies dissolved in sea-water are more than adequate and such substances are probably never limiting. Such a presumption might be valid if most of the micro-nutrient in question were readily available to absorption by plants. Many substances, such as trace metals in seawater, occur in a

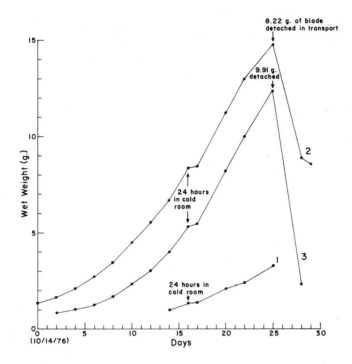

Figure 2. *Growth records from three juvenile Macrocystis sporo-phytes held in a plexiglass container attached to a mooring 6 km offshore from Newport Bay, California. The container received water continuously pumped from depths of 300 m. The plants were temporarily held in flowing bay water in a cold room on day 16 because of an equipment failure.*

variety of chemical combinations of which only a fraction may be readily available to plants. If the "available" fraction of a given micronutrient is a fairly small portion of the total, the substance could be limiting growth, yet the total quantity present might display little or no variation in its vertical distribution. Unfortunately, there are no suitable chemical analytical methods that can distinguish between the "available" and "non-available" fractions of micronutrients. The "available" fractions in different water samples might, however, be compared by employing bio-assay techniques. Juvenile *Macrocystis* sporophytes have been used in the laboratory to compare nutritional characteristics of deep and surface seawater.

BIOASSAYS WITH MACROCYSTIS

Experimental techniques for determining growth rates of juven-ile *Macrocystis* sporophytes in the laboratory have been described (6). In brief, plants were held in aquaria of 40 L capacity. Water was recirculated at a flow of 15-20 L/min. Plants were held in a narrow deep trough within the aquarium so that water flowed vigor-

ously across the blades. Media were renewed every other day in
most experiments. Plants were maintained under continuous illumination in a temperature-controlled cold room. The shape of the aquarium led to the name "deep trough," and all studies using these con-
tainers were given the designation DT, followed by the number of the
experiment as seen in Fig. 1.

In one series of experiments, some plants were held in unenriched surface water from various sources and others in deep water
pumped from depths of 200 to 300 meters. Plants did not grow in a
medium of offshore surface water and died in a few days (5). One
plant held in surface water from Newport Bay, California, grew at
6% per day, but bay water typically contains 2 to 20 times as much
NO_3 as our offshore surface samples. The average rate for young
plants in a natural kelp bed was 9% per day. Growth in deep water
was variable but maximal values of about 14% per day were obtained
in initial experiments (5, 6). More typically, however, growth in
deep water media was moderate to poor (Fig. 1).

Poor growth in deep water might be caused either by toxicity
or by inadequacy of one or more micronutrients. These two possibilities were distinguished by holding young plants in flowing deep
water. If deep water were toxic, more of the toxicant should be
available in a flowing system and plants should do poorly compared
to those in a batch-culturing system without flow. If micronutrients in the laboratory were limiting in deep water, they should become more available in a flowing system. The results showed that
all experimental plants held in flowing water from 300 m or deeper
have grown well (Fig. 2), suggesting that poor growth in the aquaria
resulted from inadequacy of one or more nutrients. Mixtures of
equal parts of surface and deep water enhanced growth rates in the
flowing system (7). When a selected group of trace elements was
added (Table III) to deep water in experiment DT-20HH (i.e., a batch-
culture type of experiment), growth improved substantially (Fig. 1).
Concentrations listed in Table III are generally one or two orders
of magnitude higher than the naturally occurring concentration in
the ocean, except for N and P. The objective was to "swamp out" the
more subtle chemical reactions in seawater so as to force higher concentrations of "available" forms of these micronutrients. Selections
of elements in Table III and their concentrations were based on published information on micronutrients as well as experiments with
sporophytes and gametophytes in this laboratory.

Information on the availability of individual elements of surface and deep water has been sought by growing plants in seawater
enriched with the nutrient solution of Table III minus one constituent at a time. N and P, however, were always adequate or were in
excess. Comparisons among the records for several seawater sources
indicated that best growth always occurred for the medium in which
Zn was omitted (Fig. 3). The Mo-deficient medium also usually supported substantial growth, although generally not as good as obtained when Zn was omitted. The Mo- and Zn-deficient solutions sometimes supported more rapid growth than the controls (Fig. 4).

Table III. *Standard nutrient solution for Macrocystis*[1].

Element	Supplied As	Concentration used, ugat/L of Seawater
Nitrogen	$NaNO_3$ or NH_4Cl	30.00
Phosphorus	K_2HPO_4	2.00
Iron	$FeCl_3 16H_2O$	1.00
Manganese	$MnCl_2 \cdot 4H_2O$	1.00
Zinc	$ZnCl_2$	1.00
Molybdenum	$Na_2MoO_4 \cdot 2H_2O$	1.00
Cobalt	$Co(NO_3)_2 \cdot 6H_2O$	0.01
Arsenic	$Na_2HAsO_4 \cdot 7H_2O$	0.30
Iodine	KIO_3, KI	10.00

[1]The Standard Nutrient Solution is used with seawater and may not contain all trace elements needed by *Macrocystis*. Missing elements would be supplied by the seawater. Zinc and molybdenum at times may not be necessary.

Possibly Zn and Mo were inhibitory at 10^{-6}M (the concentrations listed in Table III). However, other records have been obtained of plants receiving Zn and Mo enrichment at 10^{-6}M which equalled or exceeded the growth rates and survival times shown in Figs. 3 and 4. Joyce Lewin (personal communication) has observed that certain diatom species were inhibited by additions of Zn to enrichment media. In any case, Zn and Mo normally in seawater are apparently adequate to support vigorous growth by *Macrocystis*.

The media without added Fe and Co yielded variable results. Sometimes moderate growth rates occurred for periods up to 15 days. Sometimes results were poor even at the beginning of the experiment. Attempts to revive test plants by adding a missing nutrient were never successful. By the time clearly defined signs of unhealthiness appeared, damage was apparently irreversible. In some cases only a few days were required to create an irreversible condition associated with a presumed nutritional deficiency. Possibly seawater varies, so that only those enrichment solutions without Fe and Co coupled with favorable seawater would support growth.

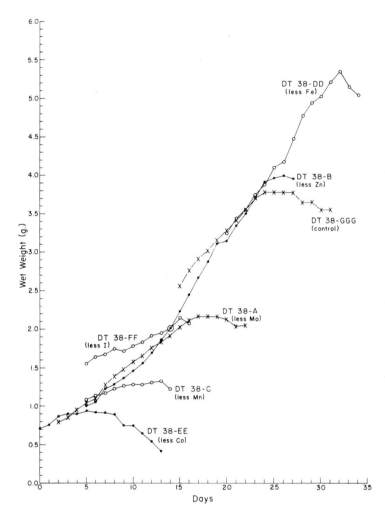

Figure 3. *Growth records from seven juvenile Macrocystis sporophytes held in seawater collected centrally from within the Pt. Loma, California, kelp bed and enriched with a standard nutrient mix (see Table III) with one element only omitted from each experiment. Triplicate plants were held in each medium except for DT-38A and DT-38B. The best of three growth records is shown for each treatment.*

Subsequent collection of seawater, usually 3-7 days later, might be unfavorable for growth. If unfavorable water were used to start a culture, poor growth could be expected immediately. Poor initial growth would be followed by irreversible changes. Such changes would prevent an improvement in growth when favorable water was used. Alternatively, if favorable water was used initially, good growth was obtained until a collection of unfavorable water was employed. Both of these patterns were observed during the experiments. Additions or omissions from nutrient solutions of elements such as Fe,

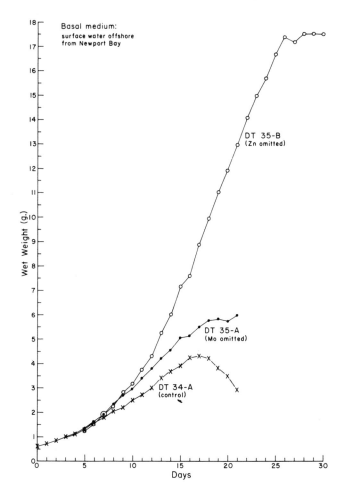

Figure 4. *Growth records from three juvenile Macrocystis sporophytes held in seawater collected about 6 km offshore from Newport Bay, California, and enriched with a standard nutrient mix (see Table III) except for omissions of elements as indicated. In this instance, omissions of zinc and molybdenum led to more rapid growth than that displayed by the control plant.*

that can form extremely fine particulates, could affect availability of other micronutrients or micro-toxicants that might absorb onto suspended particles. Additions of 10^{-6}M $FeCl_3$ to seawater sometimes does favor *Macrocystis* growth, but the availability of this element cannot be assessed at this time. Probably the availability of Co in seawater also fluctuates unpredictably from adequate to inadequate.

Five plants apiece supplied with Mn-deficient and I-deficient enrichments failed to demonstrate suitable growth. Thus the availability of these two elements appeared inadequate under the experimental conditions. Supplementing a N fertilizer with both might be

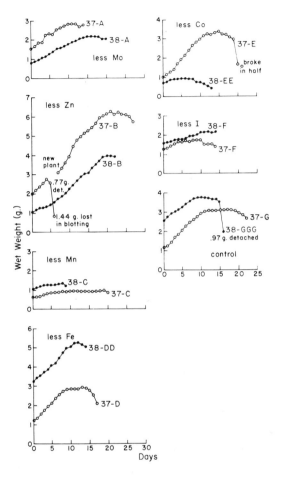

Figure 5. *Growth comparisons between juvenile Macrocystis sporophytes grown in seawater from about 300 m (DT-37 Series) and plants raised in surface water from the Pt. Loma kelp bed (DT-38 Series). Media were enriched with the standard nutrient solution (see Table III) except for omissions of single elements as indicated.*

beneficial. However an experiment should be conducted in a flowing system.

A full range of nutrient-omission experiments using offshore surface water, surface water from within a large kelp bed at Pt. Loma, and water from 300 m deep as basal media was conducted. Except for the medium without Co, growth recorded from deep water experiments was not appreciably better than that for surface waters. It may have been slightly poorer in some cases (Fig. 5).

SUMMARY AND CONCLUSIONS

This initial group of experiments suggests that large differences

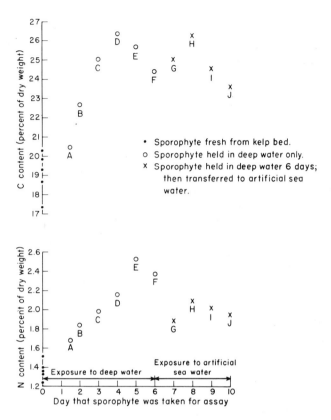

Figure 6. *Percent carbon and nitrogen contents of juvenile Macro-cystis sporophytes held in various media and sacrificed serially for analysis. Increases in both C and N after growth in nutrient-rich deep water from about 250 m suggest enhancement of photosynthesis and an accumulation of N.*

in the availability of the micronutrients between deep and surface water cannot be expected. It could be inferred that upwelling processes, important as they may be for renewing N and P, are probably considerably less influential as renewal mechanisms for other trace elements. Thus, in N-rich waters such as those which are obtained during upwelling periods or in estuaries and embayments of the Pacific Northwest, algal growth might become limited by the availability of certain micronutrients. Dissolved and suspended organic matter may strongly influence the availability of trace metals. Experiments conducted in southern California might not be valid for the Pacific Northwest because of differences in organic contents of the seawaters. It would be wise to conduct similar experiments somewhere in the Pacific Northwest, using water and plants from the region in which an ocean farming operation is contemplated.

Renewal of micronutrients other than N and P probably depends on runoff and local regeneration from decomposition processes. Advection, of course, is always necessary. Marine algae may accumulate micronutrients continuously and at fairly steady rates from water flowing across blade surfaces. If supplies are fairly dependable, there may be no need for storing reserves. Inability to accumulate reserves would explain the devastating and rapidly irreversible effects we observed in nutrient-omission experiments. Where a micronutrient is renewed by some intermittent process such as upwelling, there are advantages in a storage capability. Stored reserves could sustain growth during periods when a micronutrient such as N or P is scarce. Many algae increase their content of nitrogen when transferred to N-enriched media, which suggests a storage function. *Macrocystis* is no exception (Fig. 6). Fertilizing strategies should take these characteristics into account. Certain classes of limiting micronutrients might be dispersed continuously while others should be applied only as needed.

Caution must be used when seeking to define the nutritional quality of seawater solely by determining concentrations of inorganic N. Other micronutrients may become limiting, particularly where waters are enriched in N because of upwelling. Experiments have indicated that growth rates by juvenile *Macrocystis* sporophytes in seawater can be greatly stimulated by proper nutrient enrichment (5, 6, 7). Fertilizing programs should be able to contribute very significantly to algal biomass production and possibly make commercial operations more profitable.

ACKNOWLEDGMENTS

Grant support from the Energy Research and Development Administration and the National Science Foundation is gratefully acknowledged. The author is indebted to B. S. Anderson, L. M. Anderson, J. F. Devinny, H. C. Fastenau, J. S. Hsiao, G. A. Jackson, F. E. Jannsen, L. G. Jones, P. D. Kirkwood, J. Kuwabara, V. A. Kuga, A. L. Long, M. Martini, J. J. Morgan, V. A. Roberts, T. G. Stephan, L. M. Taylor, M. Wheeler, P. A. Wheeler, H. A. Wilcox, and A. Zirino for advice and assistance in all aspects of these studies. Certain other studies within the Marine Farm Project have received support from the American Gas Association.

REFERENCES

1. Brewer, P. G., and D. W. Spencer. 1972. Trace element profiles from the Geosecs-II test station in the Sargasso Sea. Earth Planet. Sci. Lett. 16:111-116.
2. Cheng, T. 1969. Production of kelp--a major aspect of China's exploitation of the sea. Econ. Bot. 23:215-236.
3. Jackson, G. A., and W. J. North. 1973. Concerning the Selection of Seaweeds Suitable for Mass Cultivation in a Number of Large Open-ocean Solar Energy Facilities ("Marine Farms") in order to Provide a Source of Organic Matter for Conversion to Food, Synthetic Fuels, and Electrical Energy. U. S.

Naval Weapons Center, China Lake, California, Final Report.
Contract No. N60530-73-MV176. 135 pp.

4. Lewin, R. A. 1962. Physiology and Biochemistry of Algae. Academic Press, New York. 929 pp.

5. North, W. J. 1976. Ocean Food and Energy Farm Project, Subtasks One and Two, Biological Studies of *Macrocystis pyrifera* in Upwelled Oceanic Water. Naval Undersea Center, San Diego, California, Final Report. Contract No. N00123-76-C-0313, ERDA/USN/1027/76/4, Dist. Category UC-61. 258 pp.

6. North, W. J. 1977a. Possibilities of biomass from the ocean; the marine farm project. *In:* Proceedings, Symposium on Biological Conversion of Solar Energy, University of Miami, Nov. 1976. (in press, preprint avail. from author)

7. North, W. J. 1977b. Growth factors in the production of giant kelp. *In:* Proceedings, Symposium on Clean Fuels from Biomass and Wastes, Orlando, Florida, Jan. 1977. Institute of Gas Technology, Chicago, Illinois. (in press, preprint avail. from author)

8. Provasoli, L., J. J. A. McLaughlin, and M. R. Droop. 1957. The development of artificial media for marine algae. Arch. Mikrobiol. 25:392-428.

9. Riley, J. P., and D. Taylor. 1972. The concentrations of cadmium, copper, iron, manganese, molybdenum, nickel, vanadium and zinc in part of the tropical northeast Atlantic Ocean. Deep Sea Res. 19:307-317.

10. Robertson, D. E. 1970. The distribution of cobalt in oceanic waters. Geochim. Cosmochim. Acta. 34:553-567.

11. Ryther, J. H., and W. M. Dunstan. 1971. Nitrogen, phosphorus, and eutrophication in the coastal marine environment. Science 171:1008-1013.

12. Sclater, F. R., E. Boyle, and J. M. Edmond. 1976. On the marine geochemistry of nickel. Earth Planet. Sci. Lett. 31:119-128.

13. Slowey, J. F., and D. W. Hood. 1971. Copper, manganese, and zinc concentrations in Gulf of Mexico waters. Geochim. Cosmochim. Acta. 35:121-138.

14. Spencer, D. W., D. E. Robertson, K. K. Turekian, and T. R. Folsom. 1970. Trace element calibration and profiles at the Geosecs Test Station in the Northeast Pacific Ocean. J. Geophys. Res. 75:7688-7696.

15. Stein, J. R. 1973. Handbook of Phycological Methods. Cambridge University Press, Oxford, England. 448 pp.

16. Stewart, W. D. P. 1974. Algal Physiology and Biochemistry. University of California Press, Berkeley. 989 pp.

17. Tamura, T. 1970. Marine Aquaculture, Part II. National Science Foundation PB 194051T. Also available from National Tech. Info. Serv. 639 pp.

18. Vinogradoff, A. P. 1953. The Elementary Composition of Marine Organisms. Sears Foundation Marine Research, New Haven, Connecticut, Mem. 2. 674 pp.

19. Wong, G. T. F., and P. G. Brewer. 1976. The determination of iodide in seawater by neutron activation analysis. Anal. Chim. Acta. 81-90.

-13-

Potential Yields from a Waste-Recycling
Algal Mariculture System[1]

JAMES A. DEBOER AND JOHN H. RYTHER

For centuries seaweeds have been an integral part of the
Oriental diet, but only since World War I have they become an important commodity in the Western World. In recent years they have been
used in the United States primarily for their phycocolloids (alginates, agar, and carrageenan). Agar and carrageenan, cell wall polysaccharides produced by various red algal species, are widely used
in food, pharmaceutical, textile, cosmetic, and other industries as
suspending, thickening, stabilizing, and emulsifying agents (13,
26, and Moss, this volume). The principal source of agar has been
Gelidium spp., which is harvested in Japan where the agar is extracted and exported to the rest of the world. Carrageenophytes
have been harvested from natural populations throughout the world,
dried, and shipped to factories in North America or Western Europe,
where the phycocolloid is extracted and refined for sale. Most of
the world's supply of carrageenan comes from *Chondrus crispus* (Irish
moss) populations in Eastern Canada and to a lesser extent New
England and Northern Europe.

These seaweed resources are limited in area and are now heavily
exploited. At the same time, the demand for phycocolloids is steadily increasing. The discovery that different algal species or blends
of phycocolloids from different algal species have dissimilar gelling
or emulsifying properties has led to a large number of new applications of these products. These factors together have led to screening of various species and world-wide surveys of seaweed resources
by the industry over the past two decades, in an attempt to expand

[1]This research was supported in part by NOAA Office of Sea
Grant, Department of Commerce, under Grant No. 04-6-158-44016, Contract E(11-1)-2948 from the U. S. Energy Research and Development
Administration, and the Jessie Smith Noyes Foundation, Inc. Contribution No. 4027 from the Woods Hole Oceanographic Institution.

the base of its operation. One example of such expansion is the
relatively new exploitation of the red alga *Eucheuma* in the Philippines and other parts of Southeast Asia. These resources, old and
new, are decreasing due to overharvesting (2, 27, and Doty, this
volume), pollution (17, 27, 29), and storm damage (18) to the extent
that the industry is resource limited. Attention has become focused
on cultivation as the only long-term solution.

Most studies with seaweeds have been concerned with their taxonomy, anatomy, life history, or distribution. Unfortunately, there
is very little known about the physiology and autecology of most of
these algae and even less concerning their cultivation.

CULTIVATION OF AGAROPHYTES AND CARRAGEENOPHYTES

To supplement insufficient natural supplies of agarophytes,
the Japanese initiated a seaweed cultivation program several decades
ago. One method involved scattering small fragments of *Gelidium* or
Gracilaria (11) in bays where the plants are allowed to regenerate
vegetatively. More recently the Japanese have propagated *Gracilaria*
and *Gelidium* (12) on ropes in shallow bays. *Gracilaria* culture in
Taiwan (25) has undergone a rapid expansion since its initiation in
1962. The unattached *Gracilaria* plants are grown in shallow ponds
of approximately 1 hectare, which formerly were used for milkfish
culture. *Eucheuma* farming, developed in the Philippines (7, 8, 18,
and Doty, this volume), utilizes a net-culture technique that is
similar to the cultivation of edible seaweeds in Japan. No commercial seaweed cultivation farms exist in North America[2], but several
research projects involving the cultivation of red algae have evolved
in the past decade.

Beginning in the late 1960s, a group headed by A. C. Neish
at the Canadian National Research Council Atlantic Regional Laboratory near Halifax initiated studies on the culture of unattached
Chondrus crispus in tanks containing flowing seawater. One of their
early observations was that plants of different origin grew at considerably different rates in the same tank. One clone (T-4) grew
much faster than others and attracted fewer undesired algal species
as epiphytes (14). Another important finding was that the chemical
composition of the plants could be altered by manipulation of the
culture environment. Neish and his co-workers discovered that plants
grown in unenriched seawater have a higher carrageenan content than
those grown in nitrogen-enriched seawater. If *Chondrus* grown in
nitrogen-enriched seawater was transferred to unenriched seawater,
its carrageenan content increased. The effects of several other
operating parameters were also investigated (14, 15, 23, 24).

---

[2]Atlantic Mariculture Ltd. of Grand Manan Island, New Brunswick,
Canada, uses V-shaped, air-agitated ponds as part of a commercial
Rhodymenia operation (16). Their primary objective is not cultivation, but rather to keep the plants alive until they can be
processed.

A   B

C

Figure 1. *Enclosures used in the cultivation of seaweeds at the Woods Hole Oceanographic Institution.   A - Rectangular plywood tanks.   B - Circular fiberglass tanks.   C - Concrete raceways.*

Following Neish's lead, several research groups in the U. S. have experimented with growing unattached seaweeds in suspended culture. For example, Ryther's group (3, 4, 5, 10, 19, 20, 22) in Woods Hole and in Florida has grown *Gracilaria* sp., *Neoagardhiella baileyi, Chondrus crispus, Gracilariopsis sjoestedtii, Hypnea musciformis,* and other species in tanks (Figs. 1A-B), raceways (Figs. 1C and 2), and ponds (Fig. 2). Other research teams in the U.S. have used similar tank culture methods to grow *Hypnea* (9), *Iridaea* (28), and *Gigartina* (28 and Waaland, this volume).

Figure 2. *Aluminum raceways (upper left) and PVC-lined ponds (right) used to cultivate seaweeds at the Harbor Branch Foundation in Ft. Pierce, Florida.*

Two commercial seaweed companies, Marine Colloids, Ltd., and GENU Products, Ltd., have started pilot *Chondrus* projects in Nova Scotia, with partial support from the Canadian Government, using modifications of Neish's technique. Full-scale production in both projects was delayed by the slow growth of *Chondrus* and the difficulties in control of algal contaminants (e.g., *Ulva*, *Enteromorpha*, and *Ectocarpus*) in the culture system.

A WASTE-RECYCLING MARINE-POLYCULTURE SYSTEM

Beginning about 1970, a project at the Woods Hole Oceanographic Institution developed a waste-recycling marine-aquaculture system. This system has the capacity of removing the inorganic nutrients from treated sewage effluent, prior to its discharge to the environment, and recycling these nutrients into commercially valuable crops of marine organisms.

The concept of this system is to grow unicellular marine algae in ponds in continuous flow cultures on mixtures of seawater and secondarily treated sewage effluent (Fig. 3). The algae are fed to bivalve molluscs, maintained in Nestier® trays in raceways. The

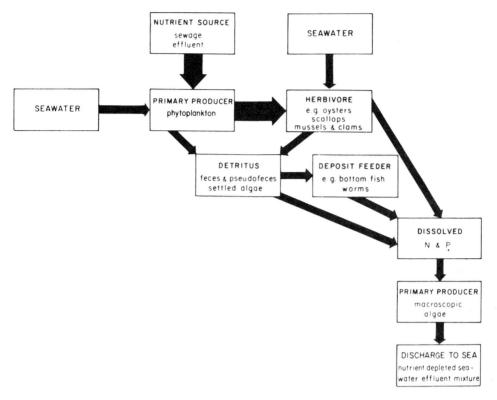

Figure 3. *Model of the Woods Hole waste recycling-polyculture system.*

algae remove the nutrients from the wastewater and the shellfish remove the algae from suspension. Finally, the effluent from the shellfish culture passes through tanks containing macroscopic algae (seaweeds). The seaweeds remove the dissolved nutrients regenerated by the animal culture before final effluent is discharged. The objective of this polyculture system is to achieve a low nutrient final effluent that will meet the standards for tertiary treatment while producing commercially valuable marine organism crops.

Two species of macroscopic red algae, *Gracilaria* sp. and *Neoagardhiella baileyi*, were grown in the Woods Hole waste-recycling seaweed-mariculture projects. Several species of *Gracilaria* harvested from natural populations serve as a major source of agar. *Neoagardhiella*, which contains iota-carrageenan (4), is not yet harvested but has potential commercial value. Species of a closely related genus, *Eucheuma*, which contain the same phycocolloid, are cultivated and harvested commercially in the Philippines (Doty, this volume). Unattached, free-floating plants were grown in concrete raceways 12.2 m (long) x 1.2 m (wide). Sloping plywood bottoms provided depths ranging from 0.6 m to 1.5 m. The seaweeds were kept in suspension by aeration from an airline at the bottom of the raceway which extended its entire length. The aeration provided for mixing,

Table I. *Culture conditions for Neoagardhiella and Gracilaria in raceways at the Woods Hole Oceanographic Institution during 1975-1976.*

| | Nutrients (µM) | | Water Flow | Density (kg wet wt/m^2) | |
Dates	ΣiN	PO_4^{-3}	1/min*	*Neoagardhiella*	*Gracilaria*
Mar 20–Apr 15	37	12	48	3.9	2.8
Apr 15–May 6	14	8	48	9.5	3.9
May 6–May 20	21	7	72	8.9	5.5
May 20–May 28	11	2	72	12.3	6.4
May 28–Jun 5	27	7	72	10.1	8.1
Jun 5–Jun 12	16	7	72	11.5	8.4
Jun 12–Jun 27	34	9	96	8.2	10.2
Jun 27–Jul 16	75	13	138	5.5	5.9
Jul 16–Aug 14	58	10	96	6.6	6.3
Aug 14–Aug 27	81	14	48–144	5.9	6.5
Aug 27–Sep 18	–	–	48–144	4.2	4.0
Sep 18–Oct 6	–	–	48–144	3.1	2.6
Oct 6–Nov 7	–	–	70	2.6	1.7
Nov 7–Nov 21	44	8	72	3.1	2.8
Nov 21–Dec 19	37	5	60–80	2.0	2.2
Dec 19–Feb 20	40	6	60–80	3.0	–
Feb 20–Mar 19	–	–	60–80	2.9	–

*48 1/min = 3 exchanges/day

gas exchange, and uniform exposure of the plants to sunlight. The seaweed cultures received the effluent from similar raceways containing various species of bivalve molluscs.

Because of major problems with the molluscan cultures during the first year of operation, considerable research was devoted to experimentation with flow rates, temperatures, shellfish stocking densities, and other factors. Therefore, the chemical and physical characteristics (Table I) of the effluent from the shellfish raceways

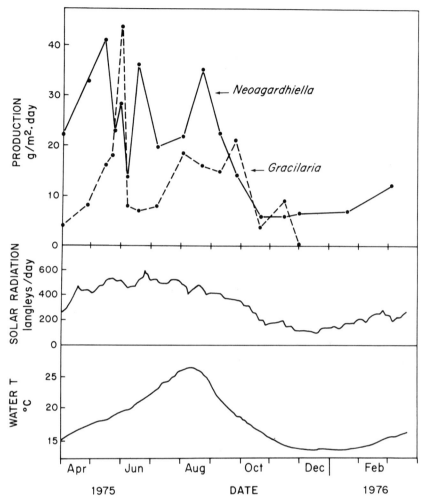

Figure 4. *Production rates of Neoagardhiella and Gracilaria grown in raceways at Woods Hole Oceanographic Institution during 1975-76.*

varied throughout the year. In spite of variations, the flow of water and the supply of nutrients were not believed to have been limiting to seaweed growth. The stocking density was varied experimentally during the year because the optimum density for the maximum rate of production in the raceways had not been previously determined. *Gracilaria* and *Neoagardhiella* were monitored from March 20, 1975, to March 19, 1976.

Production was measured by dip-netting the algae from the raceways, draining the plants in nylon mesh bags, and weighing. After weighing, the biomass was reduced to its initial weight or allowed to accumulate. Populations were weighed at intervals of about 1 week during the period of most rapid growth in spring and early summer and at longer intervals during the remainder of the year. Samples of the algae were oven dried at 60° C to determine the

relation between fresh and dry weight. Production rates or yields
were expressed in terms of dry weight which ranged between 10 and
15% of the fresh weight.

The seaweed production rates are shown in Fig. 4. *Neoagardhiella* had surprisingly high rates of production in the spring (22–41
g dry weight/$m^2 \cdot$day) and summer (20–36 g/$m^2 \cdot$day). *Gracilaria* production rates were variable, increasing from 4 g/$m^2 \cdot$day in April to
43 g/$m^2 \cdot$day in early June and decreasing again during the latter
part of the summer to 7–18 g/$m^2 \cdot$day. Production rates of both species
declined during the fall and early winter. By mid-December the *Gracilaria* production rate had fallen to zero, and the plants had deteriorated. *Neoagardhiella* remained viable (although stunted) with a
production rate of 6 g/$m^2 \cdot$day.

Gross seasonal changes in production appear to be correlated
with both water temperature and incident solar radiation. The seawater entering the bivalve mollusc cultures was heated in winter to
enable the shellfish to feed and grow, but it had cooled to as low
as 13° C by the time it entered the seaweed cultures and as low as
8° C when it left during midwinter. From June 1 to October 1, however, when the water temperatures and solar radiation were presumably
optimal for the seaweeds (3), production showed little correlation
with those variables. Low production rates then were probably due
to the high densities of seaweeds maintained in the raceways (as
high as 12.3 and 10.2 kg fresh weight/m^2 for *Neoagardhiella* and
Gracilaria, respectively).

SEAWEED MARICULTURE--OPERATIONAL CONSIDERATIONS

In addition to using seaweeds as a final step in a waste
recycling-polyculture system, they may also be grown in a one-step
waste recycling system to remove the nutrients directly from mixtures
of seawater and secondary sewage effluent. This one-step system has
been more successful than polyculture because of its simplicity.
Even so, there are several operational parameters which need examination to ensure success and to optimize yields.

One important operational consideration in the polyculture experiment was the biomass of seaweeds to be maintained to provide maximum yield per unit of area. Two experiments were conducted to determine the relationship of seaweed density to the growth rates and
production of *Gracilaria* sp. One experiment was conducted June 2–28,
1976, and the other, November 3–30. Each of six plywood tanks (Fig.
1A) was stocked with 180–4000 g fresh weight/m^2. These tanks, 2.4 m
long x 1.0 m wide x 1.2 m deep at maximum, were designed with sloping
bottoms. An airline on the bottom along the deep side of each tank
enabled the plants to be maintained in suspension and circulated by
aeration. The tanks were in a geodesic dome fitted with a vinyl
cover to retain heat during the winter. Filtered seawater heated to
18.5–21.5° C was enriched with ammonium chloride and sodium phosphate
to give a concentration of 50 µM NH_4^+ and 10 µM PO_4^{-3}. This seawater was circulated through the tanks continuously at a rate of 2
tank volumes (3650 liters) per day. Three times per week the plants

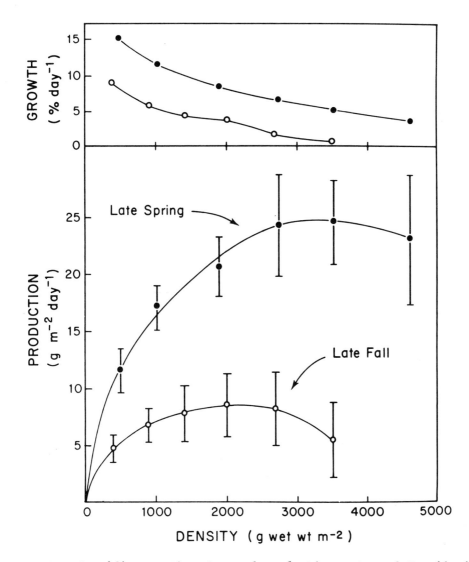

Figure 5. *Specific growth rates and production rates of Gracilaria as a function of culture density in the tanks at the Woods Hole Oceanographic Institution during 1976.*

were weighed and the density in each tank adjusted to the initial stocking density by harvesting the incremental growth. The specific growth rate (μ), equivalent to the percent increase per day, was calculated by the equation:

$$\mu = \frac{100\,(\ln {}^{N}/_{No})}{t}$$

where No is the initial biomass and N is the biomass on day t. The mean specific growth rates and production rates at the

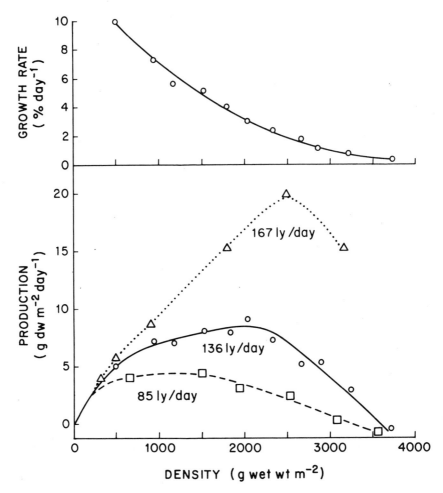

Figure 6. *Specific growth rates and production rates of Neoagardhi-ella baileyi as a function of culture density in tanks at the Woods Hole Oceanographic Institution during 1975.*

different mean densities are shown in Fig. 5. The average solar radiation was 549 langleys/day (ly/day) during the late spring experiment and 152 ly/day during the late fall experiment. The growth rate decreased exponentially with increasing culture density from 14%/day to 4%/day during the spring and from 9%/day to 1%/day in the fall. Production, or yield, which is a function of both specific growth rate and density, was highest at an intermediate density of 3000-4000 g fresh weight/m^2 in spring. In the fall, maximum production was achieved at densities of 2000-2500 g fresh weight/m^2.

Another experiment using similar methods (3) investigated the relationships among growth rate, production rate, and density in *Neoagardhiella baileyi*. The results, shown in Fig. 6, are averages over the entire 40-day experiment (Nov.-Dec., 1975), when the mean incident solar radiation was 136 ly/day. Maximum production occurred

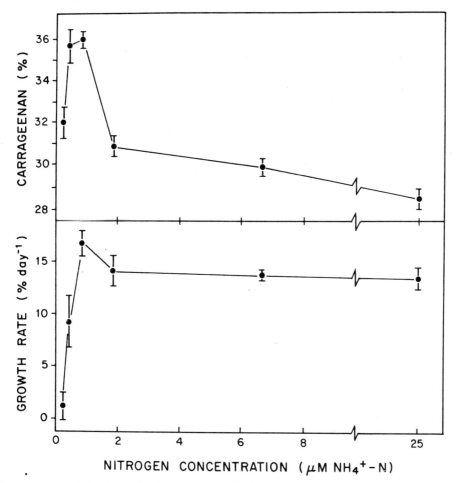

Figure 7. *Effects of nitrogen enrichment on the growth rate and carrageenan content of Neoagardhiella grown in tanks at the Woods Hole Oceanographic Institution during 1976.*

at densities of 2000-2500 g fresh weight/m^2. During brief periods (ca 1 week) of sunny weather when the solar radiation averaged 160 ly/day, productivity increased markedly to a peak of 22 g dry wt/m^2/day at a density of 2900 g fresh wt/m^2. During cloudy periods (85 ly/day) production decreased to less than 5 g dw/m^2/day at densities of 500-1500 g fresh wt/m^2.

These experiments indicate that to obtain maximum yields of *Gracilaria* and *Neoagardhiella*, the density should be maintained in the range of 1800-2800 g/m^2 during the winter, 2800-4500 g/m^2 during late spring and summer, and at intermediate densities during the remainder of the year. These relations between production and seaweed density for *Gracilaria* and *Neoagardhiella* are similar to those reported for other red algae. For example, the maximum rate of production for *Chondrus* during the summer was obtained at a population density of 5800 g fresh weight/m^2 (15). The optimum density for

Iridaea in April was approximately 2100 g fresh weight/m^2 (28).

The optimum depth of the culture system is another variable that is best determined empirically. Shacklock *et al.* (24) found that the growth of *Chondrus* was greater at a depth of 91 cm than it was at either 46 or 1800 cm. Experience with various types of culture enclosures and modes of circulation of the plants and water has shown that the optimum depth for the suspended cultures of *Gracilaria* and *Neoagardhiella* is 60–110 cm.

Another critical operating parameter in seaweed cultivation is the concentration of nutrients necessary to sustain a maximum growth rate. Ryther and Dunstan (21) found that nitrogen is the chemical nutrient most likely to limit algal growth in marine waters. Seaweed studies have also indicated nitrogen to be the critical limiting factor in the Woods Hole seaweed mariculture system. As a result, an investigation (4) was undertaken (Mar. 21–Apr. 8, 1976) to determine the concentration of ammonia at which maximum growth rate occurs in *Neoagardhiella baileyi*. Each of the six tanks described in the density experiments was stocked with 1500 g of the algae. Influent nutrient concentrations of the enriched seawater ranged from 4 to 70 μM NH_4^+ and from 1 to 14 μM PO_4^{-3}, respectively, in five of the experimental tanks, with an unenriched seawater control in the sixth tank. The continuous flow rates were equivalent to three tank volumes per day. Every 3 days the algae were weighed and the biomass in each tank adjusted to the initial stocking density. Growth rate (Fig. 7) increased with increasing nitrogen concentration up to approximately 0.8 μM NH_4^+ but remained constant at higher concentrations.

At the conclusion of the experiment described above, the carrageenan content was determined. Details of the methods and results have been previously reported (4). Carrageenan (Fig. 7) was highest at a residual nitrogen concentration of 0.8 μM NH_4^+ and was less at both higher and lower nitrogen levels. These results are similar to those of Neish and Shacklock (15) who found that *Chondrus* grown in unenriched seawater has more carrageenan than when grown in nitrogen-enriched seawater. Our results show, however, that even higher levels of carrageenan can be produced at low (0.5–1.2 μM) NH_4^+ concentrations than in unenriched seawater.

In another study[3] it was determined that the half-saturation constants for growth in *Gracilaria* and *Neoagardhiella* are approximately 0.5 μM NH_4^+ or NO_3^-. These constants are very low, demonstrating that seaweeds can utilize very low concentrations of inorganic nitrogen.

Half-saturation constants for growth in phytoplankton are

----

[3]DeBoer, J. A., H. J. Guigli, T. L. Israel, and C. F. D'Elia. Studies on the cultivation of the macroscopic red algae. I. Growth rates in cultures supplied with nitrate, ammonium, urea, or sewage effluent. (In prep.)

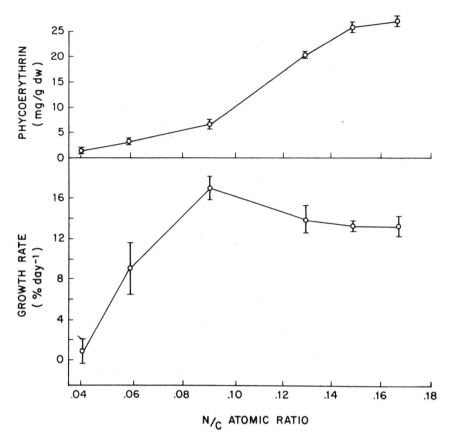

Figure 8. *Phycoerythrin content and growth rate of Neoagardhiella as a function of their analytical nitrogen: carbon ratio.*

usually in the 0.1-5 M range for inorganic nitrogen. As far as we are aware, no other studies describe the half-saturation constants for growth or uptake of inorganic nitrogen by seaweeds. This study also showed that ammonia-grown *Gracilaria* and *Neoagardhiella* exhibit higher growth rates than nitrate-grown plants, that both species show a decided preference for ammonia over nitrate in mixtures of the two, and that growth of both species is essentially equivalent whether the nitrogen source is secondary sewage effluent or chemical fertilizer.

These results indicate that for maximum biomass and carrageenan production, the inorganic nitrogen concentration should be maintained at 0.5-1.5 μM NH_4^+. Because of diurnal and other changes in the rate of nitrogen uptake by the seaweeds, maintaining optimum nutrient concentration constantly is difficult. Some simple indicator of the nutrient condition or status of the plants would facilitate large-scale culture operations. The nitrogen:carbon ratio (N:C) of the plants may serve as that indicator. Growth rate (Fig. 8) increases with increasing N:C atomic ratio up to approximately 0.85 with no increase in growth rate at higher N:C ratios. This and other studies (4) indicate that plants having an N:C ratio above 0.85 are probably not nitrogen limited but lower ratios suggest nitrogen limitation. Additional studies under a variety of environmental conditions are necessary to substantiate these findings, but initial results indicate that large deviations in the N:C ratio can be used to predict a nitrogen deficiency or surplus. Such an indicator would be extremely useful in seaweed mariculture to show when and how much to fertilize for maximum biomass and phycocolloid production.

A simple color index may be used as an indicator of nutrient status and phycocolloid content (4). Plant color in *Neoagardhiella* is due primarily to the relative proportions of chlorophyll and the red accessory pigment, phycoerythrin. Fig. 8 shows the phycoerythrin concentration as a function of the N:C ratio. Phycoerythrin concentration increased rather consistently with increasing N:C ratio. Plants with high concentrations of the red pigment appear dark reddish-brown while those with very low levels are yellow to straw colored. However, before such a simple color index of phycocolloid content and nitrogen status is considered reliable, it must be verified that other nutritional deficiencies that may not affect carrageenan levels do not influence the development of the accessory pigment. At very high light intensities, for instance, photo-oxidation of pigments occurs, so that in those circumstances pigmentation may not be a valid indicator of nutritional status and carrageenan content.

One common observation is that if the seaweed density varies greatly from the optimum or if the nitrogen concentration is higher than is necessary to support the maximum growth rate, algal contaminants often proliferate, significantly decreasing both the growth and the economic value of the cultured species. We have found that if the plants are grown near the optimum density and at optimum nutrient concentrations (1-2 μM NH_4^+ or NO_3^-), the algal contamination does not become serious. We have shown that the carrageenan content is also higher in plants grown at low nitrogen concentrations. The agar content in *Gracilaria* sp. is also higher at low nitrogen concentrations[4]. A nitrogen level to maximize both growth rate and polysaccharide content is a crucial parameter.

SEAWEED MARICULTURE--PRODUCTIVITY

The polyculture experiment described previously was conducted before the density and nutrient experiments. The densities in the raceways were at times far above those optimum for the maximum rate of production. The production rates were, therefore, probably lower than might have been obtained if optimum densities had been maintained throughout the year. Nevertheless, the mean annual dry weight production was 17 g/m^2·day or 63 metric tons/hectare·year for *Neoagardhiella* and 9 g/m^2·day or 33 t/ha·year for *Gracilaria*. These rates may

---

[4]DeBoer, J. A. Effects of nitrogen enrichment on growth rate and phycocolloid content in *Gracilaria foliifera* and *Neoagardhiella baileyi*. Submitted to Proceedings of the Ninth International Seaweed Symposium.

be higher than would be realized in a commercial enterprise because
the raceways were maintained at elevated temperatures for approximately 6 months of the year. Annual production rates based on a 5 1/2-
month growing season when the raceways were not heated (May 8 - Oct.
20, 1975) were 46 t/ha·year and 28 t/ha·year for *Neoagardhiella* and
Gracilaria, respectively (Table II).

Table II. *Comparative productivity values for cultivated seaweed
crops.*

Species	Location	Annual Production tons/hectare dry weight	Culture Method	Reference
Neoagardhiella	Massachusetts, USA	46.0	Raceway	This study
Gracilaria	Massachusetts, USA	28.0	Raceway	This study
Gracilaria	Florida, USA	46.0	Tank	10
Hypnea	Florida, USA	39.0	Tank	10
Iridaea	Washington, USA	20.0	Tank	28
Gracilaria	Taiwan	7-12.0	Pond	25
Gracilaria	Taiwan	2.0	Net	18
Gracilaria	Japan	0.4-1.3	Bay	11
Eucheuma	The Philippines	13.0	Net	18
Gelidium	Japan	1.5	Stones	18
Gelidium	Korea	1.4	Stones?	6
Porphyra	Japan	0.3-3.0	Net	18,29

The yields obtained in this study were similar to those recently reported for *Gracilaria* sp. and *Hypnea musciformis* grown
in essentially the same way but for shorter periods at the Harbor
Branch Foundation in Florida. In other small-scale, short-duration
experiments, production rates of cultured seaweeds exceed these values 2-3 fold (3, 9, 16, 28, and Waaland, this volume), but these
yields have not been substantiated by large-scale, long-term production studies. However, given favorable growing conditions it seems
realistic to expect production rates to exceed 50 t/ha/yr based on
a 5- to 6-month growing season in temperate latitudes. The yields

obtained in both the Woods Hole and Florida studies are considerably
greater than values reported in the literature for other highly
profitable seaweed crops (Table II) and are as high or higher than
many agricultural crops (5, 10).

SUMMARY AND CONCLUSIONS

Most agarophytes and carrageenophytes currently cultivated are
grown attached to ropes or nets (1, 13, and Doty and Mumford, this
volume). These methods, although profitable in some areas, are
labor intensive and are, therefore, best suited to countries with
low labor costs. The method used in the present study, growing unattached plants in raceways or ponds, is a more versatile means of
culture and one that could easily be mechanized.

We have used domestic sewage effluent as a nutrient source in
the waste-recycling polyculture systems in Woods Hole and in Florida,
but other nutrient sources could also be used. Alternative nutrient
sources include agricultural wastes from cattle feed lots or swine
farms, wastes from seafood processing plants or other food processing wastes, and wastes from open animal mariculture systems such as
penaeid shrimp farms. In all of these applications seaweeds can be
used to lower the nutrient and heavy metals content of the wastes,
enabling the discharge to meet state and federal regulations. Algae
may also be incorporated into a closed animal mariculture system,
serving as biological filters to remove the animals' toxic metabolic
wastes. Upwelled, nutrient-rich waters may also be used as a nutrient source in seaweed mariculture systems (9), and conventional commercial fertilizers will serve as adequate (though perhaps more costly) nutrient sources in seaweed monoculture operations.

In summary, cultivated seaweeds are apparently among the most
productive primary producers and prospects for seaweed cultivation
look optimistic. However, there is still a need for considerable
basic and applied research on the biology of the macroscopic marine
algae. Both biomass and phycocolloid production are undoubtedly
functions of many interrelated variables: availability of chemical
nutrients, light intensity, light quality, photoperiod, turbidity,
CO_2 availability, mixing, rate of water exchange, temperature, population density, culture depth, water quality (toxins, growth enhancement factors, pH), and the seaweeds themselves (differences
among species and physiological races, seasonality of growth, etc.).
Results of our studies on some of these variables suggest that, if
properly managed, the yields of seaweed mariculture can be substantially increased.

ACKNOWLEDGMENTS

The writers are grateful to B. E. Lapointe, R. A. Fralick,
J. P. Clarner, N. Corwin, S. Locke, and F. G. Whoriskey for their
assistance on this project.

REFERENCES

1. Bardach, J. E., J. H. Ryther, and W. O. McLarney. 1972.
    Aquaculture: The Farming and Husbandry of Freshwater and
    Marine Organisms. Wiley-Interscience, New York. 868 pp.
2. Blanco, G. J. 1972. Status and problems of coastal aquacul-
    ture in the Philippines. *In:* Pillay, T. V. R. (ed.)
    Coastal Aquaculture in the Indo-Pacific Region. Fishing
    News (Books) Ltd., London, pp. 60-67.
3. DeBoer, J. A., and B. E. Lapointe. 1976. Effects of culture
    density and temperature on the growth rate and yield of
    Neoagardhiella baileyi. In: Ryther, J. H. (ed.) Marine
    Polyculture Based on Natural Food Chains and Recycled
    Wastes. Woods Hole Oceanographic Institution, Tech. Rep.
    WHOI-76-92, Pap. No. 9. NTIS No. PB261 939/3GA. 281 pp.
4. DeBoer, J. A., B. E. Lapointe, and C. F. D'Elia. 1976. Effects
    of nitrogen concentration on growth rate and carrageenan
    production in *Neoagardhiella baileyi. In:* Ryther, J. H.
    (ed.) Marine Polyculture Based on Natural Food Chains and
    Recycled Wastes. Woods Hole Oceanographic Institution,
    Tech. Rep. WHOI-76-92, Pap. No. 10. NTIS No. PB261 939/
    3GA. 281 pp.
5. DeBoer, J. A., B. E. Lapointe, and J. H. Ryther. 1977. Pre-
    liminary studies on a combined seaweed mariculture-tertiary
    waste treatment system. *In:* Proceedings, Eighth Annual
    Workshop in San Jose, Costa Rica, January. World Mariculture
    Society, Louisiana State University, Division of Continuing
    Education, Baton Rouge, Louisiana. (in press)
6. Department of Fisheries, Republic of Korea. 1970. Present
    status and problems of coastal aquaculture in the Republic
    of Korea. *In:* Pillay, T. V. R. (ed.) Coastal Aquaculture
    in the Indo-Pacific Region. Fishing News (Books) Ltd.,
    London, pp. 48-51.
7. Doty, M. S. 1973. Farming the red seaweed, *Eucheuma,* for
    carrageenans. Micronesica 9:59-71.
8. Doty, M. S., and V. B. Alvarez. 1975. Status, problems, ad-
    vances and economics of *Eucheuma* farms. Mar. Technol. Soc.
    J. 9:30-35.
9. Haines, K. C. 1975. Growth of the carrageenan-producing tropi-
    cal red seaweed *Hypnea musciformis* in surface water, 870 m
    deep water, effluent from a clam mariculture system, and in
    deep water enriched with artificial fertilizers or domestic
    sewage. *In:* Tenth European Symposium on Marine Biology.
    1:207-220.
10. Lapointe, B. E., L. D. Williams, J. C. Goldman, and J. H. Ryther.
    1976. The mass outdoor culture of macroscopic marine algae.
    Aquaculture 8:9-21..
11. Lin, M. N. 1970. The production of *Graciliaria* in Yuanchang
    Reservoir, Putai, Chiayi. Reports of Fish Culture. Chinese-
    American Joint Committee on Rural Reconstruction Fisheries
    Series No. 9, pp. 30-34.

12. MacFarlane, C. I. 1968. The Cultivation of Seaweeds in Japan and Its Possible Application in the Atlantic Provinces of Canada. Canadian Department of Fisheries, Ind. Dev. Serv., Ottawa. 96 pp.

13. Mathieson, A. C. 1975. Seaweed aquaculture. Mar. Fish. Rev. 37:2-14.

14. Neish, A. C., and C. H. Fox. 1971. Greenhouse Experiments on Vegetative Propagation of *Chondrus crispus* (Irish Moss). National Research Council of Canada, Atl. Reg. Lab. 12. NRCC No. 12034. 35 pp.

15. Neish, A. C., and P. F. Shacklock. 1971. Greenhouse Experiments (1971) on the Propagation of Strain T4 of Irish Moss. National Research Council of Canada, Atl. Reg. Lab. 14. NRCC No. 12253. 25 pp.

16. Neish, I. C. 1976. Role of mariculture in the Canadian seaweed industry. J. Fish. Res. Board Can. 33:1007-1014.

17. Okazaki, A. 1971. Seaweeds and Their Uses in Japan. Tokai University Press, Tokyo. 165 pp.

18. Parker, H. S. 1974. The culture of the red algal genus *Eucheuma* in the Philippines. Aquaculture 3:425-439.

19. Prince, J. S. 1974. Nutrient assimilation and growth of some seaweeds in mixtures of seawater and secondary sewage treatment effluents. Aquaculture 4:69-79.

20. Ryther, J. H. 1977. Preliminary results with a pilot plant waste recycling-marine aquaculture system. *In:* D'Itri, F. M. (ed.) Wastewater Renovation and Reuse. Marcel Dekkert Co., Inc., New York, pp. 89-132.

21. Ryther, J. H., and W. M. Dunstan. 1971. Nitrogen, phosphorous and eutrophication in the coastal environment. Science 171:1008-1013.

22. Ryther, J. H., J. C. Goldman, C. E. Gifford, J. E. Huguenin, A. S. Wing, J. P. Clarner, L. D. Williams, and B. E. Lapointe. 1975. Physical models of integrated waste recycling-marine polyculture systems. Aquaculture 5:163-177.

23. Shacklock, P. F., D. Robson, I. Forsyth, and A. C. Neish. 1973. Further Experiments (1972) on the Vegetative Propagation of *Chondrus crispus* T4. National Research Council of Canada, Atl. Reg. Lab. 18. NRCC No. 13113. 22 pp.

24. Shacklock, P. F., D. R. Robson, and F. J. Simpson. 1975. Vegetative Propagation of *Chondrus crispus* (Irish Moss) in Tanks. National Research Council of Canada, Atl. Reg. Lab. 21. NRCC No. 14735. 27 pp.

25. Shang, Y. C. 1976. Economic aspects of *Gracilaria* culture in Taiwan. Aquaculture 8:1-17.

26. Silverthorne, W., and P. E. Sorenson. 1971. Marine algae as an economic resource. Mar. Technol. Soc. Ann. Conf. 7:523-533.

27. Suto, A. 1974. Mariculture of seaweeds and its problems in Japan. *In:* Shaw, W. N. (ed.) Proceedings, First US-Japan Meeting on Aquaculture. NOAA Tech. Rep. NMFS CIRC-388,

pp. 7-16.
28. Waaland, J. R. 1976. Growth of the red alga *Iridaea cordata* (Turner) Bory in semi-closed culture. J. Exp. Mar. Biol. Ecol. 23:45-53.
29. Wildman, R. 1974. Seaweed culture in Japan. *In:* Shaw, W. N. (ed.) Proceedings, First US-Japan Meeting on Aquaculture. NOAA Tech. Rep. NMFS CIRC-388, pp. 97-101.

- 14 -
Potential Yields of Marine Algae—
with Emphasis on European Species

TORE LEVRING

This chapter will deal mainly with the benthic vegetation of Europe, its composition, productivity, and commercial use. The coastlines to be treated here extend from the Arctic Sea south to the Strait of Gibraltar and include those of inland seas such as the Baltic and Mediterranean. Most shores of western Europe are rocky and constitute a suitable substratum for a well developed marine algal vegetation. Many of these shorelines are exposed to a heavy sea.

THE EUROPEAN ALGAL FLORA

The flora is naturally not the same in the north as in the south. Temperature is the major controlling influence (Table I). It is, therefore, important to consider that the west coast of Europe is affected by the Gulf Stream and experiences temperatures higher than normal for that latitude.

The classic attempt to divide the algal flora of western Europe into geographical groups was made by Børgesen and Jónsson (5). They distinguished the following five groups: arctic, subarctic, boreal-arctic, cold-boreal, and warm-boreal. Additions and adjustments have since been made to this system. The character of the flora of a certain area is often determined by calculating the percentage of the total number of species which belong to each of these groups. It is obvious that species with a wide distribution are normally eurythermal, and those with a limited one, stenothermal. It is clear that the flora gradually changes from arctic to warm-boreal as one proceeds southward (Table II).

The influence of the tide on coastal vegetation is well known. The littoral zone, normally characterized by regular submerging and draining twice in about 24 hours, contains several species of commercial importance. The difference between the tide levels in Europe, as elsewhere, is slightly unequal. Normally the tide is

Table I. *Average water temperatures in different regions of Europe given in degrees C.*

Region	Warmest month	Coldest month
Spitzbergen	$+ 6^{\circ}$ C	$- 2^{\circ}$ C
Northern Norway	10°	1°
Western Norway	$14 - 15^{\circ}$	$2 - 3^{\circ}$
Western Sweden	$17 - 19^{\circ}$	$1 - 3^{\circ}$
British Isles, west	$15 - 18^{\circ}$	$8 - 9^{\circ}$
British Isles, east	$13 - 15^{\circ}$	$4 - 5^{\circ}$
Western France	$16 - 17^{\circ}$	$8 - 9^{\circ}$
Portugal	17°	13°

1.0 - 3 m, but in some places it is different because of local conditions. The extreme tides of the bay between Normandy and Brittany in France are famous. The difference of level reaches 14 m at spring tide (St. Malo) and at neap tide is 6 m. On the Isle of Man, between Ireland and England, the spring tide is about 9 m and the neap tide, 5 m. The south point of Norway (Lindesnes) has no tides at all. On the Swedish west coast like in the Baltic the tide is almost negligible (0.1 - 0.3 m). The sea level changes observed here (0.4-0.5) are mainly due to winds.

One factor which is not normally considered in connection with the geographical distribution of marine algae is day-length. We do not know much about photoperiodism in marine algae or how this phenomenon influences distribution and development, but recent studies have shown that it does. It is obvious that photoperiodic reactions are involved in certain stages of the life history of several species (10, 14 , 22). It may be a factor to consider in connection with large-scale culture.

From the North of Europe to the South the day-length differs considerably in different seasons. At Spitzbergen, for instance, there is complete darkness for 4 months of the year and 24 hours of daylight for 4 months. The corresponding figure for North Norway is 2 months. Scotland, Denmark, and South Sweden have about 6 hours of daylight at midwinter and 18 at midsummer. Southern Europe has 9 and 15, respectively.

The sketch (Fig. 1) by Chapman (7) gives the outlines of the zonation on an average West European rocky shore. Various *Fucaceae* dominate the littoral. In the uppermost part *Pelvetia canaliculata* is found. It is followed by *Fucus spiralis*. The main part of the littoral is occupied by *Fucus vesiculosus* and *Ascophyllum nodosum*. In areas with exceptionally great tides the development of the *Ascophyllum* communities can be enormous. In the lower part of the

Table II. *The number of marine algal species in western Europe given as the total and percent of total in the three major algal divisions.*

Group	Spitz-bergen	White Sea	Norway, Arctic	Norway, western	Sweden, western	Baltic, south	British Isles	France, western	Por-tugal
CHLOROPHYTA									
Number of Species	19	41	35	86	79	55	116	75	60
Percent of Total	19.6	22.0	16.4	22.8	23.7	47.0	18.8	15.4	14.8
PHAEOPHYTA									
Number of Species	47	81	104	133	119	37	179	137	98
Percent of Total	48.4	43.0	49.0	35.3	35.6	30.0	29.0	28.1	24.3
RHODOPHYTA									
Number of Species	31	66	74	158	136	28	321	275	246
Percent of Total	32.0	35.0	35.0	41.9	40.7	23.3	52.2	56.5	60.9
Total species	97	188	213	377	334	120	616	487	404

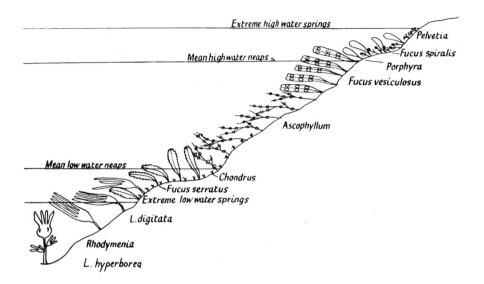

Figure 1. *Typical zonation on a European rocky shore (after Chapman, 7).*

littoral is a *Fucus serratus* belt, and around the low water line *Laminaria digitata* appears. All these algae can be collected by hand at low tide. *Laminaria hyperborea* is normally always submerged.

A great number of small algae are also found in the littoral and sublittoral. *Chondrus crispus* and sometimes *Gigartina stellata* form communities in the lower part of the littoral. The diagram illustrates normally developed vegetation as it appears in western Norway, Great Britain, Ireland, and northwestern France. Very exposed or very sheltered localities differ. With very strong exposure, most of the fucoids disappear.

The flora changes gradually between the Arctic Sea and Portugal (Table III). At Spitzbergen the littoral vegetation is poorly develope due to the ice. The tide is slightly less than 1 m. Of the major West European algae *Ascophyllum* and *Laminaria hyperborea* are lacking. Others, like *Alaria grandifolia* and *Laminaria digitata* and *saccharina* are very well developed.

Iceland has a rich and abundant algal flora. The southern part of the island has a warm-boreal character. The rocks are covered with belts of *Pelvetia, Fucus spiralis, vesiculosus,* and *distichus.* Broad belts of *Ascophyllum* often dominate. Considerable amounts of the carrageenan-yielding species *Chondrus crispus* and *Gigartina stellata* are also found.

The west coast is influenced by Atlantic water, the east and north coasts by arctic water. In the south and west the seawater never freezes, in the other parts only rarely. The vegetation and the chemical composition of several species have recently been studied by Munda (23, 25, 26, 27). *Laminaria hyperborea, Alaria esculenta,* and *Saccorhiza polyschides* are common and well developed.

Table III. *The occurrence and harvest of commercially valuable algal species in Europe*[1].

Species	Spitz-bergen	Ice-land	Norway	Den-mark	W. Ger-many	British Isles	France	Spain	Por-tugal	Note
Pelvetia canaliculata		o	o	o	o	o	o	o		
Fucus spiralis		o	o	o	o	o	o	o	o	
vesiculosus		o	o	o	o	o	o	o	o	
distichus		o	o	o		o	o			
serratus		o	o	o	o	o	o	o	o	
Ascophyllum nodosum		o +	o +	o	o	o	o	o	o	
Bifurcaria bifurcata						o	o	o	o	
Halidrys siliquosa		o	o	o	o	o	o	o		
Cystoseira sp.							o	o	o	
Sargassum sp.							o	o	o	
Laminaria digitata	o	o	o +	o	o	o +	o +	o	o	
hyperborea		o	o +	o	o	o +	o	o		
saccharina	o	o	o		o	o	o	o	o	
Saccorhiza polyschides			o			o	o	o	o	
Alaria	o	o	o	o		o +	o			
Porphyra umbilicalis		o	o	o	o	o +	o	o	o	Azores o +
Pterocladia capillacea							o +	o	o	
Lithothamnion calcareum						o	o +	o	o	
Furcellaria fastigiata			o	o +	o	o	o +	o +	o +	USSR Baltic o +
Chondrus crispus		o	o	o	o	o	o	o +	o +	
Gigartina stellata etc.		o	o	o	o	o	o	o +	o +	
Phyllophora	o	o	o	o	o	o	o	o		
Rhodymenia palmata	o	o	o	o	o	o +	o	o	o	

[1]Those locations where the alga occurs are designated by o; where it is also harvested, the designation is +.

Chemical analysis of several of the most common littoral species showed a remarkably high protein content in some.

According to Haug and Myklestad (13), the area covered with rockweed and kelp in Norway is as large as that of cultivated land. The vegetation has been studied by the Norwegian Institute of Seaweed Research. These investigations form one of the few serious attempts made to evaluate the standing crop of an area rich in seaweeds (2, 3, 4, 12, 32).

Laminaria hyperborea is the dominant alga along the exposed coast. Printz (32) estimated the weight of this species to comprise 80-90% of the total algal vegetation. It dominates in the sublittoral zone to a depth of 12 - 15 m. Densities of up to 30 kg/m^2 have been recorded. The mean density is estimated at 7.2 kg/m^2, with a regeneration time of 3 - 4 years. It corresponds to a mean yearly yield of 2.4 - 1.8 kg/m^2 according to Svendsen (33). Drift-weed is cast ashore in enormous quantities, especially in the southwest part of Norway.

Together with *Laminaria hyperborea*, *Ascophyllum nodosum* is the most harvested species in Norway. It grows in great quantities just below *Fucus vesiculosus* and prefers protected localities. The maximum density has been calculated to 26 kg/m^2, and the mean to 5.2 kg/m^2. The coast of Sweden, eastern Denmark, and the Baltic are special types. There is practically no tide, the littoral zone is only 0.4 - 0.5 m. Futhermore, the salinity of the sea water is reduced. On the Swedish west coast it is about $20 - 30^0/oo$ and in the Baltic less than $8^0/oo$. As salinity decreases, the algae become fewer. A great number of the Atlantic species disappear. From a commercial point of view, however, there is at least one valuable species, *Furcellaria fastigiata*, which is harvested in quantities. It is collected in Denmark and the USSR (formerly Lettland). This species, which is common to most parts of Europe, occurs in parts of the Baltic and between the Danish Isles as a loose form which only reproduces vegetatively.

The only place of the North Sea coast of West Germany where bottom conditions are suitable for a normal algal vegetation is Heligoland. The vegetation corresponds to the sketch in Fig. 1. The quantities of seaweeds, however, are not large enough for any commercial utilization. The growth conditions in the North Sea have recently been reported in some detail by Lüning (18, 19).

The algal vegetation of the British Isles has been studied by many phycologists. The sketch for this area (Fig. 1) was originally made by Chapman. In places with great tides, e.g., the Isle of Man, the littoral zone extends vertically with heavy algal growth as a result. The same is true in northwestern France, especially in the bay between Normandy and Brittany.

On the Atlantic shores of Spain and Portugal the fucoids of northwestern Europe gradually disappear. The littoral becomes dominated by other brown algae and various red ones: *Chondrus, Gigartina* species, *Gelidium* species, and *Pterocladia*. In the sublittoral zone *Saccorhiza polyschides, Laminaria digitata,* and *hyperborea* form communities.

THE USE OF SEAWEEDS

Only a moderate number of the algal species growing along the coast of Western Europe are commercially useful. In certain areas some species are used as food for humans or cattle. This is a very old tradition. *Porphyra umbilicalis,* known as laver, is used for human consumption in Wales. In Ireland, at least on the west coast, air-dried *Rhodymenia* is eaten. *Ascophyllum* in earlier days was used as pig feed in Norway and elsewhere. The use of *Porphyra* and *Rhodymenia* is, however, of little commercial importance.

According to the Icelandic sagas, several algae were eaten as long ago as A.D. 961, especially in times of starvation. Sheep still graze on fresh *Rhodymenia, Alaria,* and other algae. It is also known that in Scandinavia seaweeds were burned for salt production during the Middle Ages.

Farmers in coastal districts have long used seaweeds as fertilizer. In many places in Europe they still do. Seaweeds which have drifted ashore are collected and stored in stacks. Today some algal products are manufactured as growth stimulants for certain horticultural plants. These products are interesting and probably should be evaluated in a scientific manner. For production of seaweed meal, *Ascophyllum* is the main raw material in Europe. This is an important industry in Norway, where products are of a high quality. The total output there is about 15,000 tons a year. Meal is also made in Iceland and Ireland. For the alginate industry, *Laminaria hyperborea* is the main raw material. In Norway *L. digitata* and to some extent *Ascophyllum* are also of importance. *Laminaria* is harvested for this purpose in Great Britain, Ireland, France, and Norway. Part of the raw material is collected as drift-weed or cut by hand at low tide. Several experimental harvesting machines have been constructed, but most of them have turned out to be useless on open coasts. One, however, is used successfully in Norway (Fig. 2).

Chondrus, Gigartina, Gelidium, and *Pterocladia* are used for the manufacture of agar and carrageenan. They are all harvested by hand. Of special interest is the use of *Furcellaria* in Denmark and along the Baltic shore of the USSR.

QUANTITATIVE SURVEY METHODS

Good knowledge of the different seaweed producing areas is essential, but sound quantitative studies of seaweed growth rates and yields are in fact rather scarce. There are only a few areas which have been investigated thoroughly in this respect: Scotland, Nova Scotia, the southwest coast of Norway, and the California kelp beds. It is important to distinguish between biomass (or standing crop) on one hand and production or potential yield (harvest) on the other. There is also a distinction between theoretically feasible harvest estimates and those which are economically practical.

Rough seas can make rocky localities with a very rich algal growth inaccessible for collectors or divers. Cold uninhabited areas may have a great biomass, but collecting and transport costs may be far too high to make harvesting economically feasible.

Figure 2. *Large seaweed dredge designed for fishing boats of 20 meters. Its capacity is 1,000 kg of Laminaria. Its net weight is about 500 kg (after Svendsen, 33).*

Some algae, like *Fucus, Cystoseira,* or other fucoids, may be easy to harvest but have little practical value.

Different methods have been developed for surveying the algal beds. Littoral beds must be examined in a different way from the sublittoral ones. The methods employed depend on the actual species and the topography of the sea bottom. The first detailed studies of algal resources were those by Cameron (6). The kelp beds of the Pacific coast of North America were charted from small vessels. During the last world war some other methods were tried in Great Britain and will be discussed later.

Littoral weeds (rockweeds) can be estimated by weighing cut
samples and determining the areas occupied by the different species.
For calculations of total seaweed quantities the following two methods have been used: 1) multiply the mean density (weight per unit
area) by the area of the beds, or 2) multiply the mean weight per
m shoreline by the length of the shoreline. The second method is
preferable as only the shoreline length has to be measured, which
normally can be done on a map. However, it is difficult to obtain
an accurate estimate of the weight per m of shoreline. Aerial photographs may be of great help. The structure of the *Ascophyllum* zone
has been treated statistically using such methods by Baardseth (3).

At the Marine Botanical Institute in Göteborg considerable interest has been directed towards algal growth and seaweed resources
of the world. Dr. G. Michanek, in particular, has devoted many
years to such investigations (16, 21).

HARVESTS COMPARED TO POTENTIAL YIELDS IN WESTERN EUROPE

The main areas of seaweed harvesting are Iceland and Norway
(*Ascophyllum, Laminaria*); Denmark (*Furcellaria*); Scotland, Ireland,
and France (*Ascophyllum, Laminaria, Chondrus, Gelidium*); and Spain
and Portugal (*Gelidium*).

Iceland

Iceland's seaweed resource is one of the largest in the world,
but very little is used today. According to old sagas several seaweeds were eaten. *Rhodymenia* is rarely collected for human consumption, but sheep still graze on it and fresh *Alaria*. Production of
seaweed meal from *Ascophyllum* started in 1960 for sale on the domestic market as a supplementary cattle food.

The algal vegetation of Iceland has been studied recently by
Munda (23, 24, 26, 27). A study of the chemical composition of the
most common littoral algae shows the content of protein (in some
cases notably high), fat, mannitol, and alginic acid. Some of these
algae could possibly be used as raw material for industrial purposes.
The density for some important communities was also calculated in
kg/m^2 wet weight by Munda (25): *Chordaria* 3.0 kg, *Fucus vesiculosus*
up to 7.0 kg, *F. distichus* up to 5.6 kg, *Ascophyllum* up to 7.7 kg,
Porphyra umbilicalis up to 2.3 kg, *Gigartina stellata* up to 5.6 kg,
and *Rhodymenia palmata* up to 3.6 kg.

USSR (Arctic)

The Arctic coast of the USSR has a rich algal vegetation. The
density of the *Ascophyllum* beds of the Murman littoral is reported
to be up to a wet weight of 28 kg/m^2. The weight of the rockweeds
is estimated to at least 500,000 metric tons. The sublittoral is
dominated by *Laminaria digitata* and *saccharina*, with an estimated
wet weight of 5 - 6 million tons. The algal beds of the White Sea
appear to be larger than those of the Murman peninsula. The amount

of *Laminaria* is reported to be 800,000 tons, *Fucus* 25,000 tons, and
Zostera 400,000 tons (38, 39).

Norway

        The algal vegetation of the Norwegian coast is very rich.   It
is one of the few areas of the world where quantitative studies of
the seaweeds have been carried out on a large scale.   These investigations have been assessed through the Norwegian Institute of Seaweed Research.   According to Haug and Myklestad (13) the area covered
by seaweeds is as large as the area of cultivated land in Norway.
The most harvested species are *Ascophyllum, Laminaria digitata,* and
hyperborea.   Gigartina was earlier collected in small quantities,
but this harvesting has ceased entirely.
        Ascophyllum occurs in large quantities on littoral rocks, es-
pecially in sheltered localities.   Normally this species forms a
belt immediately below the *Fucus vesiculosus* belt.   Below *Ascophyllum*
lies *Fucus serratus*.   Baardseth (4) has studied these communities
quantitatively.   The weight ratio between the three communities of
Ascophyllum : Fucus vesiculosus : Fucus serratus was found to be
100 : 39 : 41.   The density of the *Ascophyllum* belt varied between
0 and 26 kg/m^2 wet weight.   The mean value is calculated at 5.2
kg/m^2.   Baardseth estimates the fresh weight of the entire stock of
Ascophyllum along the Norwegian coast to 1.8 million tons.
        Laminaria hyperborea is the dominant sublittoral species down
to a depth of 12 to 15 m.   It is estimated to represent 80 - 90 percent by weight of the total algal vegetation of Norway (31).   The
mean density has been calculated at 7.2 kg/m^2 fresh weight, but
values up to 30 kg/m^2 have been recorded.   The regeneration time has
been estimated at 3 - 4 years (32), which would give a yearly yield
of 1.8 to 2.4 kg/m^2.   Harvesting every third year gives a high, perhaps the maximum, profit from the beds (12).   Drift-weeds, especially
in the southwestern part of Norway, are also harvested.   After autumn
storms countless thousands of tons are washed ashore.   The occurrence
of drift-weeds has been especially studied by Printz (31).
        Laminaria hyperborea has been commercially harvested from the
bottom with special dredges since 1964.   Together with *L. digitata*
and *Ascophyllum* it is used for alginate manufacturing.   According
to Jensen (15) about 50,000 tons of fresh *Ascophyllum* is processed
yearly, producing about 15,000 tons of seaweed meal.   An estimated
1.8 million tons of fresh *Ascophyllum* is available in Norway.   An
annual harvest of 10 percent could produce 50,000 metric tons of
seaweed meal (4).   The potential yield of the brown algal vegetation
can be estimated at 210,000 tons and the actual harvesting at 75,000
tons.

Sweden

        The algal flora is fairly rich, especially on the west coast.
However, the vegetation becomes reduced due to the lowered salinity
and the lack of tides encountered in the Baltic.   The only investigation of hard-bottom algal communities on the Swedish west coast

was made by Gislén (11) in 1930 and should be repeated. The average densities in kg/m^2 fresh weight were *Ascophyllum* 15.3, *Halidrys* 10.0, *Fucus serratus* 8.4, *Laminaria hyperborea* 5.7, *L. digitata* 8.0, *L. saccharina* 3.6, *Furcellaria* 3.2, *Corallina* 3.0, *Chordaria* 2.3, and *Ulva* 2.0.

Seaweeds have long been used by coastal populations for fertilizer and fodder. *Fucus* was used to cover the ridges of thatched houses and *Zostera* was used as straw for the cattle and as building insulation. Today, seaweeds are scarcely employed except in small amounts for fertilizer. The comparatively small quantities of seaweeds on the Swedish coast and the high wages make the raw material prohibitively expensive. During the last war, a small industry started south of Göteborg for production of seaweed meal. It seemed promising, but closed after a few years.

Denmark

The first investigation concerning the productivity of benthic marine vegetation was carried out in Danish waters (30). A special dredge, the Petersen grab, was designed for this research, which dealt with the production of eelgrass (*Zostera*) on soft bottoms and the study of the relation between food availability and marine animal stocks. Production, calculated as kg/m^2 dry weight, was found to be 1.92 for maximum growth, 1.12 for medium growth, and 0.54 for minimum growth. This corresponds to a yearly production in Danish waters of 8 million tons dry weight or 24 million tons wet weight. *Fucus* also grows abundantly, but this species is not harvested (17).

The red alga *Furcellaria fastigiata* is common and widely distributed in Europe. Normally it occurs from 2 to 30 m attached to hard bottom, stones, or other algae. It is also found in Danish waters loose lying in protected areas with sandy bottoms. In places it is almost pure and free from other algae. For more than 30 years it has been trawled for commercial purposes. In the early 1940s the yearly harvest was a few thousand tons but increased to 20 - 30,000 tons by the 5 companies involved in harvesting. This led to heavy overharvesting and finally a depredation of the stock. Only one of the original 5 companies still exists, but the crop has more or less recovered. The trawling is now done in a satisfactory way and the beds are controlled. The potential sustained yield is estimated at 11,000 tons per year. The 1975 harvest was 10,700 tons.

Furcellaria is used as raw material for "Danagar." Chemically it is often known as furcellaran, which is not a true type of agar.

USSR (Baltic coasts)

The quantity of attached *Furcellaria* has been estimated at 80 - 90,000 tons. That which is loose lying on soft bottoms is judged to be near 150,000 tons. The *Furcellaria* found on the Baltic shore is one of the most important red algal resources of the Soviet Union. The potential yield is estimated at about 50,000 fresh weight tons and the actual harvest at 25,000 tons per year.

West Germany

On the Baltic coast (184 km) the *Fucus* biomass has been estimated at 40,000 tons or 217 tons/km. However, because the vegetation is spread out, density is only 1-2.5 kg/m^2 fresh weight.
Harvesting would probably not be profitable.

Hard bottoms on the North Sea coast are found essentially only
on Heligoland. However, the area is not large enough for commercial
utilization. Lüning (18, 19) found the following standing crops:
Fucus serratus 3.7 kg/m^2 fresh weight at 0 m, *Laminaria digitata*
5.6 kg/m^2, and *L. hyperborea* with a maximum of 11.1 kg/m^2 at 2 m.
Dense vegetation was found only between 1.5 and 4 m.

United Kingdom

During World War II sources of raw material became important
and available supplies were inventoried. This inventory included
a preliminary investigation of the coastline from Dorset west to
Lands End, north along the west coast, around Scotland, and south to
Yorkshire. The purpose was to find the coastal beds of rockweed
and large stands on the bottom. A more detailed survey was planned
later. Scotland was made the object of one of the most extensive
assessments of seaweed populations ever carried out. This research
was organized and carried out by the Scottish Seaweed Association
and the Institute of Seaweed Research.

Several survey methods were used. Some estimates were made by
observation, either by walking along the coast or in shallow water
under bright sunlight from the bow of a boat. Calibrated grapnel
devices were deployed from boats either with or without the aid of
echo-sounders. Echo-sounders were found to be 90 percent reliable,
under favorable conditions, in determining the boundaries of stands.
Aerial photography ultimately proved to be the most successful
method. In the littoral zone, an estimated 180,000 tons of rockweeds were distributed over 870 km with a density of at least 4 kg/m^2.
Seventy percent of the population grow on the Outer Hebrides and
22 per cent on the Orkney Islands. The dominant species is *Ascophyllum nodosum*.

The survey of the sublittoral kelp was carried out between 1946
and 1955. Along a coastline of over 8,500 km representing 8,000 km^2
of kelp-supporting bottom, 10 million tons wet weight of *Laminaria*
were estimated to be growing in water from 0 to 18 m in depth. The
mean density was 1.25 kg/m^2. According to Walker (35, 36, 37, 38)
4 million tons were found in densities (average 3.7 kg/m^2) along a
2,700 km coastline with 1,000 km^2 of bottom productive enough to be
of significance.

According to estimates made by the Institute of Seaweed Research
in 1956 in Scotland, it was possible that from the 10 million tons of
Laminaria (mainly *L. hyperborea*) growing in Scottish water, 1 million
tons wet weight, equivalent to 200,000 tons dry weight, could be harvested each year. Complete recovery of the harvested area would be
guaranteed by such sustained yield practices. *Laminaria hyperborea*
is dominant in most places, constituting 80 - 100 percent of the

Table IV. *Estimates of European seaweed consumption compared with potential yields of species of two major algal divisions given in metric tons wet weight for 1975.*

Region	Actually harvested PHAEOPHYTA (Brown algae)	Actually harvested RHODOPHYTA (Red algae)	Potential PHAEOPHYTA (Brown algae)	Potential RHODOPHYTA (Red algae)
Iceland	700	-	100,000	2,000
USSR-White Sea	5,000*	-	400,000*	2,000*
Baltic Sea	-	25,000*	-	50,000*
Norway	75,000	-	210,000	2,000
Sweden	-	-	5,000	-
Western Germany	-	-	3,000	20
Denmark	-	10,700	-	11,000
United Kingdom	26,200	-	20,000[1] 1,000,000[2]	150
Ireland	25,000	-	75,000	700
France	51,300	8,900	225,000	10,000
Spain	1,000	59,000	50,000	75,000
Portugal	-	23,500	25,000	46,000
Total	185,000	100,000 - 125,000	2,000,000	200,000

*National data are not authenticated.
[1]rock weed
[2]kelp

Laminariaceae. *L. digitata* reaches 0 - 5 percent, and *L. saccharina,* 0 - 15 percent. In some places *Saccorhiza polyschides* reached 1 percent.

    Gigartina and *Chondrus* are found in harvestable quantities from Cornwall to Scotland in the west and on the south coast of Scotland and north coast of England in the east. About 100 tons have been harvested but operations seem now to have ceased (20).

Ireland

    On the west coast of Ireland there is a rich algal vegetation, with beds extending a considerable distance from shore. Unfortunately, the waters make harvesting by boat impossible and the shores are often

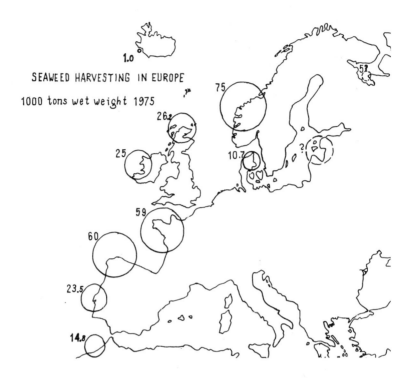

Figure 3.   *Seaweed harvesting in Europe.*

inaccessible from land.

  Chondrus crispus (Irish moss), which occurs in quantity, is
mainly harvested by hand at high tides (34).   *Laminaria hyperborea*
and limited amounts of *L. digitata* are mainly collected as drift-
weed and serve as raw material for the alginate industry.   The harvest varies from 1 to 3,000 tons.   *Ascophyllum* and *Fucus* are mainly
cut by hand and used with drift-weed for production of seaweed meal.
The standing crop of *Ascophyllum* has been estimated at 150,000 tons
wet weight.

  Saccorhiza polyschides occurs in large quantities.   The mean
density in the Cork area is 8.8 kg/m^2 (29).   This is more than twice
as much as in Norway, Scotland, and Spain.

France

  Brittany and Normandy are the two main areas for seaweed utilization in France (Atlantic coast).   *Laminaria, Fucus,* and *Himanthalia*
are used as fertilizer in Brittany.   Audoin and Pérez (1) made a
study of the *Laminaria* communities along 100 km of the Calvados
coastline.   The number of *Laminaria digitata* plants per m^2 was about
25, compared to 45 in Brittany.   For an area 17 km^2 they estimated
the standing crop to be 100,000 tons.   The potential yield of brown
algae can be estimated at 225,000 tons.   There is commercial harvest-
ings of *Chondrus crispus*, which is common on the rocky shores of the

Table V. *The approximate percentages of potential seaweed yields actually harvested in various regions of Europe.*

Region	Phaeophyta	Rhodophyta
Iceland	0.7	0.0
USSR-White Sea	1.2	0.0
Baltic	0.0	50.0
Norway	35.0	0.0
Sweden	0.0	0.0
Western Germany	0.0	0.0
Denmark	0.0	98.0
United Kingdom	26.0	0.0
Ireland	33.0	0.0
France	23.0	89.0
Spain	2.0	78.0
Portugal	0.0	51.0

same area. The estimated potential is 100,000 tons per year.

Lithothamnion calcareum, known only in France, occurs in large quantities in banks, known as "maerl bottoms," on the coast of Brittany. The product, originally used in the fields to improve crops, is now dredged commercially. The final product is sold often under the trade name "Goemar." The product contains various trace elements and organic growth substances and has a neutralizing effect on acid soils. It is extremely porous and contains calcium and magnesium in the ratio 10:1. It is also used in the hardening of acid drinking water.

Spain

Agar is the main product of the Spanish seaweed industry which ranks as the second largest in the world, exceeded only by that of Japan. *Gelidium, Gigartina stellata,* and *Chondrus crispus* are used as raw materials. Most are collected on beaches after autumn and winter storms. From May to October weeds are collected by hand at low tide or harvested by frogmen. Some *Fucus* is used for production of animal fodder.

Portugal

The algal flora of Portugal is rich. *Gelidium sesquipedale* and other species (*G. attenuatum, latifolium,* and *pusillum*) are harvested for agar. On the Azores, *Pterocladia capillaceae* is the main raw material. For carrageenan manufacturing *Chondrus crispus*

Table VI. *Worldwide potential sustained yields and actual harvests of two divisions of marine algae given in thousands of tons wet weight per year. The percent of the potential harvest represented by actual harvests is also given.*

Area	Rhodophyta Potential yields	Recent harvests[1]	%	Phaeophyta Potential yields	Recent harvests[1]	%
Arctic Sea	-	-	-	-	-	
NW Atlantic	100	35	35.0	500	6	1.2
NE Atlantic	200	100	50.0	2,000	185	10.7
WC Atlantic	(10)	-		1,000	1	0.1
EC Atlantic	50	10	20.0	150	1	0.7
Mediterranean & Black Sea	1,000	50	5.0	50	1	2.0
SW Atlantic	100	23	23.0	2,000	75	3.7
SE Atlantic	100	7	7.0	100	13	13.0
W Indian Ocean	120	4	3.3	150	5	3.3
E Indian Ocean	100	3	3.0	500	10	2.0
NW Pacific	650	545	82.0	1,500	825	56.0
NE Pacific	10	-		1,500	-	
WC Pacific	50	20	40.0	50	1	2.0
EC Pacific	50	7	14.0	3,500	153	4.4
SW Pacific .	20	1	5.0	100	1	1.0
SE Pacific	100	30	30.0	1,500	1	0.7

[1]Recent levels of harvests are based upon estimates for 1971–73.

and various *Gigartina* species are harvested. Seaweed collection is restricted to the second half of the year, in order not to interfere with the reproductive cycle of *Gelidium* and *Pterocladia*. There are also brown algae (*Laminaria digitata, L. hyperborea, Saccorhiza polyschides,* and *Himanthalia elongata*) in such quantities that harvesting could be possible.

Madeira has no algal industry, although there are large beds of *Cystosira abies marina* and *Pterocladia*. At least a modest utilization might be possible.

Estimates of European seaweed consumption and potential yields

are grouped together in Table 4 and Fig. 3. The figures are based
on the FAO Yearbook of Fishery Statistics, 1975 (9), supplemented
by estimates from Michanek (21) and Naylor (28). The corresponding
percentage figures for major sources have been calculated in Table 5.

WORLD-WIDE SEAWEED UTILIZATION

The world's seaweed resources are still not exploited fully.
Increased harvesting is dependent on economic factors such as the
demand for products and the cost of harvesting and processing. The
global production of red algae could probably be increased by at
least 50 percent. Utilization of the brown algae biomass could
probably be increased 20 times. An estimate of the potential yield
contrasted with the actual harvests of seaweeds in the world is
found in Table VI.

One means of satisfying the demand for algal raw material is
by large-scale culture of selected species. Cultivation has long
been practiced in Japan and is now being introduced in other areas.
There is no commercial cultivation in Europe. A group of marine
biologists in France (31) proposed to introduce *Macrocystis*, and
some preliminary studies have been carried out. The idea is not
new. Similar plans were advanced in Great Britain after the last
war, but were never realized. The project was regarded as too hazardous, because of possible injurious effects on fishing in the
North Sea. For the same reason there has been very strong resistance to the present French project. However, new uses for seaweeds
certainly will be found which will increase the demand for raw material. The importance of marine plants to the world-wide economy
seems destined to increase.

ACKNOWLEDGMENTS

I am very much indebted to Dr. G. Michanck of the Marine
Botanical Institute, Göteborg, Sweden, for valuable information
and discussions in the preparation of this paper.

REFERENCES

1. Audoin, J., and R. Pérez. 1970. Cartographie des populations
    de Laminaires des côtes francaises de la Manche orientale.
    [Chart of the Laminaria populations from the French coast
    of the eastern Channel area.] Sciences Pêche 194:1-11.
2. Baardseth, E. 1954. Kvantitative Tare-undersøkelser i Lofoten
    og Salten Sommeren 1952. [Quantitative Seaweed Researches
    in Lofoten and Salten in the Summer 1952.] Norwegian In-
    stitute of Seaweed Research, Oslo, Norway, Rep. 6. 47 pp.
3. Baardseth, E. 1955. A Statistical Study of the Structure of
    the *Ascophyllum* Zone. Norwegian Institute of Seaweed
    Research, Oslo, Norway, Rep. 11. 34 pp.
4. Baardseth, E. 1970. Synopsis of Biological Data on Knobbed
    Wrack, *Ascophyllum nodosum* (Linnaeus) Le Jolis. Food and

Agriculture Organization of the United Nations, FAO Fish. Synops. 38. Rev. 1.

5. Børgesen, F., and H. Jónsson. 1905. The Distribution of the Marine Algae of the Arctic Sea and the Northernmost Part of the Atlantic. Botany of the Faeröes. 3. Appendix. 28 pp.

6. Cameron, F. K. 1912. A Preliminary Report on the Fertilizer Resources of the United States. U.S. Senate, 62nd Congress, 2nd Session, Washington, D.C., Doc. 190. 290 pp.

7. Chapman, V. J. 1943. Zonation of marine algae on the seashore. Proc. Linn. Soc. London 154:239-253.

8. Chapman, V. J. 1948. Seaweed resources along the shores of Great Britain. Econ. Bot. 2:363-378.

9. Food and Agriculture Organization of the United Nations. 1976. Yearbook of Fishery Statistics. Vol. 40. Catches and Landings, 1975. 417 pp.

10. Føyn, B. 1955. Specific differences between northern and southern European populations of the green alga *Ulva lactuta* L. Pub. Staz. Zool. Napoli. 27:261-270.

11. Gislén, T. 1930. Epibioses of the Gullmar Fjord. 2. Marine sociology. Kristinebergs Zoologiska Station 1877-1927. K. Svenska Vet. Akad. Handl. 4. 380 pp.

12. Grenager, B. 1953. Kvantitative Undersøkelser av Tare-forekomster på Kvitsøy of Karnøy 1952. [Quantitative Studies on Seaweed Occurrences at Kvitsøy and Karnøy 1952.] Norwegian Institute of Seaweed Research, Oslo, Norway, Rep. 3. 53 pp.

13. Haug, A., and S. Myklestad. 1960. Aktuelle problemer i norsk tang- og tareforskning. [Actual problems in Norwegian sea-weed research.] Tekn. Ukebl. 107(35):1-12.

14. Hygen, G. 1948. Fotoperiodiske reaksjoner hos alger. [Photo-periodic reactions in algae.] Blyttia 6:1-6.

15. Jensen, A. 1966. Carotenoids of Norwegian Brown Seaweeds and of Seaweed Meals. Norwegian Institute of Seaweed Research, Oslo, Norway, Rep. 31. 138 pp.

16. Levring,T., H. A. Hoppe, and O. J. Schmid. 1969. Marine Algae, a Survey of Research and Utilization. Cram, De Gruyter, and Co., Hamburg, West Germany. 421 pp.

17. Lund, S. 1941. Tangforekomsterne i de Danske Farvande og Mulighederne for deres Udnyttelse. [The occurrence of sea-weeds in Danish waters and the possibilities for their uses.] Dansk Tidskr. Farm. 15(6):158-174.

18. Lüning, K. 1969. Standing crop and leaf area index of the sublittoral *Laminaria* species near Helgoland. Mar. Biol. 3:282-286.

19. Lüning, K. 1970. Tauchuntersuchungen zur Vertikalverteilung der sublitoralen Helgoländer Algenvegetation. [Diving studies on the vertical distribution of the sublittoral at Heligoland.] Helgol. Wiss. Meeresunters 21:271-291.

20. Marshall, S. M., L. Newton, and A. P. Orr. 1949. A Study of Certain British Seaweeds and Their Utilization in the

Preparation of Agar. Her Majesty's Stationary Office, London, England. 184 pp.

21. Michanek, G. 1975. Seaweed Resources of the Ocean. Food and Agriculture Organization of the United Nations, FAO Fish. Tech. Pap. 138. 127 pp.

22. Müller, D. G. 1962. Uber Jahres- und Lunarperiodische Erschein- ungen bei einigen Braunalgen. [On the yearly and lunarperio- dically appearances of some brown algae.] Bot. Mar. 4:140- 155.

23. Munda, I. 1969. Differences in the algal vegetation of two Icelandic fjords, Dyrafjördur (West Iceland) and Reydarfjördur (East Iceland). *In:* Margalef, R. (ed.) Proceedings, Sixth International Seaweed Symposium. Dirreción General de Peace Maritama, Ruiz de Alarcón, Madrid, Spain, pp. 255-261.

24. Munda, I. 1970a. Algological investigations around the coast of Iceland in 1963-68. Natturufroeðingurinn 40:1-25. (in Icelandic with English summary)

25. Munda, I. 1970b. A note on the densities of benthic algae along a littoral rocky slope in Faxaflói, Southwest Iceland. Nova Hedwigia 19:535-550.

26. Munda, I. 1972. On the chemical composition, distribution and ecology of some common benthic marine algae from Iceland. Bot. Mar. 15:1-45.

27. Munda, I. 1975. Hydrographically conditioned floristic and vegetation limits in Icelandic coastal waters. Bot. Mar. 18:223-235.

28. Naylor, J. 1976. Production, Trade and Utilization of Seaweeds and Seaweed Products. Food and Agriculture Organization of the United Nations, FAO Fish. Tech. Pap. 159. 73 pp.

29. Norton, T. A. 1970. Synopsis of Biological Data on *Saccorhiza polyschides*. Food and Agriculture Organization of the United Nations, FAO Fish. Synops. 83:(pag. var.)

30. Petersen, C. G. J. 1913. Havets Bonitering 2. [Quantitative Studies on the Sea Bottom.] Rep. Danske Biol. Stn. 21. 41 pp.

31. Perez, R., J. Couespeldu Mesnil, Y. Colin, L. le Fleur, and H. Didon. 1973. Etude sur l'opportunité d'introduire l'algue *Macrocystis* sur le littoral Français. [Studies on the possi- bility of introducing the alga *Macrocystis* on the French coast.] Rev. Trav. Inst. Pêches Marit. 37:307-361.

32. Printz, H. 1957. Norges Forekomster av Drivtang og Drivtare. [The Occurrence of Drifting Seaweeds in Norway.] Norwegian Institute of Seaweed Research, Oslo, Norway, Rep. 18. 50 pp.

33. Svendsen, P. 1972. Some observations on commercial harvesting and regrowth of *Laminaria hyperborea*. Fiskets Gang. 22:448- 460. (in Norwegian with English summary)

34. Valéra, M. de. 1958. A Topographical Guide to the Seaweed of Galway Bay with Some Brief Notes on Other Districts on the West Coast of Ireland. Dublin, Ireland. Institute of In- dustrial Research and Standards. 36 pp.

35. Walker, F. T. 1947. Sublittoral seaweed survey. 1. Development of the view box -- spring grab technique for sublittoral weed

survey. 2. Survey of Orkney, Scapa Flow. 3. Survey of Orkney, Bay of Firth. J. Ecol. 35:166–185.

36. Walker, F. T. 1953. Summary of seaweed resources of Great Britain. *In:* Proceedings, First International Seaweed Symposium, 1952, Edinburgh, Scotland, Pergamon Press, London, England, pp. 91–92.

37. Walker, F. T. 1954. Distribution of Laminariaceae around Scotland. J. Cons. Explor. Mer. 20:160–166.

38. Walker, F. T. 1958. Some ecological factors conditioning the growth of the Laminariaceae around Scotland. Acta Adriat. 8(13):1–8.

39. Zenkevitch, L. 1963. Biology of the Seas of the USSR. Allen and Unwin, London, England. 955 pp.

- 15 -
Survey of Chemical Components
and Energy Considerations

GLENN W. PATTERSON

CARBOHYDRATES

The algal compounds used by man in the largest amounts and for
the greatest length of time are the carbohydrates. Carrageenan is
a good example. Carrageenan is the cell wall polysaccharide found in
a number of red algae. The species most used in commercial production are *Chondrus crispus* and *Gigartina stellata*. Actually carrageenan is not a single compound but in a mixture of polysaccharides,
each of which contains closely related α-D or β-D galactose units
which are sulfated to various degrees. Carrageenans were originally
classified on the basis of their solubility in potassium chloride
solutions (30). Kappa-carrageenan, precipitated by potassium chloride, consists of 1,3 linked β-D-galactose 4-sulfate and 1,4 linked
3,6-anhydro-α-D-galactose (2, 4, 37). Lambda-carrageenan is not
precipitated by potassium chloride and is considered to be composed
of alternating units of 1,3 linked β-D-galactose 2-sulfate and 1,4
linked α-D-galactose 2,6-disulfate (Fig. 1). Now it appears that
the composition of carrageenan is much more complex. For instance,
κ-carrageenan also apparently contains units of 1,4 linked α-D-
galactose 6-sulfate, 1,4 linked α-D-galactose 2,6 disulfate, and
various others as well (51). It is not known whether these units
represent rarely encountered units in the κ-carrageenan chain or
whether they represent a chain of highly sulfated units which is
itself a minor constituent of the κ-carrageenan fraction.

As with the κ-carrageenan fraction, the structure of κ-carrageenan is idealized to some extent, since a number of various sulfated galactose units are found in the fraction in addition to the
two primary ones. Further clarification of the structure of κ- and
λ-carrageenan, as well as other carrageenans that have been named,
is dependant on methods to separate carrageenan fractions and to
unambiguously degrade the molecule.

271

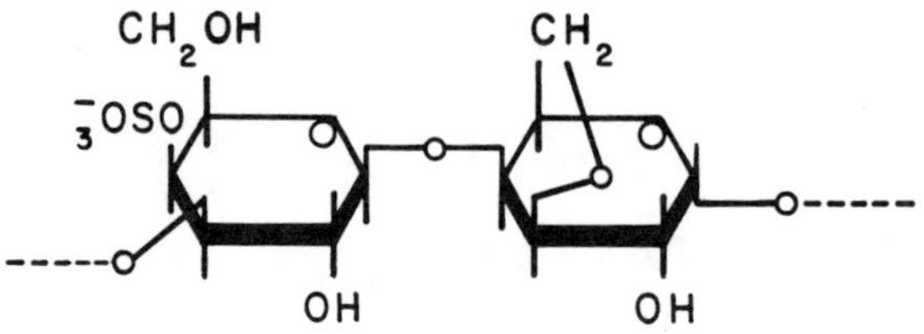

κ – CARRAGEENAN

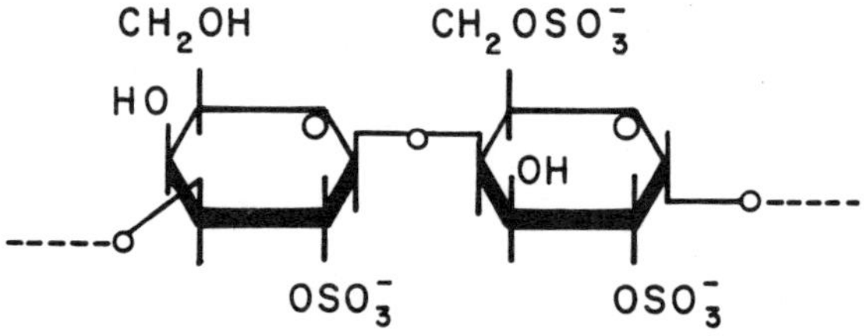

λ CARRAGEENAN

Figure 1.   *Partial structures of κ- and λ-carrageenan.*

Carrageenan has properties which make it attractive for numerous applications. It stabilizes emulsions (puddings), improves
smoothness and consistency (ice cream), and is an excellent suspending agent for cocoa (chocolate milk). Many of the uses of carrageenan are rather new, and demand for it has resulted in new industries in many parts of the world. It is now apparent that the demand for carrageenan will continue to outstrip the supply in the
foreseeable future.

Agar is one of the most valuable products derived from algae.
It is obtained primarily from the red algae *Gelidium* and *Gracilaria,*
although several other red algae have also been used for agar production. Agar is a complex carbohydrate which is not degradable by
most living organisms. It can be fractionated to yield two products,
agarose and agaropectin (3). Agarose is a neutral linear polysaccharide containing 1,3 linked β-D-galactose units attached to
1,4-linked 3,6 anhydro-α-L-galactose units (Fig. 2). Sometimes a
β-D-galactose unit is replaced by a unit of 6-0-methyl-D-galactose.
The frequency of occurrence of 6-0-methyl-D-galactose units varies
from 1% to 20% of the total agarose depending on the particular seaweed from which it was obtained. Thus, agarose itself has slightly
different properties depending on its origin.

Agaropectin is also composed of galactose residues, but in
agaropectin these residues frequently contain acidic groups such as
sulfates, pyruvates, and glucuronic acid. There is no agreement on
whether agaropectin is composed of a single polysaccharide or several

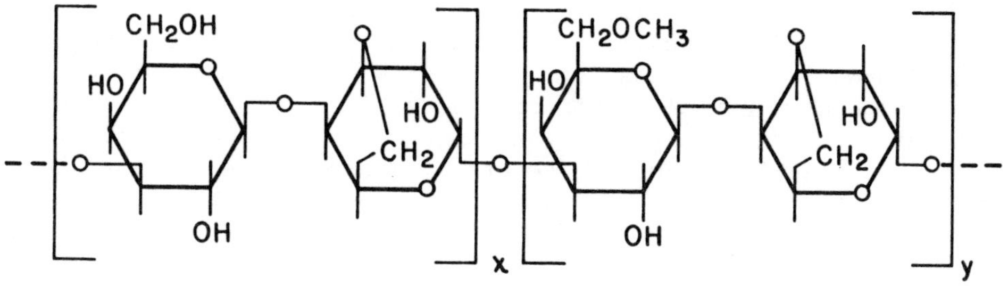

PARTIAL STRUCTURE OF AGAROSE

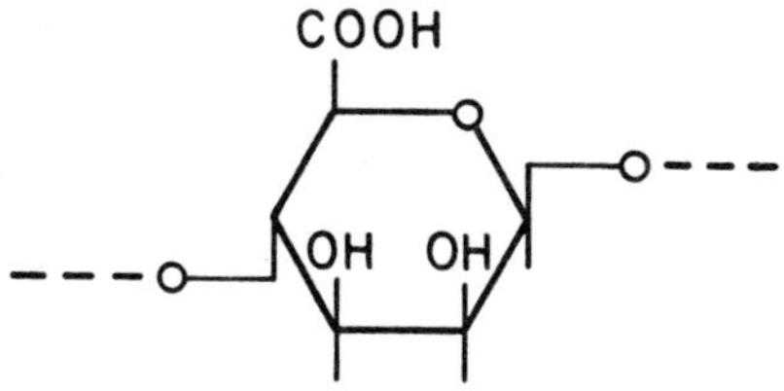

B – I, 4 LINKED – D – MANNURONIC
ACID

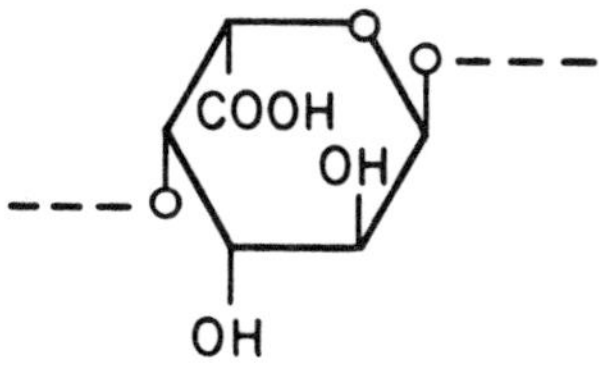

I, 4 LINKED L – GULURONIC ACID

Figure 2.   *Components of the structure of agar and alginic acid.*

(9).   The specific structure of agaropectin cannot be known without more careful analytical work.

Although agar has numerous commercial uses, all of them are overshadowed by its well-known employment in the preparation of plate cultures in microbiological work.   Due partially to the decline in the production of agar-producing algae in Japanese waters and its increasing cost, carrageenan and alginic acid are replacing agar in some products.   Carrageenan has largely replaced agar in the manufacture of pet food and alginic acid substitutes for agar in the making of dental impressions (5).   A relatively new development is the purification of agarose from agar for use in electrophoresis, gel filtration, and as a molecular sieve for separation of substances of a high molecular weight.

The third major algal polysaccharide of commerce is alginic acid.   The same problems of structure are present that are encountered with carrageenan and agar--difficulty of achieving purity and lack of unambiguous means of degradation.   Although early work indicated that alginic acid was a polymer of β-1,4 linked-D-mannuronic acid (14), it is apparent now that alginic acid is composed of both β-1,4 linked-D-mannuronic acid and β-1,4 linked L-guluronic acid (Fig. 2).   The ratio of mannuronic acid to guluronic acid varies in different brown algae such as *Laminaria*, *Ectocarpus*, *Macrocystis*, and *Ascophyllum*.   The alginic acid molecule apparently contains blocks of D-mannuronic acid residues, blocks of L-guluronic acid residues, and in portions of the molecule, the two acids alternate (13).

Alginic acid is usually used commercially as salts of sodium, calcium, ammonium, or potassium. Alginates are now used in scores of food products, especially frozen foods, desserts, salad dressing, and bakery products. Alginates suspend cocoa in chocolate milk, reduce ice crystal formation in ice cream, and serve as an emulsifying agent or a consistency-improving agent producing smoothness in a product as carrageenan does. The development of a propylene glycol alginate (45) gave a product which is stable under acidic conditions and opened a new application for alginates--the suspension of pulp in fruit drinks. Agar and carrageenan are not acid stable and cannot be used for this purpose. New applications for alginic acid arise regularly and demand is sharply increasing.

A polysaccharide prepared in Denmark from *Furcellaria fastigata* is called furcellaran. It resembles κ-carrageenan chemically, although it appears to have a branched structure and the sulfate content is generally lower. The properties of furcellaran are intermediate between those of agar and carrageenan. Furcellaran can be used in many of the food applications of other products (5).

A number of brown algae contain polysaccharides rich in L-fucose. These polysaccharides are now termed fucoidan and are found in *Fucus, Ascophyllum, Pelvetia,* and other brown algae. Fucoidan apparently varies in composition and quantity from species to species, and definitive chemical work on it is still incomplete.

Many other polysaccharides of almost endless variety apparently are present in seaweeds. Although man is utilizing some of them for agar, carrageenan, alginic acid, etc., there is no assurance that the most useful chemical structures in this group of compounds have been discovered. More work is needed on chemical structure, alteration of structure, and separation of fractions of mixed polysaccharides. Determination of factors leading to variation in composition and advances in the production and harvest of these plants will produce a larger and more productive marine plant industry. There is every indication that demand will outstrip supply.

LIPIDS

Only recently has a significant amount of structural work been accomplished on the fatty acids of algae. The algae contain the usual fatty acid chains with an even number of carbon atoms, but the green, red, and brown seaweeds, in particular, also produce significant quantities of 18-, 20-, and 22-carbon fatty acids with 3, 4, or 5 double bonds which are "essential" fatty acids in the mammalian diet (50, 55). These acids are also effective in lowering blood cholesterol. Twenty-carbon polyunsaturated fatty acids such as these are biological precursors in the synthesis of prostaglandins, an extremely potent new class of drugs. Prostaglandins are 20-carbon, oxygenated, unsaturated fatty acid derivatives which were first isolated from the genitals and seminal fluid of man and sheep (Fig. 3). Their possible medical applications include such areas as fertility

$$CH_3(CH_2)_4(CH=CHCH_2)_4(CH_2)_2COOH$$

ARACHIDONIC ACID

$$CH_3(CH_2)_4(CH=CHCH_2)_3(CH_2)_5COOH$$

EICOSATRIENOIC ACID

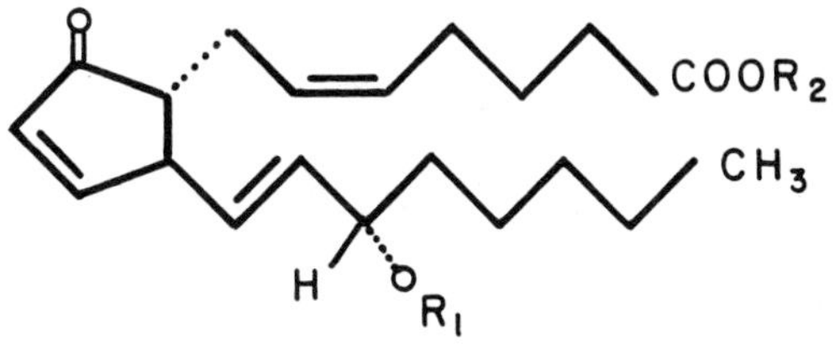

COOR$_2$

CH$_3$

H   R$_1$

(15S) = PGA$_2$

R$_1$ = H or HCO
R$_2$ = H or CH$_3$

Figure 3. *Unsaturated fatty acids and prostaglandins from marine
sources.*

control, labor induction and abortion, and reproductive biology. They
occur in mammals in such small quantities, however, that other sources
have been sought for prostaglandins for experimental research. A gorgonian, *Plexaura homomalla,* has recently been found to be a rich
source of 15-epi-PGA$_2$, which, although not active, can be converted
chemically to the active PGA$_2$ (52, 53). Although prostaglandins have
not been reported in marine algae, their biochemical precursors are
present. The possibilities of the presence of prostaglandins or substances convertible to prostaglandins in marine algae is worthy of
research.

   In an energy-conscious world, living plants are now being examined
as a source of hydrocarbons to replace our historical fossil sources.
In the last 10 years a number of new and interesting hydrocarbons have
been identified from marine algae. In *Ascophyllum nodosum,* the polyunsaturated hydrocarbon, all cis-3,6,9,12,15,18-heneicosahexaene has
been isolated (54). In *Ascophyllum* this compound apparently occurred
exclusively in the reproductive parts. This C$_{21}$ hexaene and its biosynthetic precursor, the corresponding C$_{22}$ hexaenoic acid, are found
together in a number of green and brown seaweeds (54). Although the
C$_{22}$ acid is relatively abundant, the hydrocarbon is usually found in

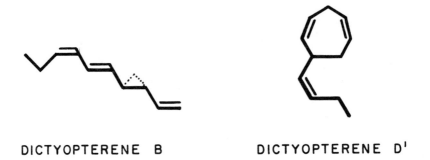

DICTYOPTERENE B   DICTYOPTERENE D'

Figure 4.   *Structure of Dictyopterene B and D' from marine algae.*

quantities too small to be considered of economic interest as an
energy source.

   Most living organisms contain a series of n-paraffins ranging
from 15 to 33 carbon atoms in length.   In some higher plants the concentrations of these hydrocarbons can be of economic interest.   In
algae, these odd-chain hydrocarbons are present but only in small
quantities.

   One more possibility of using algae to produce hydrocarbons
should be considered.   The freshwater green alga *Botryococcus* produces, as over 10% of its dry weight, a series of polyunsaturated,
high-molecular-weight hydrocarbons containing from 27 to 31 carbon
atoms.   Under conditions which are poor for growth, this organism
goes into a resting stage (6, 26) where up to 70% of the weight of
the organism is made up of these hydrocarbons.   Similar hydrocarbon
synthesis in marine algae has not been reported, but neither the
economic potential nor the biosynthetic capabilities of such organisms should be underestimated.

   A number of new hydrocarbons have been isolated from brown algae
in recent investigations.   Dictyopterene B (Fig. 4) and several related C_{11} unsaturated hydrocarbons have been isolated from species
of the brown alga *Dictyopteris* (32, 40).   During this same time period, Müller discovered and isolated the sex attractant produced by the
female gametes of the brown alga *Ectocarpus siliculosus* (33).   Mass
culturing of the algae produced enough of the attractant to identify
it as an unusual C_{11} cyclic compound referred to as dictyopterene D'
(35) (Fig. 4).   Chemical conversion of dictyopterene B to dictyopterene D' has been accomplished (1, 25).   More recent work with *Fucus*
indicates that besides a similar sex attractant, there is also a hydrocarbon, apparently having a lower molecular weight (15, 34).
These reports are just the beginning of our chemical understanding
of the reproductive process in algae.   Such hormone-like compounds
in algae may also affect the animals feeding on them.   Paffenhöfer
has speculated that the sex ratios of adult *Calanus helgolandicus*
may depend on the amount of heneicosahexaene (21:6) in the diet of

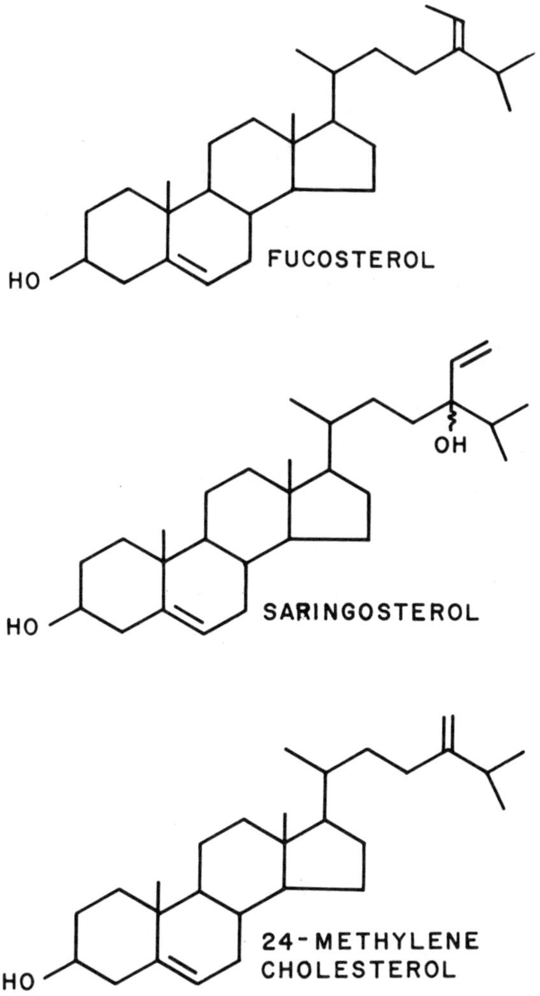

FUCOSTEROL

HO

SARINGOSTEROL

OH

HO

24-METHYLENE
CHOLESTEROL

HO

Figure 5. *Sterols of brown algae.*

the immature copepods.   Increasing ratios of males were found using
diets with increasing amounts of 21:6 (38).
     The sterols of many marine algae have been examined in recent
years.   Nearly all of the red algae contain cholesterol as their
principal sterol and are the only large group of plants to do so
(39).   Many brown algae contain 24-methylene cholesterol or saringosterol (Fig. 5), although fucosterol is the principal sterol in
all species examined (39).   These sterols were thought to produce
the hypocholesterolemic action in chicks which were fed a diet containing them (42).   The unsaponifiable fraction from clams containing these sterols also lowered blood cholesterol (41).   The hypocholesterolemic action of sitosterol, a terrestrial plant sterol,
was not as pronounced.   Sterols of marine plants may be important

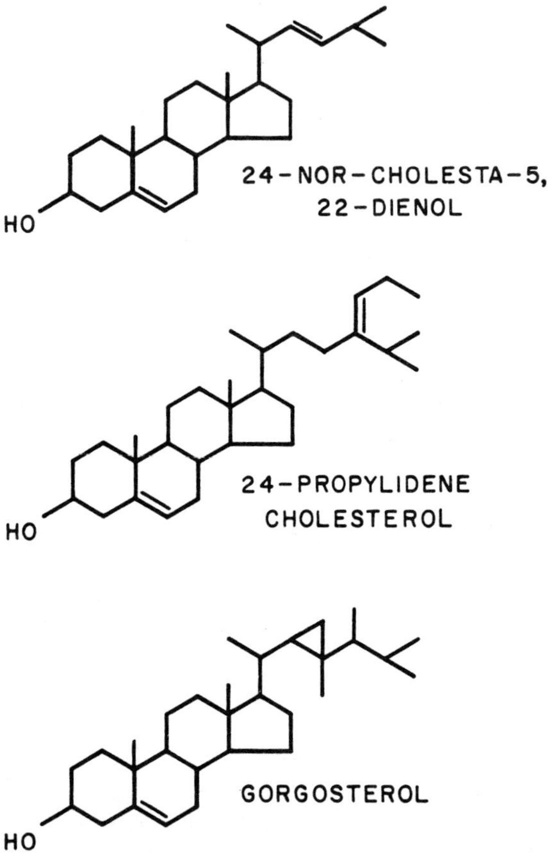

24-NOR-CHOLESTA-5, 22-DIENOL

24-PROPYLIDENE CHOLESTEROL

GORGOSTEROL

Figure 6.   *Unusual sterols from the marine environment.*

in the diet of marine invertebrates because many of them have been
demonstrated to be incapable of sterol synthesis (49).   Sterols are
considered indispensable to the life of virtually every living being.
Mariculture ventures should proceed with these possible requirements
in mind.   Sterols with extremely unusual structures have been isolated from marine organisms in recent years.   Sterols with unprecedented side chains containing 7 carbons (24-nor-cholesta-5,22-dienol),
11 carbons (24-propylidene cholesterol), and a cyclopropane ring
(gorgosterol) (Fig. 6) have been isolated from marine plants and
animals (12, 17, 18).   Although these unusual structures are now
frequently found in marine organisms, attempts to demonstrate their
biosynthesis have failed (11).   The biochemical or physiological
significance of such unusual structures is unknown.

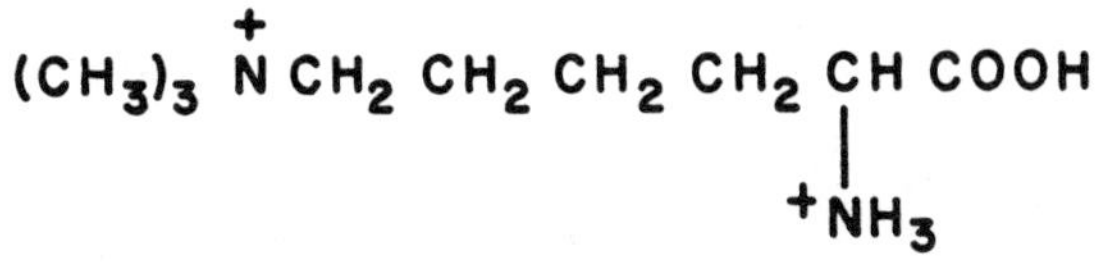

$$(CH_3)_3 \overset{+}{N} CH_2\ CH_2\ CH_2\ CH_2\ \underset{\underset{+NH_3}{|}}{CH}\ COOH$$

LAMININE

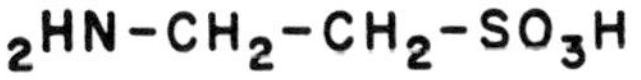

$$_2HN-CH_2-CH_2-SO_3H$$

TAURINE

$$_2HN-\overset{\overset{NH}{\|}}{C}-\overset{H}{N}-CO-\overset{H}{N}-(CH_2)_3-\underset{\underset{NH_2}{|}}{CH}-COOH$$

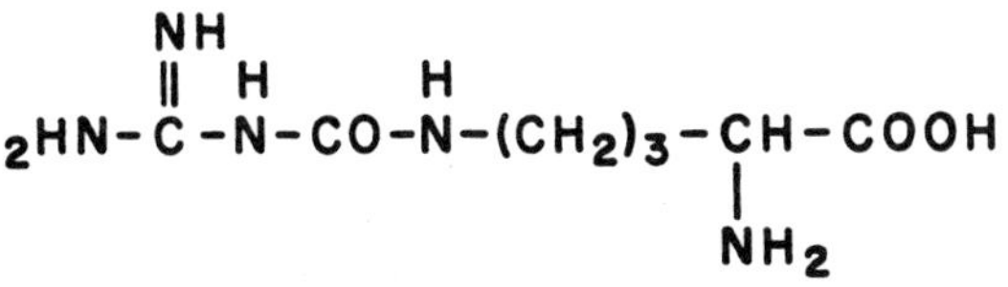

GIGARTININE

Figure 7. *Structures of laminine, taurine, and gigartinine.*

NITROGEN COMPOUNDS

Because of the large accumulations of such carbohydrates as agar, carrageenan, alginic acid, and others in seaweeds, protein composition is necessarily lower on a percentage basis than in the micro-algae. Neither the amino acid composition nor the feeding value to animals of the protein of these algae has been sufficiently studied. When one considers the large amount of seaweed material available as a by-product of algal polysaccharide production, it appears that more information in this area would be valuable. For instance, two new amino acids, gigartinine (Fig. 7) and the closely related gongrine, were isolated from the red alga *Gymnogongrus flabelliformis* (22, 23). A survey of 25 species of algae (24) showed that gigartinine was widely distributed in the red algae. Our knowledge of essential amino acid requirements of marine animals lags far behind that of terrestrial animals. Amino acids which are unusual from the terrestrial point of view but are widely distributed in marine plants may play an important role and could even be essential amino acids in the diet of marine animals.

Several brown algae have been used in the Orient as hypotensive

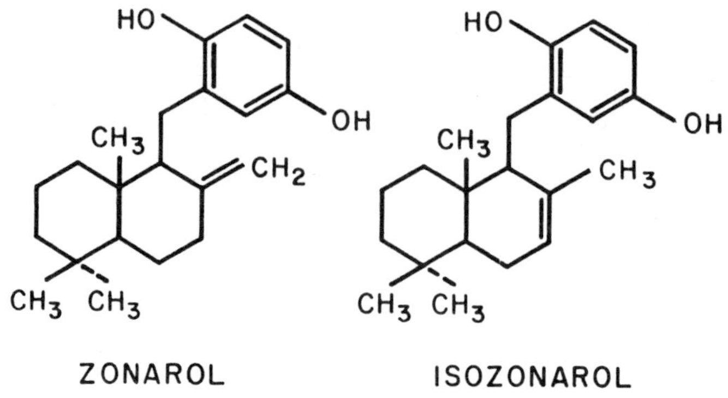

Figure 8. *Structures of zonarol and isozonarol.*

drugs for many years. The active principle, laminine, is widely
distributed in the Laminariaceae (47, 48) (Fig. 7).
    Taurine and several other related compounds have been isolated
from the red algae, *Ptilota, Porphyra,* and *Gelidium* (29).

ANTIBIOTICS

    One of the most promising potential uses of seaweeds is in the
production of antibiotics. Antibiotic properties of algae have been
known for decades, but as yet none have reached commercial use.
Dictyopteris zonarioides produces two components which are reported
to be highly fungitoxic although apparently not antibacterial (10).
They are the isomeric isoprenoid hydroquinones, zonarol and isozonarol (Fig. 8). The corresponding quinones were also isolated and
all of these compounds were effective against fungal plant pathogens.
    The red algae *Digenia* of the Middle East and *Chondria* of Japan
have long been used as anthelmintics. Murakami purified and identified the active compound from *Digenea* as L-α-kainic acid (36). The
active principle from *Chondria* is domoic acid (Fig. 9), an analog of
L-α-kainic acid (8, 46).
    The intestine of the penguin contains a potent antibacterial
substance identified as acrylic acid ($CH_2CH\ COOH$), whose origin was
traced through the food chain to marine algae (43).
    Several bromine-containing compounds have recently been iso-
lated from the red alga *Laurencia*. The first of these compounds
was isolated from *L. glandulifera* and was named laurencin (19)
(Fig. 10). Closely related compounds were isolated from *L. nipponica*
(20, 21). These unusual compounds are cyclic ethers containing one
or more bromine atoms and an acetylenic bond. Although a number of
brominated natural products have shown activity as antibiotics, little
yet is known about the medicinal or other values of laurencin and related compounds.
    Many bromophenols are suspected to have a close connection with
antibiotic properties in some algal extracts. Cragie and coworkers

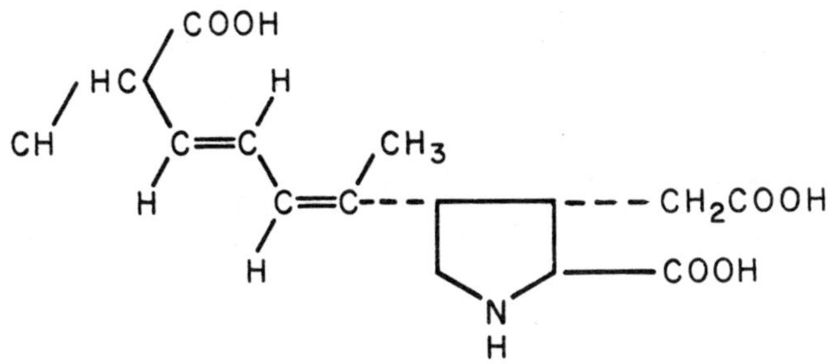

DOMOIC ACID

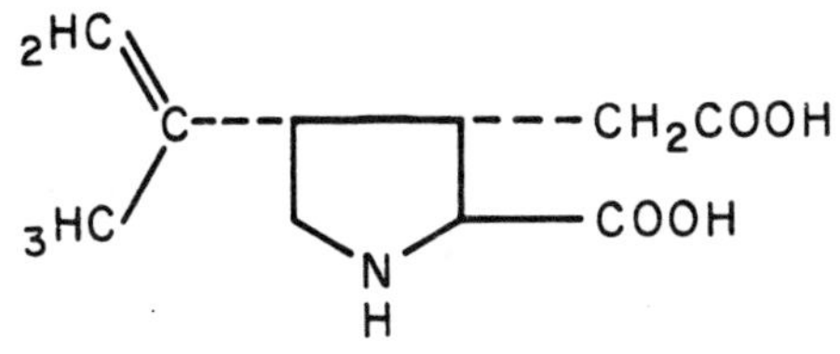

L-α KAINIC ACID

Figure 9. *Structures of domoic acid and L-α kainic acid.*

(7) have isolated 3,5-dibromo-p-hydroxybenzyl alcohol from *Odontha-lia* and *Rhodomela* and dipotassium 2,3-dibromo-benzyl alcohol-4,5-disulfate (Fig. 10) from *Polysiphonia lanosa* (16).

Several tannin-like substances which have antibiotic properties occur in *Sargassum*. These substances, called sarganins, have not been chemically identified (31).

Polyphenol extracts from *Sargassum* contain antifouling properties which, although they are apparently not of commercial quality, may affect the growth of epiphytic organisms (44).

ENERGY

One of the promising new developments in the use of algae for commercial purposes has been in the field of energy. Nearly all commercial uses of algae take advantage of some chemical component abundant in the algae. With rapid depletion of fossil hydrocarbons and sharply increasing energy prices, it is expedient to examine marine plant biomass in terms of its energy value. In this new concept of algal utilization, methane, produced by anaerobic fermentation of *Macrocystis*, is considered to be the primary product (28). Alginic acid, mannitol, inorganic salts, and animal feeds are among the by-products.

Complete modern analyses of the components of *Macrocystis* are not yet available, but the major components which may be commercially

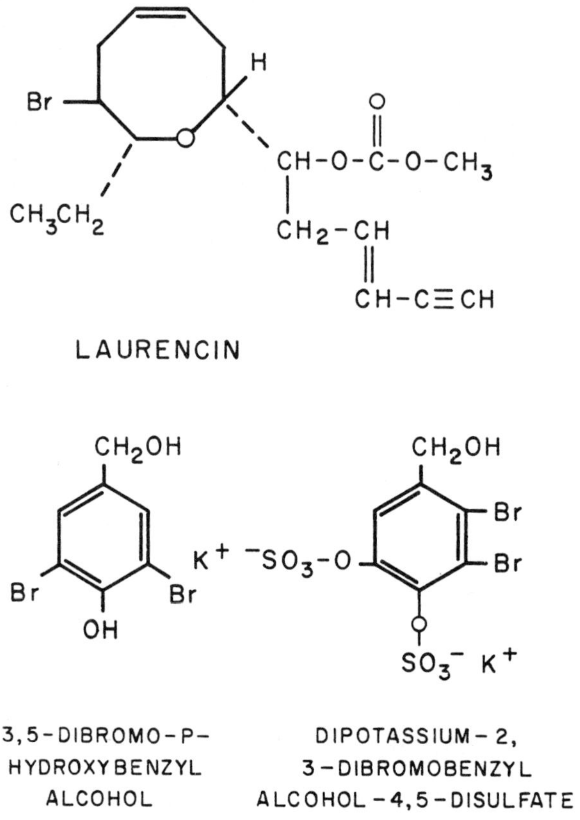

LAURENCIN

3,5-DIBROMO-P-   DIPOTASSIUM-2,
HYDROXYBENZYL   3-DIBROMOBENZYL
ALCOHOL   ALCOHOL-4,5-DISULFATE

Figure 10. *Some halogenated compounds from marine algae.*

utilized are listed below along with their maximum concentrations
on a dry weight basis: mannitol (14%), alginic acid (24%), laminarin (21%), and potassium chloride (4%). Other components include
protein (13%), fats (0.5%), cellulose (8%), fucoidan (2%), and several inorganic salts. The chemical composition of these algae fluctuates drastically, so a flexible scheme to take advantage of this
variability is necessary.

The carbon/hydrogen ratio of *Macrocystis* is about 6:1 which is
a favorable ratio for conversion to liquid fuel. The overall energy
content of *Macrocystis* is about 4,400 BTU/lb which is comparable to
salad vegetables such as lettuce and celery (27). While this value
is only about 1/3 that of soft coal, it is 6 times that of oil shale.
The higher growth rate of *Macrocystis* gives it superior potential to
many terrestrial plants for biomass conversion.

A kelp processing plant could operate as follows: raw kelp is
chopped and pressed at 625 psi, removing up to 83% of the fresh
weight. From this juice mannitol, other soluble carbohydrates, and
salts such as potassium chloride can be recovered. An attractive
option is to treat the pressed cake with Na_2CO_3 for extraction of
alginic acid before preparation for anaerobic fermentation. Anaerobi

fermentation with yeast to produce ethanol is feasible, but fermentation producing methane is favored. Carbon conversion of this process is estimated to be near 50%. Sludge from the fermentation process can be used as a feed supplement or as fertilizer.

Such a process has the following advantages: (1) market demand for these products is high, (2) there are many viable alternate methods of processing, (3) fermentation technology is well developed, (4) dry plant material is not necessary, and (5) the process is potentially profitable even with plant material known for its variable quality.

Process data are now being analyzed prior to design of a pilot plant (27).

SUMMARY AND CONCLUSIONS

In summary, commercial utilization of algae for specific chemicals is limited primarily to the carbohydrates, agar, carrageenan, and alginic acid. Other carbohydrates such as mannitol, laminarin, furcellaran, fucoidan, and related compounds show promise but are not used widely in commerce at present. The essential fatty acids and sterols may be important if algae are used as feed in mariculture. They both have potential as starting materials in the synthesis of other chemicals. Research to determine the presence of prostaglandins in seaweed would be useful. Direct production of hydrocarbons by algae does not appear to be sufficient to be of commercial value, but little research has been directed toward screening important seaweeds for hydrocarbons or determining if large-scale hydrocarbon production can be induced as in *Botryococcus*.

Laminine, domoic acid, and L-α-kainic acid have been shown to be of medicinal value, and algae containing them have been used for medicinal purposes for years, but commercial development of these drug sources has not occurred. No antibiotics are now commercially produced from algae, but zonarol and isozonarol look promising. Other antibiotic principles have been isolated from a number of algae, but testing of them is incomplete. Perhaps the most exciting new development is the possibility of recovering economically competitive methane or other energy-rich compounds from various varieties of seaweed. Such processes not only could produce needed energy but would help to make possible economic recovery of many other algal products.

REFERENCES

1. Ali, A., D. Saranthakis, and B. Weinstein. 1971. Synthesis of the natural product (±)-Dictyopterene B. J. Chem. Soc., London, Chem. Com., Sec. D, p. 940.
2. Anderson, N. S., T. C. S. Dolan, and D. A. Rees. 1968. Carrageenans Part III. Oxidative hydrolysis of methylated *k*-carrageenan and evidence for a masked repeating structure. J. Chem. Soc., London, Sec. C, pp. 596-601.

3. Araki, C. 1966. Some recent studies on the polysaccharides of agarophytes. *In:* Young, E. G., and J. L. McLachlan (eds.) Proceedings, Fifth International Seaweed Symposium, 1965. Pergamon Press, Oxford, pp. 3-17.

4. Araki, C., and S. Hirase. 1956. Partial methanolysis of the mucilage of *Chondrus crispus* Holmes. Bull. Chem. Soc. Japan. 29:770-775.

5. Booth, E. 1975. Seaweeds in industry. *In:* Riley, J. P., and G. Skirrow (eds.) Chemical Oceanography Vol. IV. Academic Press, New York, pp. 219-267.

6. Brown, A. C., B. A. Knights, and E. Conway. 1969. Hydrocarbon content and its relationship to physiological state in the green alga, *Botyrococcus braunii*. Phytochemistry 8:543-547.

7. Craigie, J. S., and D. E. Gruenig. 1967. Bromophenols from red algae. Science 157:1058-1059.

8. Daigo, K. 1959. Studies on the constituents of *Chondria armata* II. Isolation of an anthelmintical constituent. Yakugaku Zasshi 79:353.

9. Duckworth, M., K. C. Hong, and W. Yaphe. 1971. The agar polysaccharides of *Gracilaria* species. Carbohyd. Res. 18:1-9.

10. Fenical, W., J. J. Sims, P. Radlick, and R. M. Wing. 1972. *In:* Abstract of papers, Third Food-Drugs from the Sea Conference, University of Rhode Island. Marine Technology Society, 1030 15th St., N.W., Washington, D.C. 20005, p. 17.

11. Ferezou, J. P., M. Devys, J. P. Allais, and M. Barbier. 1974. Sur le sterol a 26 atomes de carbone de l'algue rouge *Rhodymenia palmata*. [On the C-26 sterol from the red alga, *Rhodymenia palmata*.] Phytochemistry 13:593-598.

12. Hale, R. L., J. LeClercq, B. Tursch, C. Djerassi, R. A. Gross, Jr., A. J. Weinheimer, K. Gupta, and P. J. Scheuer. 1970. Demonstration of a biogenetically unprecedented side chain in the marine sterol, gorgosterol. J. Am. Chem. Soc. 92: 2179-2180.

13. Haug, A., B. Larsen, and O. Smidsrød. 1967. Studies on the sequence of uronic acid residues in alginic acid. Acta Chem. Scand. 21:691-704.

14. Hirst, E. L., J. K. N. Jones, and W. O. Jones. 1939. The structure of alginic acid, Part I. J. Chem. Soc., London, pp. 1880-1885.

15. Hlubucek, J. R., J. Hora, T. P. Toube, and B. C. L. Weedon. 1970. The gamone of *Fucus vesiculosis*. Tetrahedron Lett. 5163-5164.

16. Hodgkin, J. H., J. S. Craigie, and A. G. McInnes. 1966. The occurrence of 2,3-dibromobenzyl alcohol 4,5-disulfate, dipotassium salt, in *Polysiphonia lanosa*. Can. J. Chem. 44:74-78.

17. Idler, D. R., P. M. Wiseman, and L. M. Safe. 1970. A new marine sterol, 22-*trans*-24-norcholesta-5,22-dien-3β-ol. Steroids 16:451-461.

18. Idler, D. R., L. M. Safe, and E. F. MacDonald. 1971. A new C-30 sterol (Z)-24-propylidenecholest-5en-3β-ol (29-methyli-

sofucosterol).   Steroids 18:545-553.

19.   Irie, T., M. Suzuki, and T. Masamune.   1965.   Laurencin, a constituent from *Laurencia* species.   Tetrahedron Lett. 1091-1099.

20.   Irie, T., M. Izawa, and E. Kurosawa.   1968.   Laureatin, a constituent from *Laurencia nipponica* Yamada.   Tetrahedron Lett. 2091-2096.

21.   Irie, T., M. Izawa, and E. Kurosawa.   1968.   Isolaureatin, a constituent from *Laurencia nipponica* Yamada.   Tetrahedron Lett. 2735-2738.

22.   Ito, K., and Y. Hashimoto.   1965.   Occurrence of γ-(guanylureido) butyric acid in a red alga, *Gymnogongrus flabelliformis*. Agr. Biol. Chem. 29:832-835.

23.   Ito, K., and Y. Hashimoto.   1966.   Gigartinine: a new amino-acid in red algae.   Nature 211:417.

24.   Ito, K., K. Miyazawa, and Y. Hashimoto.   1966.   Distribution of gongrine and gigartine in marine algae.   Nippon Suisan Gakkaishi 32:727-729.

25.   Jaenicke, L., T. Akintobi, and D. G. Miller.   1971.   Synthesis of the sex attractant of *Ectocarpus siliculosus*.   Angew. Chem., Int. Ed. Engl. 10:492-493.

26.   Knights, B. A., A. C. Brown, E. Conway, and B. S. Middleditch. 1970.   Hydrocarbons from the green form of the fresh water alga *Botyrococcus braunii*.   Phytochemistry 9:1317-1324.

27.   Leese, T. M. 1975.   Personal communication from a paper presented to the 141st Annual Meeting of American Association for the Advancement of Science, New York City, entitled Kelp farm product conversion.

28.   Leese, T. M.   1976.   The conversion of ocean farm kelp to methane and other products.   *In:* Symposium Proceedings, Clean Fuels from Biomass, Sewage, Urban Refuse and Agricultural Wastes. Institute of Gas Technology, 3424 S. State St., Chicago, Illinois 60616.

29.   Lindberg, B.   1955.   Methylated taurines and choline sulfate in red algae.   Acta. Chem. Scand. 9:1323-1326.

30.   Mackie, W., and R. D. Preston.   1974.   Cell wall and intercellular region polysaccharides.   *In:* Stewart, W. D. P. (ed.) Algal Physiology and Biochemistry.   University of California Press, Berkeley, pp. 40-85.

31.   Martinez-Nadal, N. G., L. V. Rodriguez, and C. Casillas.   1964. Isolation and characterization of sarganin complex, a new broad-spectrum antibiotic isolated from marine algae.   Antimicrob. Agents Chemother., pp. 131-134.

32.   Moore, R. E., J. A. Pettus, Jr., and M. S. Doty.   1968.   Dictyopterene A., an odoriferous constituent from algae of the genus *Dictyopteris*.   Tetrahedron Lett. 4787-4790.

33.   Müller, D. G.   1968.   Versuche zur charakterisierung eines Sexual-Lockstoffs bei der Braunalge *Ectocarpus siliculosus*. Planta 81:160-168.

34.   Müller, D. G.   1972.   Chemotaxis in brown algae.   Naturwissenshaften 59:166.

35. Müller, D. G., L. Jaenicke, M. Donike, and T. Akintobi. 1971.
        Sex attractant in a brown alga: chemical structure. Science
        171:815-817.

36. Murakami, S., T. Takemoto, and Z. Shimizu. 1953. Studies on
        the effective principles of *Digenea simplex* Aq. I. Separation
        of the effective fraction by liquid chromatography. Yakugaku
        Zasshi 73:1026.

37. O'Neill, A. N. 1954. Degradative studies on fucoidan. J. Am.
        Chem. Soc. 76:5074-5076.

38. Paffenhöfer, G. A. 1970. Cultivation of *Calanus helgolandicus*
        under controlled conditions. Helgol. wiss. Meeresunters
        20:346-359.

39. Patterson, G. W. 1971. The distribution of sterols in algae.
        Lipids 6:120-127.

40. Pettus, J. A., Jr., and R. E. Moore. 1970. Isolation and
        structure determination of an undeca-1,3,5,8-tetraene and
        dictyopterene B from algae of the genus *Dictyopteris*. J.
        Chem. Soc. London, Chem. Comm., Sec. D, pp. 1093-1094.

41. Reiner, E., D. R. Idler, and J. D. Wood. 1960. A hypocholes-
        terolemic factor in marine sterols. Can. J. Biochem.
        Physiol. 38:1499-1500.

42. Reiner, E., J. Topliff, and J. D. Wood. 1962. Hypocholesterol-
        emic agents derived from sterols of marine algae. Can. J.
        Biochem. Physiol. 40:1401-1406.

43. Sieburth, J. McN. 1961. Antibiotic properties of acrylic acid,
        a factor in the gastrointestinal antibiosis of plar marine
        animals. J. Bacteriol. 82:72-79.

44. Sieburth, J. McN., and J. T. Conover. 1965. *Sargassum* tannin,
        an antiobiotic which retards fouling. Nature 208:52-53.

45. Steiner, A. B. 1947. Manufacture of glycol alginates. U.S.
        Pat. 2,426,125. Abstr. in Office. Gaz. U.S. Patent Office
        601:495.

46. Takemoto, T., and K. Daigo. 1958. Constituents of *Chondria
        armata*. Chem. Pharm. Bull. 6:578-580.

47. Takemoto, T., K. Daigo, and N. Takagi. 1964. Studies on the
        hypotensive constituents of marine algae I. A new basic
        amino acid "laminine" and the other basic constituents iso-
        lated from *Laminaria angustata*. Yakugaku Zasshi 84:1176.

48. Takemoto, T., K. Daigo, and N. Takagi. 1964. Studies on the
        hypotensive constituents of marine algae II. Synthesis of
        laminine and related compounds. Yakugaku Zasshi 84:1180.

49. Teshima, S., and A. Kanazawa. 1976. Comparison of the sterol
        synthesizing ability in some marine invertebrates. Mem.
        Fac. Fish. Kagashima Univ. 25:33-39.

50. Wagner, H., and P. Pohl. 1964. Zur Kenntnis der Polyen-
        fettsaüren von Meeresalgen. Naturwissenschaften 51:163-
        164.

51. Weigl, J., J. R. Turvey, and W. Yaphe. 1966. The enzymatic
        hydrolysis of κ-carrageenan. *In:* Young, E. G., and J. L.
        McLachlan (eds.) Proceedings, Fifth International Seaweed
        Symposium. Pergamon Press, Long Island City, New York,

and Oxford, England, pp. 329-332.

52. Weinheimer, A. J., and R. L. Spraggins. 1969. The occurrence of two new prostaglandin derivatives (15-epi-PGA$_2$ and its acetate, methyl ester) in the gorgonian *Plexaura homomalla*. Tetrahedron Lett. 5185-5188.

53. Weinheimer, A. J., and R. L. Spraggins. 1970. Two new prostaglandins isolated from the gorgonian *Plexaura homomalia* (Esper). Chemistry of coelenterates XVI. *In:* Youngken, H. W., Jr. (ed.) Food-Drugs from the Sea. Marine Technology Society, Washington, D.C., pp. 311-314.

54. Youngblood, W. W., M. Blumer, R. L. Guillard, and F. Fiore. 1971. Saturated and unsaturated hydrocarbons in marine benthic algae. Mar. Biol. 8:190-201.

55. Youngken, H. W., Jr., and Y. Shimizu. 1975. Marine drugs: chemical and pharmacological aspects. *In:* Riley, J. P., and G. Skirrow (eds.) Chemical Oceanography. Academic Press, Inc., New York 10003. 4:269-317.

- 16 -
Potential Pharmaceutical Products
GEORGE H. CONSTANTINE

The development of pharmaceutical products has progressed from the art of the tribal medicine man preparing secret potions, to the discoveries of cause and effect relationships of diseases (Koch, Pasteur, Lister), to the conquering of previously fatal or debilitating diseases (polio, small pox, etc.), and currently to the design of medicinal agents at the biochemical or molecular level. Many of the pharmaceutical products now available were discovered by fortunate accidents or suggested by the use of plants in folklore, or were developed by chemical modifications of previously known active compounds. One need only recall the chance isolation of penicillin by Fleming, or the discovery that extracts of *Catharanthus* (periwinkle), being evaluated for hypoglycemic activity, had the capability of altering the growth of certain neoplasms, or the recognition that certain antimicrobial sulfonamide derivatives had hypoglycemic properties and subsequently could be used in the treatment of diabetes, to show that these approaches may continue to be valid.

In more recent times, the approach to drug discovery has become more systematic. The risk and cost of chance discoveries may outweigh their potential benefit. Modern research proceeds in a more orderly fashion, primarily modifying known chemical entities to yield compounds of either greater potency and/or less toxicity. This has resulted in a myriad of pharmaceuticals which in many instances surpass the qualities of their predecessors, but unfortunately in some cases do not. A few examples of such drugs include 1) psychotherapeutic agents since the discovery of the phenothiazine nucleus, 2) the modifications of the penicillin nucleus to yield agents which are as effective orally as the injectables, 3) the modification of the steroid nucleus to yield hormones and other agents to mimic our own body processes, and 4) the potent agents currently utilized for treating hypertension.

In some areas of research involving chemical modification techniques, the chemically modified components have yielded a host of effective and useful components. The cephalosporin antibiotics are a typical example of chemical proliferation. The organism secreting an antibiotic was discovered by Brotzu off the coast of Sardinia in 1945. Subsequent isolation of cephalosporin C by Chain, Florey, and Abraham in 1954 made available an entire new group of potentially useful antibiotics. Since that time there have been no fewer than 3000 compounds derived from this nucleus and 12 of these are currently marketed in this country.

Chemical modification has repeatedly provided more effective and safer drugs, but there is still need to search for new basic starting molecules upon which additional structures can be built. The search for new and more effective chemical entities with biologic activity has taken a variety of routes. Not only is there a quest for compounds from natural sources but the place where these compounds have an effect in the human body must be determined. Research on the activity of drugs in the body has greatly expanded in the past decade. The study of pharmacokinetics, pharmacodynamics, drug receptors, and biotransformations will greatly influence the development of new drugs. Some information concerning development and evaluation of new drugs is useful before considering the future of marine products and specifically algal products in pharmacy.

DRUG DEVELOPMENT

There are constraints placed upon the pharmaceutical industry in the development of a new product. If interesting activity is discovered in an extract or mixture of components, isolation and purification of the active principle must be accomplished. Such isolations are often lengthy and tedious, though isolation techniques have advanced significantly in the past decade.

After isolation and identification, chemical and physical characterization is required. Technology has advanced so that a complete structural formula often can be determined within months.

Prior to initial studies of a chemical entity in humans, the full pharmacological spectrum and its pharmacokinetic characteristics (absorption, fate, excretion) must be thoroughly explored in lower animals of several species. An Investigational New Drug (IND) application must be submitted to the Food and Drug Administration (FDA) prior to administering a new drug to man. This document contains the results of lengthy preclinical tests in animals, the composition and properties of the drug, studies of the drug's metabolism, and the procedures and tests for production and quality control.

Following submission of the IND, initial clinical studies in healthy human volunteers may begin. These may be followed by controlled clinical trials if the initial results are favorable. The extent and duration of these trials varies. If clinical trials are successful, selected practicing physicians will receive the drug for testing in patients. Since no single agent can be tested on all types of patients or under all conditions, these trials may continue into the period of general clinical use. Therefore, agents are introduced in this country in which toxicities appear only after extensive long-term use. Consequently, the development of an entirely new drug for use in fertility control, for example, may be prohibitive due to the costs and the long testing time required for certification.

After all trials are complete, the manufacturer can submit a New Drug Application (NDA) to the FDA. In 1973, five NDA's were approved for new chemical entities. It took an average of 6.6 years to receive this approval (34).

The pharmaceutical industry is not alone in its difficulty in dealing with increased constraints. Since the promulgation of the Toxic Substance Control Act (Public Law 94-469, October 11, 1976), agro-chemical development may also be impeded. If one envisions drug development for veterinary utilization, one must also be aware that not only is the FDA responsible for approval of such products, there are U.S. Department of Agriculture standards which must be met.

These constraints paint a bleak picture for future drug development. However, they have not yet caused the pharmaceutical industry to reduce their research and development budgets (Table I). Note the investment and the amount placed into basic research.

Table I. *U.S. applied R & D expenditures on pharmaceuticals, by product class, 1973-1974* *

Ranking Per Year 1974	1973	Product Class	Percentage Share of Applied R&D Dollar 1974	1973	Percent Change 1973/1974
1	1	Central nervous system	18.5%	17.4%	+ 6.3
2	2	Anti-infectives	17.8	15.8	+ 12.7
3	4	Cardiovasculars	15.1	15.4	+ 18.0
4	3	Neoplasms	14.0	15.4	− 9.1
5	5	Digestive	6.2	6.0	+ 3.3
6	9	Respiratory	4.8	3.9	+ 23.1
7	8	Biologicals	4.5	4.0	+ 12.5
8	7	Dermatologicals	3.5	4.1	− 14.6
9	10	Vitamins	2.1	2.0	+ 5.0
a	6	Diagnostics	--	5.1	--
		Other	6.1	5.8	+ 5.2
		Veterinary preparations	6.9	7.2	− 4.2
		Veterinary biologicals	.5	.5	0.0
			100.0%	100.0%	

[a]*Diagnostic products were omitted in the 1974 survey.*
*From Prescription Drug Industry Fact Book (29)

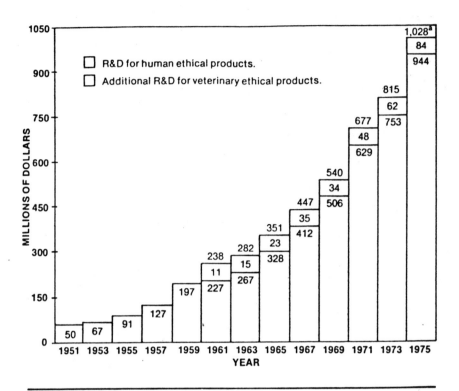

ªBudgeted.
Sources: Published PMA survey reports.

Figure 1. *Research and development expenditures for ethical products 1951-1975 (odd numbered years) (29).*

There are still many diseases to be conquered and hopefully these investments will ensure their conquest (Fig. 1 and Table II).

DRUGS FROM THE SEA

It is interesting that some of the earliest printed data on the use of substances from the sea for the biomedical properties involved the use of various seaweeds for dropsy, menstrual difficulties, gastro-intestinal disorders, abscesses, and cancer. However, man has utilized the marine environment as a source of very few medicinal products, compared to the total of those found on land. The search for drugs from the sea, in fact, is a relatively recent undertaking.

In 1960, one of the first symposia was held for people of diverse disciplines to discuss their research interests in compounds derived from marine organisms (27). Halstead's 3-volume compendia published from 1965 to 1967 greatly assisted those who were interested in studying marine animal toxins, since this is always a useful indicator of potential medical use (15). In 1968, Schwimmer and Schwimmer paralleled this effort by reporting the toxic

Table II. *Ten leading causes of death in the U.S., 1900 and 1975*[*]

Cause of Death	Rate of Death (per 100,000 persons)
1900	
Influenza and pneumonia	202.2
Tuberculosis	194.4
Gastroenteritis	142.7
Diseases of the heart	137.4
Cerebral hemorrhages and vascular lesions affecting central nervous system	106.9
Chronic nephritis	81.0
All accidents	72.3
Cancer and other malignant neoplasms	64.0
Certain diseases of early infancy	62.6
Diphtheria	40.3
1975	
Diseases of the heart	338.6
Malignant neoplasms	174.4
Cerebrovascular diseases	91.1
All accidents	46.9
Influenza and pneumonia	26.3
Diabetes mellitus	16.8
Cirrhosis of the liver	15.0
Arteriosclerosis	13.7
Certain diseases of early infancy	12.5
Suicide	12.4

*From Prescription Drug Industry Fact Book (29)

properties of marine and fresh water algae (35).

A number of symposia have subsequently been held dealing with marine natural product subjects. Four, sponsored by the Marine Technology Society, on "Food-Drugs from the Sea" were held in 1967, 1969, 1972, and 1974 (30). The fifth is planned for September 1977. A symposia by the American Society of Pharmacognosy at Oregon State University also contributed to this expanding interest (33). More recently, the Gordon Research Conferences have established a Marine Natural Products section to their schedule. The second was held in June 1977 in Santa Barbara. The IX International Seaweed Symposium in August 1977 will also contain a special symposium, "Marine Algae in Pharmaceutical Science."

Texts and review articles dealing with marine natural products are greatly benefiting those who seek entrance into this exciting research area. Baslow's text Marine Pharmacology (3) updates Halstead's earlier work, Martin and Padilla's Marine Pharmacognosy (23) presents an in-depth look at the actions of marine toxins at the cellular level, Scheuer's The Chemistry of Marine Natural

Products (32) reviews the chemistry of those agents known to 1972, and most recently the contributions of Baker and Murphy's Compounds from Marine Organisms (2) has updated the information. Review articles by Premuzic (28), Faulkner and Anderson (9), and Ruggieri (31) have kept track of the current literature.

Specific reviews in the literature dealing with bioactive marine algal products are more limited. Dawson's text Marine Botany (6) has only a brief mention of medicinal uses of marine algae. The symposium "Plants in the Development of Modern Medicine" did present information, current to 1968, of drugs from marine plant sources (7). Gentile's review of algal toxins (12) may assist in our understanding of potentially useful products. Aside from these, most of the data are in journals dealing with the isolation of algal compounds. These reports appear in chemical, algal, pharmacological, microbiological, or medical journals. The data generated often require further biologic assay or evaluation. This requires interdisciplinary expertise and more than the modest funds now allocated for such research.

Dr. Patterson in this volume has dealt with the classes of organic compounds in algae. A discussion of those compounds which are biologically active may be helpful.

Antimicrobial Agents

Many feel that a sufficient armamentarium of antimicrobial agents already exists. However, there are gaps in this arsenal. Not only are many of the once useful agents powerless against new resistant strains of microorganisms, such as *N. gonorrheae* and *Ps. aeruginosa*, but there are totally new microbial diseases emerging (39).

Antimicrobial assay for evaluating activity is one of the few relatively simple, short methodologies in drug research. Usually within 48 to 72 hours one can determine whether an extract possesses activity. Unfortunately, such a simple assay system is unavailable for other research, e.g., cardiovascular, psychotherapeutic, etc. However, these tests are only a preliminary indicator of activity and must be verified.

Antibiotic assays may be conducted against a variety of different organisms, both human and nonhuman pathogens, and may be used to evaluate anything from mixtures in a crude water extract to pure compounds. One difficulty associated with this methodology, substantiated by Hornsey and Hyde (20), pertains to the seasonal variability in the concentration of antimicrobial agents; positive or negative results may depend upon the time of collection.

It is interesting to note that this search for newer agents has yielded a great number of halogenated terpenoidal derivatives (10, 36, 37). Many publications have dealt with isolation and identification of these compounds (e.g., 4, 13, 41), the majority of which have antibiotic activity. However, the risks of toxicities associated with such products may outweigh their benefits and this information cannot be extrapolated from these results.

Lipid and Cholesterol Depressants

It has been reported that the intravenous injection of laminarin sulfate from *Laminaria digitata* and of carrageenan from *Chondrus crispus* significantly reduces serum lipids and prevents the development of atherosclerosis (25). More recent reports show that orally administered algal polysaccharides (22) or dried algae (1) could reduce serum cholesterol in rats. The latter report hypothesizes that the acidic polysaccharides form a lyophilic colloid which prevents the absorption of cholesterol from the intestinal tract. Guven's group (14) has recently demonstrated that the lipolytic effect of *Cystoseira barbata* is not due to polysaccharides but to polypeptides. He has separated two such components and shown one to be lipolytic and the other hypoglycemic.

Synergists of Inorganic Ion Metabolism

Rats treated with carrageenan from *Hypnea japonica* have demonstrated an increased uptake of calcium into the bone and slightly decreased serum calcium (16). This may be of potential benefit to persons suffering from osteoporosis or similar disorders, although the number of animals used in this study was not reported, nor was there a statistical evaluation of the data.

Early reports in the literature indicated that radioactive strontium can form a gel with alginates in the gastrointestinal tract, thereby preventing its absorption (38). Subsequent investigations have shown that alginates with higher concentrations of guluronic acid rather than mannuronic acid have a greater affinity for divalent cations and a higher selective binding capacity for strontium than calcium (17). It has been proposed that alginates can be used for the prevention and treatment of radioactive strontium poisoning.

Overdoses of barium, cadmium, and zinc have also been effectively treated in rats by administering algal polysaccharide, albeit simultaneous administration (40). However, this methodology does indicate a possibility for removing such pollutants from contaminated water supplies.

Other Reported Uses in Medicine

Chapman (5) reports on many isolated instances of algal utilization in medicine under the category, minor uses. These include their use as a source of iodine in treatment of certain goitrous conditions; the use of kainic acid, isolated from *Digenia simplex*, for the treatment of certain parasitic worms (24); and the effect of laminarin from *Laminaria digitata* to prolong coagulation time in laboratory animals (18).

Yamamoto has recently reported a non-dialyzable fraction of *Sargassum fulvellum* that has activity against the Sarcoma 180 tumor system (41). Ito has subsequently demonstrated that a non-dialyzable fraction from *Sargassum thunbergii* is effective against Ehrlich ascites carcinoma (21).

Nakazawa has reported in a series of papers other anti-tumor effects of aqueous extracts of marine algae. The latest report deals with an acidic polysaccharide effective against Ehrlich ascites and Sarcoma 180 (26).

Compounds Which May Develop into Human Use

Steroids - Research on the existence of novel steroid molecules in marine organisms has been conducted with emphasis primarily on those from marine invertebrates (holothurogenins, etc.). Among the algae, it appears that the green and red algae contain steroids usually found in terrestrial plants whereas the brown algae have some unique compounds. The need for steroid precursors in the pharmaceutical industry at this time is satisfied by the relatively inexpensive diosgenin available from the *Dioscorea* species. However, should there be some economic or political upheavals, another source of steroid precursors from marine algae is available.

Fatty acids - The fatty acid constituents of marine algae have been most recently investigated by Hayashi *et al.* (19). Most of the emphasis of this work to date has been primarily because of biosynthetic, geochemical, or chemotaxonomic interests. As a byproduct of a marine algal industry, the fatty acids may provide an inexpensive dietary component, not only for man but for animals.

SUMMARY AND CONCLUSIONS

The marine environment has yielded a number of interesting, exciting, new, and potentially useful components. Although this presentation deals only with those components of algal origin, a great deal needs to be learned about the marine environment. Many so-called marine animal natural products are now known to be derived from the food source which marine algae supply and which are modified biochemically by the animals. Not only should further investigation provide direct answers to our immediate needs but it should provide a better understanding of biochemical and cellular processes.

REFERENCES

1. Abe, S., and Takashi, K. 1972. The effects of seaweeds on cholesterol metabolism in rats. *In:* Nisizawa, K. (ed.) Proceedings, Seventh International Seaweed Symposium, Wiley, New York, pp. 562-565.

2. Baker, J.T., and V. Murphy. 1976. Compounds from Marine Organisms. CRC Press, Cleveland, Ohio, p. 216.

3. Baslow, M.H. 1969. Marine Pharmacology. Williams and Wilkins, Baltimore, Maryland, p. 286.

4. Burreson, B.J., R.E. Moore, and P. Roller. 1975. Haloforms in the essential oil of the alga *Asparagopsis taxiformis*. Tetrahedron Lett., pp. 281-284.

5. Chapman, V.J. 1970. Seaweeds and Their Uses. Methuen, London, England, p. 304.

6. Dawson, E.Y. 1966. Marine Botany. Holt, Rhinehart and Winston, New York, p. 371.

7. Der Marderosian, A. 1972. Drug plants from the sea. *In:* Swain, T. (ed.) Plants in the Development of Modern Medicine. Harvard Univ. Press, Cambridge, pp. 209-233.

8. Djerassi, C. 1969. Prognosis for the development of new chemical birth control agents. Science 166:468-473.

9. Faulkner, D.J., and R.J. Anderson. 1974. Natural products chemistry of the marine environment. *In:* Goldberg, D. (ed.) The Sea, Vol. 5. J. Wiley, New York, pp. 679-714.

10. Fenical, W., *et al.* 1973. Zonarol and isozonarol, fungitoxic hydroquinones for the brown seaweed *D. zonairordes*. J. Org. Chem. 38:2383-2386.

11. Fenical, W., and O. McConnell. 1975. Simple antibiotics from the red seaweed *Dasya pedicellata* var. *Stanfordiana*. Phytochemistry 15:435-436.

12. Gentile, J.H. 1971. Blue-green and green algal toxins. *In:* Kadis, S., A. Ciegler, and S.J. Ajl (eds.) Microbial Toxins Vol. III, Algal and Fungal Toxins. Academic Press, New York, pp. 27-66.

13. Glombitza, K.W., *et al.* 1975. Fucole, polyhydroxyolisophenyle aus *Fucus vesiculosus*. Phytochemistry 14:1403-1405.

14. Guven, K.C., *et al.* 1974. Lipolytic and hypoglycemic compounds from *Cystoseira barbata*. Arzneim.-Forsch. 24:144-147.

15. Halstead, B.W. 1965-67. Poisonous and Venomous Marine Animals of the World, Vol. 1-3. U.S. Government Printing Office, Washington, D.C.

16. Hang, K.C., *et al.* 1972. Effect of sulfated galactan from *Hypnea japonica* on bone metabolism in the rat. Can. J. Physiol. Pharmacol. 50:784-790.

17. Haug, A. 1967. Strontium-calcium selectivity of alginates. Nature (London) 215:757.

18. Hawkins, W.W., and V.G. Leonard. 1958. The physiological activity of laminarin sulfate. Can. J. Biochem. Physiol. 36:161-170.

19. Hayashi, K. 1974. Component fatty acids of acetone soluble lipids of seventeen species of marine benthic algae. Nihon Suisan-Gakkai-Shi (Biol. Abst. 75-02257) 40:609-617.

20. Hornsey, I.S., and D. Hyde. 1976. The production of antimicrobial compounds by British marine algae. II. Seasonal variation. Br. Phycol. J. 11:63-67.

21. Ito, H., and M. Sigiura. 1976. Anti-tumor polysaccharide fraction from *Sargassum thunbergii*. Chem. Pharm. Bull. (Tokyo) 24:1114-1115.

22. Ito, K., and Y. Tsuchiya. 1972. The effect of algal polysaccharides on the depressing of plasma cholesterol levels in rats. *In:* Nisizawa, K. (ed.), Proceedings, Seventh International Seaweed Symposium, Wiley, New York, pp.558-561.

23. Martin, D.F., and G.M. Padilla. 1973. Marine Pharmacognosy. Academic Press, New York, p. 317.

24. Murakami, S., *et al.* 1953. Studies on the effective principles of *Digenea simplex*. Yakagaku Zasshi (J. Pharm. Soc. Jpn.) 73:1026-1028.

25. Murata, K. 1962. The effects of sulfated polysaccharides obtained from seaweeds on experimental atherosclerosis. J. Gerontol. 17:30-36.

26. Nakazawa, S., *et al.* 1976. Anti-tumor effect of water extracts of marine algae. III. *Codium pugniformis*. Chemotherapy (Tokyo) 24:448-450.

27. Nigrelli, R.F. (ed.) 1963. Biochemistry and pharmacology of compounds derived from marine organisms. Ann. N.Y. Acad. Sci. 90:615-950.

28. Premuzic, E. 1971. Chemistry of natural products derived from marine sources. Fort. Org. Natur. 21:417-488.

29. Prescription Drug Industry Factbook '76. 1976. Pharmaceutical Manufacturers Association, Washington, D.C.

30. Proceedings of the Food-Drugs from the Sea Conferences, 1967, 1969, 1972, 1974. Marine Technology Society, Washington D.C.

31. Ruggieri, G.D. 1976. Drugs from the sea. Science 194:491-497.

32. Scheuer, P.S. 1973. The Chemistry of Marine Natural Products. Academic Press, New York, p. 201.

33. Schwarting, A.E. (ed.) 1964. Marine Biomedicinals Symposium. Lloydia 32:407-484.

34. Schwartzman, D. 1976. Innovation in the Pharmaceutical Industry. Johns Hopkins University Press, Baltimore, p. 399.

35. Schwimmer, D., and M. Schwimmer. 1968. Medical aspects of phycology. *In:* Jackson, D.F. (ed.) Algae, Man and the Environment. Syracuse University Press, Syracuse, New York, pp. 279-358.

36. Sims, J.J., *et al.* 1972. Marine natural products III, johnstonol, an unusual halogenated epoxide from the red algae, *Laurencia johnstonii*. Tetrahedron Lett., pp. 195-198.

37. Sims, J.J., *et al.* 1973. Marine natural products IV, prepacifenol, a halogenated epoxy sesquiterpene and precursor to pacifenol from the red algae *Laurencia filiformis*. J. Am. Chem. Soc. 95:972.

38. Skoryna, S.C., *et al.* 1964. Studies on inhibition of intestinal absorption of radioactive strontium. Can. Med. Assoc. J. 91:285-288.

39. Spencer, D.J. 1961. Emerging diseases of man and animals. Ann. Rev. Micro. 26:465-487.

40. Tanaka, Y. 1972. Application of algal polysaccharides as *in vivo* binders of metal pollutants. *In:* Nisizawa, K. (ed.) Proceedings, Seventh International Seaweed Symposium, Wiley, New York, pp. 602-604.

41. Waraszkiewicz, S.M., and K.L. Erickson. 1975. Halogenated
      sesquiterpenoids from the Hawaiian marine algae. Tetra-
      hedron. Lett., pp. 281–284.
42. Yamamoto, I., *et al*. 1974. Anti-tumor effect of seaweeds I.
      Anti-tumor effect of *Sargassum fulvellum* and *Laminaria*.
      Japan J. Exp. Med. 44:543–546.

- 17 -

Essential Considerations for Establishing
Seaweed Extraction Factories

JAMES R. MOSS

One possibility for utilizing certain red and brown seaweeds is in the production of widely used colloids such as algin,
carrageenan, and agar. The economic feasibility of establishing a
colloid extraction plant depends on market availability, stability,
future growth potential, and assured seaweed availability at competitive costs. Capital requirements and profitability are among other
important considerations. This chapter will furnish information and
suggested guidelines to aid in developing the data necessary to determine whether an extraction factory will be economically feasible.

Probably, the first question any prospective investor will ask
is whether seaweed extraction factories are part of a growing industry. No one wishes to invest in a dead or nongrowth industry.
Other questions are whether the industry is stable and whether the
growth and profit records of other companies in the same general
type of business look attractive. Also, what are the predictions
for future growth? A quick review of certain historical data will
go far towards answering these essential questions.

HISTORY

The growth and development of the seaweed extraction industry
can be seen from the estimated world production of seaweed extractions shown in Fig. 1.

Agar was the first seaweed extractive to achieve commercial
status. Shortly after 1900 it was introduced for nonfood uses,
especially in bacteriological plating media, which is probably the
first significant employment of seaweed extractives for other than
food. Before World War II, Japan monopolized agar production with
an estimated volume exceeding 2,000 tons annually. World War II
stopped Japanese exports, causing severe shortages which resulted
in small agar processing factories being started in several other
countries.

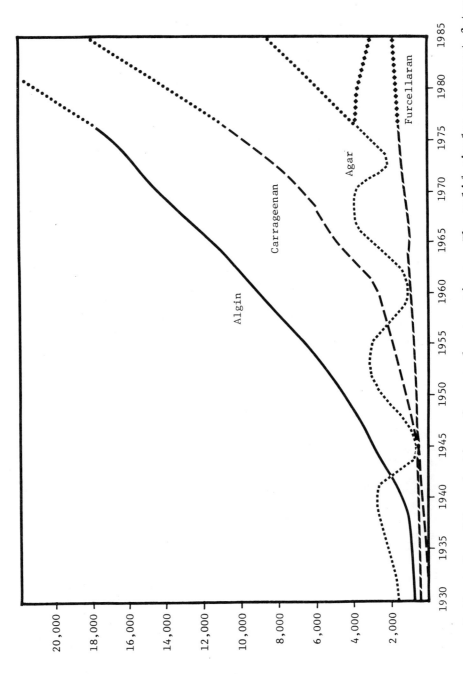

Figure 1. *Estimated world production of seaweed extractions. The solid circles represent future production assuming controlled culture of seaweeds; the solid diamonds represent future production based on wild sources only. The ordinate is in metric tons.*

By 1950 the demand for agar and production had recovered to a point where available seaweed supplies were overburdened. This resulted in overharvesting of known seaweed populations which, in turn, reduced succeeding harvests. As a consequence, the price of food grade agar increased out of proportion to its value, causing a rapid decline in sales. Eventually seaweed supply, sales demand, and prices returned to a rational balance, whereupon the destructive cycle started again. Up and down cycles occur every few years, discouraging the commercial producer as well as the consumer. Production represented in Fig. 1 for the years 1970–1976 is more accurate than earlier data and illustrates the most recent of these cycles. Cyclic production declines cause significant numbers of agar manufacturers to experience bankruptcy. Other manufacturers must periodically exist without profits. Agar customers become discouraged with an undependable source and some abandon its use permanently.

Algin was the second seaweed extract to reach commercial status. In the late 1920s and early '30s a small processing factory was started in San Diego, which later became the Kelco Company. After a slow start the company began to develop efficient *Macrocystis* harvesting methods, extraction processes, and consumer markets for ammonium, sodium, calcium, and potassium alginates. Many new products and uses resulted in rapid growth shortly before, during, and after World War II. After Kelco began to flourish, other algin extraction factories appeared in Maine, England, Norway, France, India, Japan, and Oxnard, California. It is reported that additional factories are now under construction in Brazil and Argentina. As Fig. 1 shows, world production has steadily and consistently increased, and there is no apparent reason why the present growth should slow or stop.

Carrageenan made its first commercial appearance as an extractive from *Chondrus crispus* during the mid-1930s. For several years it was produced by one company to stabilize chocolate milk. This inspired the development of new and refined products and improved extraction techniques. World War II also stimulated the growth of sales of carrageenan. By 1950 there were three producing factories in New England. In 1959 two of the U.S. carrageenan factories merged to become Marine Colloids, Inc. The other U.S. factory is located in Portland, Maine, and is owned by Stauffer Chemical Co. The success of the U.S. factories encouraged the establishment of additional ones in Denmark, France, Spain, and Japan. Reports are that new factories are under construction in Argentina, Brazil, and Norway.

Furcellaran is another seaweed extractive which has been commercialized, mainly by Danish manufacturers. The growth of the industry was limited in Denmark by an inadequate supply of seaweed. Sales volume is, therefore, now relatively small in comparison with algin, agar, or carrageenan.

SIZE AND DISTRIBUTION OF SEAWEED EXTRACTION FACTORIES

A potential investor in a new seaweed extraction factory will benefit from the informaion estimated in Table I. Almost without exception, seaweed extraction factories are located in maritime countries

Table I. *1976 estimated world production of seaweed extractives by countries and product types.*

Producing Country	No. of Factories	Algin (Tons)	Carrageenan (Tons)	Agar (Tons)	Furcellaran (Tons)
Canada	1	1,200	*	400	
Argentina	*	*	*	*	
Brazil	*	*		200	
Chile	2	200		**	
China	Unknown	**			
Denmark	3		2,300	100	1,500
England	2	6,500	100	?	
France	3	1,200	2,000	100	
India	1	200			
Japan	10–20(?)	?	300	2,000	
Korea	1 or 2			200	
Mexico	1			100	
Morocco	1			250	
Norway		3,000		100	
New Zealand	1			100	
Philippines	1			400	
Portugal (& Azores)	2			800	
Spain	8 to 10		1,000	100	
United States	5	5,500	4,800		
USSR	Unknown	**	?	**	
TOTAL PLANTS	43				
ESTIMATED TOTAL TONS (2,200 lbs.)		17,800	10,500	4,800	1,500
ESTIMATED TOTAL POUNDS		38,200,000	23,100,000	10,560,000	3,300,000
ESTIMATED AVERAGE SELLING PRICE (Per Pound)		$2.30	$2.80	$4.50	$2.20
ESTIMATED TOTAL ANNUAL SALES (World Production) ($207,360,000)		$87,860,000	$64,680,000	$47,520,000	$7,260,000

*Factories under construction
**Unconfirmed reports indicate extraction factories have been established

having a well-developed and diversified industrial base. Most producing countries also have a well-developed home market for their
extractive. Each country supplies a substantial portion of the basic
seaweed requirements and reagents to its own extraction factories.
Between 30 and 35 relatively small factories produce agar as compared
with 8 to 10 algin factories and 9 or 10 carrageenan factories.

THE MARKET

Production volumes for each nation shown in Table I are estimated, because manufacturers jealously guard their sales and distribution figures. Consumption of the various extractives on a national
basis is also very difficult to estimate because, generally, major
producing countries are major consumers of their own products. For
example, it is estimated that Japan normally uses more than 3,000
tons of agar and 1,000 tons of carrageenan annually. The United
States and Canada normally use more than 500 tons of agar, 4,000
tons of algin, and 3,500 tons of carrageenan annually. The European
Common Market countries are believed to use more than 2,500 tons of
algin and 2,500 tons of carrageenan. The differences between these
estimates and the production figures shown in Table I are due to the
quantities exported or imported. One can also conclude from Table
I that seaweeds containing extractives are widely distributed among
the maritime countries. This table suggests predictions about the
competitive situation and the size of the world market for each
extractive.
The uses of seaweed extractives are discussed fully in available literature and will not be reviewed here. Suffice it to say,
that the profile of uses for each extractive is very broad. Because of
the diverse properties of seaweed extractives, it seems highly unlikely
that present or future synthetic products could greatly reduce sales
or impede future growth of those extractives taken from seaweeds that
are now available in adequate supply. There is vigorous competition
among algin manufacturers for algin customers. However, algin rarely
competes for carrageenan or agar uses. Neither do carrageenan or
agar compete for algin uses. Each seaweed extractive has its special
and distinctive chemical and physical properties which channel it into
specific uses.
The market for agar requires special comment. Agar is an excellent colloid with unique and valuable properties. Its sales potential
is estimated to be as great as that of either algin or carrageenan.
However, unless someone can work out the technology for growing agar
seaweeds (particularly species of *Gelidium*) under controlled conditions, the writer predicts that production and sales will permanently
decline. It is possible that other colloids will replace all but the
bacteriological applications which worldwide consume probably less
than 1,000 tons of agar annually.

SEAWEED SOURCES

It will be seen from Fig. 1 that carrageenan production experienced a slowdown in the mid- and late 1960s because of shortages of

raw material. The problem was largely solved by 1971, when a technology developed for the cultivation or sea-farming of *Eucheuma
cottonii*, an excellent source for kappa-carrageenan. Similar technology is now being employed for *Eucheuma spinosum*, the principal
source for iota-carrageenan.

It is pertinent to note that 10 to 12 years were required to
work out a technology for growing *Eucheuma* under controlled conditions.
Credit for the original concept and the leadership for the program
which developed this technology belongs to Dr. Maxwell S. Doty of the
University of Hawaii. The program was originally and principally sponsored by Marine Colloids, with substantial support furnished by the
U.S. Office of Sea Grants, the University of Hawaii, and the Government
of the Philippines. It was initiated in 1961. The first small but
significant harvest occurred in 1970. Rapid expansion of *Eucheuma*
culture followed, and substantial and increasing tonnage is being exported from the Philippines and adjacent areas.

A similar breakthrough with *Chondrus crispus* and other carrageenan seaweeds is anticipated in the near future. Consequently, carrageenan manufacturers may now look forward to freedom from source restrictions and continued growth of the carrageenan industry.

The algin manufacturers have also experienced problems, some of
which should be examined. It is well-known that productivity of *Macrocystis* and other large species of kelp fluctuates. The intrusion
of abnormally warm water currents reduces kelp growth and algin yield.
Pollution, sea urchins, and storms are also serious problems. Although economical controlled growth of brown seaweeds has not been
achieved, work is in progress and success seems likely. The alginate
industry has time to solve this problem because substantial sources
of unused wild brown seaweeds remain available. As most of these untapped sources are located in relatively remote areas, costs for harvesting and delivering to existing manufacturing plants may be a
problem.

In considering the commercial potential for brown seaweed, it
should be kept in mind that the requirements of the large U.S. and
English manufacturers exceed their local sources. It is also reported that the Japanese, who use kelp for human food, have exceeded their
local supplies. Consequently, these countries are establishing supplemental sources and may be potential markets of imported dried kelp.
Nevertheless, the sales of extractives from brown algae should continue
to expand.

INFORMATION REGARDING SEAWEED RESOURCES

The impact of seaweed resources on sales growth and industrial
stability is now clearly evident. In considering and determining the
economic feasibility of an extraction factory, it is emphasized that
*all other problems are strictly secondary to that of obtaining an
adequate supply of the appropriate algae.* If an adequate, dependable,
and economic source exists, other requirements can probably be solved
satisfactorily.

What information concerning seaweed resources must the prospectiv

investor have to enable sound judgments on their economic exploitation? Where is this information available and how should it be evaluated? A program to address such questions is suggested in the following outline. Explanatory comments and suggestions which further
explore the information desired and indicate ways in which it may be
developed are included. For the most part, the questions raised will
apply equally to seaweeds growing wild along coasts and to those produced under controlled conditions.

The outline is confined to the major questions; however, additional problems will surely arise which can be approached within the general context of the outline.

1. *The Algal Resource*
   a. Are boundaries of proposed harvesting areas clearly defined?
      Have exclusive harvesting rights been obtained from appropri-
      ate state and federal agencies?
   b. Have major seaweed beds been mapped and evaluated for densi-
      ty, species composition, and ease of harvesting?
   c. How serious are storms? Are the seaweed beds damaged by
      pollution, disease, or grazers such as sea urchins?
   d. At what point does overharvesting occur? Several representa-
      tive beds should be scientifically overharvested to determine
      this point. To the fullest extent possible, the characteris-
      tics of the life cycle of the species under consideration
      should be determined.

2. *Factory Siting*
   a. Is a convenient harbor available with rail and highway con-
      nections? Does the harbor have available a suitable factory
      site? Will local, state, and federal regulations permit use
      of harbor and factory site for purpose intended? It should
      be noted that numerous city, state, and federal licenses and/
      or permits are generally required to harvest and produce sea-
      weed extractives.
   b. Are large quantities of cheap fresh water available? Are
      power, labor, and facilities for substantial waste disposal
      available? Are tax concessions possible?

3. *Harvesting*
   a. Has the best method of harvesting been worked out, in light
      of related costs, including that of delivering the wet weed
      to a drying point?
   b. Every seaweed manufacturer has experienced unexpected and
      major increases or decreases in his seaweed harvest. What
      is the range of variation? It is suggested the minimum har-
      vest experience be used as the standard in computing the
      productivity of a seaweed source. The size of the inventory
      of dried seaweed carried over from the prior year can influ-
      ence a decision on this point. Seaweed colloids remain quite
      stable in well-dried seaweeds, which can be stored without
      difficulty for several years.

4. *Processing and Production*
   a. Methods and costs of drying each seaweed species should be
      determined. Sun-drying and mechanical drying should be

Table II. Estimated international shipments of dried seaweeds - 1976.

Extractive	Exporting Country	Dry Tons Exported	$/Dry Ton-FOB Exporting Country*	Value to Exporting Country	Dry Tons Imported				
					England	France	Denmark	Japan	U.S.
ALGIN									
Kelps	Chile	9,000	$150-160	$1,350,000	3,000			4,000	2,000
	Argentina	1,000	150-160	155,000				1,000	
	South Africa	3,000	150-160	465,000	2,000			1,000	
	Miscellaneous	7,000	150-160	775,000	5,000			2,000	
Ascophyllum	Canada	?	130-140						
	Norway	?							
Totals				$2,640,000	10,000			8,000	2,000
CARRAGEENAN									
Chondrus crispus	Ireland	500	$500-550	$ 250,000	250	250			
	Canada (Maritimes)	9,000	500-550	4,500,000		500	4,000		
Other *Gigartina*	Mexico	600	500-550	300,000					4,500
	Peru	400	500-550	200,000					600
	Morocco	300	500-550	150,000		300			400
	Portugal	500	500-550	250,000		500			
	Korea	900	500-550	450,000				900	
Iridaea	Chile	3,600	450-500	1,800,000		?			3,600
E. cottonii	Philippines & Adjacent areas	4,500	270-300	1,260,000		1,000	1,000	1,000	1,500
E. Spinosum	" " "	2,500	320-350	850,000		700	700	400	700
Hypnea	Brazil	?	400-425	?					?
Totals				$10,010,000	250	3,250	5,700	2,300	11,300
AGAR									
Gelidium	Korea	600	$1,000-1,200	$ 600,000				400	200
	Miscellaneous	500	1,000-1,200	500,000				300	200
Gracilaria	Argentina	2,000	500-600	1,100,000				2,000	
	Chile	2,000	500-600	1,100,000				2,000	
	Taiwan	500	?-?	?				500	
	Brazil	?	?-?	?					
	Miscellaneous	1,000	?-?	400,000				1,000	
Totals				$ 3,800,000				6,200	400
TOTAL				$16,450,000					

All weights shown in metric tons.
*Ocean freight normally ranges between $80 & $95 per ton.

Table III. *The composition of representatives of algal genera employed in the production of algin, carrageenan, and agar.*

	Approximate Tons Wet per Metric Ton Dry	%Extractive Yields--Wet	%Extractive Yields--Dry
ALGIN SEAWEEDS			
Macrocystis			
Nereocystis	8 - 9	2.0 - 2.5	16 - 22
Laminaria	5 - 6		18 - 23
Ascophyllum	4 - 5	4.0 - 6.5	16 - 25
CARRAGEENAN SEAWEEDS			
Gigartina			
Iridaea	4 - 5	7.0 - 7.7	32 - 38
Eucheuma	8 - 9	4.0 - 4.5	32 - 38
AGAR SEAWEEDS			
Gelidium		Only dry	
Gracilaria		seaweeds	17 - 25
Pterocladia		used	

evaluated. Is there a good sun-drying area available?

b. From time to time the colloidal content of representative samples of seaweed should be extracted and evaluated. In such studies, the variability in properties of the extractive yield should be ascertained and related to season, water temperature, and other variables. Comparisons should also be made with yields from seaweeds which competitors are harvesting and using.

c. The prospective investor in an extraction factory will ask for cost comparisons between cultivated seaweeds and wild seaweeds. Based on current evidence, it appears that seaweeds produced under controlled conditions will cost as much or more than seaweeds growing wild. Wild seaweed should continue to compete with the cultivated product and will be needed for many years to come.

d. Tables II and III furnish certain yield, price, and market data which should serve as guides in determining whether seaweeds in new locations may be harvested and sold at a profit.

e. Specific production problems facing processors of brown seaweeds (Phaeophyta) can be reviewed.

Macrocystis is conventionally and economically harvested with large mechanical reapers. Because of the volume harvested, the cost per ton of seaweed is quite low. Southern California and Mexico weather permits *Macrocystis* to be harvested from 300 to 330 days per year. The wet kelp is delivered directly to the extraction factory. Factories

extracting wet kelp are not faced with the costs of drying,
packaging, and storing dry kelp. Furthermore, using modern
storage techniques for wet kelp, such factories can operate
340 to 350 days per year, 24 hours per day, seven days a
week

Wet *Macrocystis* and similar broad-frond seaweeds yield
44-50 pounds of alginic acid per metric ton. Under normal
kelp harvesting conditions, the unextracted alginic acid
that is delivered in wet kelp will range from 15 to 20 cents
per pound. The cost of unextracted alginic acid in dry
kelp will range between 40 and 50 cents per pound.

Table II also shows that most algin manufacturers are
currently supplementing their wet kelp supplies with imported dry kelp. This raises several interesting possibilities for those who are considering establishing an
algin extraction factory.

Although the preceding comments are specific to the production of alginates from *Macrocystis* and other large kelp
growing along the coast of Lower California and Mexico, it
should be noted that approximately the same situation holds
for the production of alginate from *Ascophyllum*. A substantial percentage of the world's alginate production is
now derived from *Ascophyllum*, because the cost of the basic
alginic acid in the delivered wet seaweed is competitive
with that of *Macrocystis* and other brown seaweeds.

Anyone considering the establishment of a new algin manufacturing operation should first compare the raw material
costs with those of existing major manufacturers. Increasing the number of days in which wet kelp or *Ascophyllum* is
available decreases the costs of the raw material and operating expenses. There have been two major failures of algin
extraction factories in areas of heavy kelp growth because
adverse weather conditions greatly reduced the number of
days when harvesting is possible.

f. The specific production problems facing processors of red
    algae (Rhodophyta) are associated with both production and
    marketing of agar and carrageenan.

    The only question facing agar producers is whether the
    seaweeds are available. If a sufficient quantity is harvest-
    able on a dependable basis, the market is relatively certain
    because of the current shortages and high prices commanded
    by the extractive. Table IV suggests the minimum investment
    required for an extraction factory.

    A much different situation exists for carrageenan manufac-
    turers who must extract a variety of red seaweeds in order
    to obtain the kappa, iota, and lambda configurations now in
    commercial demand. As the sources for each of these com-
    pounds are rarely found in one location, extraction factor-
    ies import dried seaweeds from almost every quarter of the
    globe. A manufacturer who has a portion of his dried red
    seaweeds supplied from local sources will have obvious

Table IV. *Estimated minimum extractive production and investments required for profitable plant operation.*

	ALGIN Minimum Range		CARRAGEENAN Minimum Range		AGAR Minimum Range	
Tons (Metric) Production	700-1,000		450-750		75-125	
Pounds	1,540,000 - 2,200,000		900,000 - 1,650,000		165,000 - 275,000	
Price per lb.	$2.30-$2,80		$2.70-$3.20		$4.50-$5.00	
Sales	$3,530,000 - $5,060,000		$3,100,000 - $5,100,000		$780,000 - $1,300,000	

		Tons Seaweed Required*		Tons Seaweed Required*		Tons Seaweed Required*
Price/Ton		Wet	Dry	Wet	Dry	Dry
$160	Macrocystis	32,-44,000	3,4-4,900			
	Nereocystis	32,-44,000	3,4-4,900			
140	Ascophyllum	15,-22,000	3,4-4,900			
500-600	Chondrus crispus			5,8-9,700	1,4-2,300	
500-600	All Gigartina			5,8-9,700	1,4-2,300	
400-425	Iridaea			5,8-9,700	1,4-2,300	
350-500	Eucheuma			10,0-16,700	1,2-2,100	
	Gelidium					375-625
	Gracilaria					375-625

Categories and Initial Investments Required (1977 Costs) to Achieve Given Production Goals for the 3 Extractives

Land, Buildings & Utilities;
Harvesting Equipment &
Organization; Engineering;
Product Development; Equip-
ment; Start-up Costs; Raw
Materials & Reagents;
Finished Product Inventory;
Accounts Receivable;
Operational Cash;
Miscellaneous

ALGIN	CARRAGEENAN	AGAR
700 Tons=$4,500,000	450 Tons=$4,500,000	75 Tons=$550,000
1,000 Tons=$6,500,000	750 Tons=$6,750,000	125 Tons=$850,000

*It is recommended that seaweed sources exceed these estimated tonnages by 40 to 50 percent as protection from adverse harvesting variations.

advantages over a competitor who must import all of them
from other countries. For example, if sufficient quantities of *Iridaea* and *Gigartina* can be harvested and delivered
to a centrally located factory, production of carrageenan
might be feasible. The question is, are they sufficiently
available in an area, and, if so, can they be harvested
economically? Also, could they be produced under controlled
conditions at competitive costs? The marine science depart-
ments of various universities are engaged in intensive research to develop the crucial culture technology.

Experience indicates that there is a relatively low-risk method
for developing the extensive information suggested in the foregoing
outline. Specifically, it is recommended that a 3- to 4-year pilot
program of harvesting and selling dried seaweeds be developed. Demonstrating for suitable periods of time that the seaweeds of a given
area can be profitably harvested, dried, and marketed will automatically answer most questions concerning feasibility. Established extraction factories ought to purchase seaweed produced under a program
of this type. By the end of a 3-year pilot period, the quantity and
quality of seaweeds will be known, information on yearly variations
in quantities will be available, and harvesting and drying methods
and costs will have been developed. The fact that another manufacturer considers the raw material worth buying will be a major item
in any feasibility study. A working knowledge of the area will assist
in the selection of the best factory site if the decision to construct
a new factory is made. Labor availability and costs will be known.
Licenses, permits, and other legal problems will be known and can be
solved.

However, it is recommended that a pilot program be started slowly. Perhaps a target of 100 to 200 tons of dried seaweed for the first
year would be practical, after which production can double each succeeding year. For wild sources of red seaweeds, a small harvesting
organization will need to be equipped with a few skiffs and appropriate underwater harvesting gear. For kelp, a small cutter barge should
be sufficient. Generally, some type of hot air dryer will be required
for drying. Packaging can be accomplished by chopping and bagging or
baling.

An estimate of the dried seaweed exports from producing countries
to manufacturing countries appears in Table II. Included in this
table are the approximate current prices per ton, FOB in the exporting
country. Obviously, a brisk trade for dried red and brown seaweeds
has been established.

MANUFACTURING AND PROFITS

A question which must be addressed is, what is the minimum size
required for a self-sustaining extraction operation? One can extrapolate some information from Table I. Individuals knowledgeable in
extraction engineering suggest, with some qualifications, the estimates shown in Table IV. These figures, including the minimum investment as well as that required per pound of product annually, are

subject to many variables. Limiting production to one or two basic
extracts would substantially reduce the investment. Whether the final
product will be of industrial or pharmaceutical quality will make a
difference. Location of the factory and the amount of raw material
and finished product inventory can make great differences in the size
of investment. The money needed to carry the business until it meets
operating expenses also can weight the amount of the total investment.

It is beyond the scope of this chapter to describe the specific
steps of each extraction process. This information is usually available from engineering specialists or consultants. The number of companies listed in Table I is convincing proof that others have found
solutions to the technical problems.

Manufacturers of both algin and carrageenan are currently making
and marketing more than 150 products, each having different chemical
and physical characteristics. Carrageenan can be used to illustrate
this. As mentioned earlier, there are three principal fractions used
for commercial purposes, these being κ-, ι-, and λ-carrageenan. Each
of these may be extracted as is or as calcium, sodium, potassium, or
ammonium carrageenate. Each of the latter products may be modified
to a high, medium, or low gel strength or viscosity. They may also
be produced as coarse, medium, or fine mesh. Finally, they may be
cross-blended to obtain special properties and thus result in innumer-
able combinations. Each product may be roll dried or precipitated
with alcohol, with an associated modification of properties. It is
conservative to estimate that there are over 150 products tailored
to specific commercial utilization. Essentially the same situation
exists for algin. However, agar is usually sold in only two basic
grades, food grade and pharmaceutical grade. The manufacturer's
ability to modify properties of colloidal extractives is one of
the strongest arguments for their increased consumption in the future.
However, each time capability is expanded to produce a different product, capital requirements also increase.

Capital presently invested in the major seaweed extraction companies has been generated largely from profit. The continued expansion and growth of the algin and carrageenan markets also indicate
that these industries are profitable. Because most seaweed extraction operations do not publish annual operating statements, exact
profit figures are not available. It is estimated that net profits
for algin and carrageenan range between 7% and 15% of sales. However,
no prognosis regarding long-term agar profits can be made with absolute safety.

SUMMARY AND CONCLUSIONS

Before a judgment can be made on whether a new seaweed extraction factory may be economically feasible, data must be accumulated
showing the seaweed source to be adequate, stable, harvestable, competitive in costs, and available on a controlled or exclusive basis.
A procedural outline involving minimum financial risk has been suggested for developing this data.

Extraction technology and engineering know-how for all seaweed extractives are available or can be developed. Markets are available and growing for competitively priced products. These markets can be classified easily by uses or customers. Profits for carrageenan and algin industries appear to range at acceptable levels. A stable seaweed source for agar is imperative if this product is to achieve its market potential and prevent serious investment hazards.

- 18 -
Marine Plant Production and
Utilization—A Systems Perspective

C. DAVID McINTIRE

This chapter addresses problems associated with marine
plant reproduction and biomass management by means of a systems perspective. It deals with the dynamics of plant processes as isolated
systems and as subsystems within higher levels consisting of a variety of biotic and abiotic components. The properties of ecological
systems relative to different strategies of exploitation and management of marine plant communities can be better understood by involving mathematical modeling in community analyses.

Most professional biologists would include among the major
unifying concepts in modern biology: (1) the continuity of life
through the processes of heredity and evolution and (2) the hierarchical organization of biological systems. In an examination of
the problems associated with managing the marine plant resources,
the time frame relates to physiological and ecological factors rather
than to those which are evolutionary. However, the latter may enter
the calculations in unexpected ways. Therefore, this chapter confines
the discussion to biological organization manifest through contemp-
orary, hierarchical systems operating on an ecological time scale,
i.e., a relatively short period of time during which the genetic
constitution of the constituent populations does not change appreciably. Furthermore, the approach is from an ecological perspective.
Biochemical and physiological, as well as social and economic considerations, are not emphasized.

A system is usually defined as an assemblage of components
joined in regular interaction or interdependence, or holistically,
as an orderly working totality. Traditionally, ecological systems
are organized into hierarchical schemes of individual organisms,
populations, communities, and the entire ecosystem. The latter category is somewhat inconsistent with the others, as it presumably contains both abiotic and biotic components. Moreover, concepts associated with these levels of organization become more abstract as the

level of complexity increases. Most ecologists are probably more comfortable with the ecological theory of populations than with theoretical concepts associated with communities or ecosystems.

Beginning students in biology are fond of the cliche: "The whole is greater than the sum of its parts." The statement in this form is ambiguous, but it becomes much more meaningful when restructured in the form of a question: Can behavior of the whole be predicted from a study of the parts in isolation? Can the production and distribution of plant biomass off the coast of the Pacific Northwest be predicted from an investigation of individual species by laboratory culture methods? Systems science now indicates that the behavior of the whole cannot be understood by studying the parts in isolation--unless coupling variables are carefully identified and their integrity maintained. To help understand plant production dynamics at the population and community levels of organization, more experimental studies are needed, which are motivated by ecological questions and designed to test hypotheses generated by observational data obtained from the field. For example, an understanding of competitive interactions for nutrients may eventually require studies of cultures of mixed species or more imaginative approaches that face the difficult problem of biological interaction.

Each biological system can be examined holistically on the basis of its behavioral characteristics or mechanistically in terms of its coupled subsystems. Subsystems can be handled conceptually and analytically in the same manner by a further sorting into their subsystems. As a result, hypotheses and theoretical concepts can be generated relative to different levels of resolution, the relevance of which depends on a set of well-defined objectives. Communication on ecological matters, and in the related area of resource management, ultimately depends on explicit expressions at given resolution levels relative to biological organization, space scale, and time scale.

EXPLOITATION OF PLANT BIOMASS IN ECOLOGICAL SYSTEMS

At this point the taxonomic position(s) of the marine plant biomass is not very well defined. Do the biochemical products of interest occur in macroalgae or microalgae? Will harvest be from natural or artificial substrates? What is the proper time for harvest considering biological and economic factors? Will exploitation of plant biomass destroy, limit, or enhance production of consumer organisms of commercial or aesthetic value? Answers to these questions depend on the ecological properties of the system under exploitation, the methods, and the management.

Examples of alternative management strategies are the following:

1. Complete exploitation of the product without regard for the rest of the system.

2. Management of the system for maximum sustained yield of the product without regard for the rest of the system.

3. Management of the system for maximum sustained yield of the product with protection of other organisms of commercial value.

4. Management of the system for maximum sustained yield of the product with preservation of the diversity and stability of the entire system.

The first strategy, while popular when resources and energy were unlimited, is no longer compatible with the social and economic climate. The second approach suggests the early approach in fisheries when there was a single product of interest. The problem becomes the selection of an economically feasible scheme to manage the species of interest at a biomass level that maximizes production. Production is defined as the net elaboration of new living tissue in a unit of time (assimilation or photosynthesis minus respiration), regardless of whether that tissue survives to the end of that time (13). In an exploited stock, the biomass that optimizes production is usually much less than the maximum biomass the system will support without exploitation. For example, in the case of logistic growth, maximum sustained yield is supported by a biomass equal to one-half the carrying capacity of the system. The second strategy can lead to conflicts of interest, particularly if the management of the stock of interest threatens other populations of commercial or aesthetic value. The second of the strategies might seem most appropriate for harvest of plant biomass from artificial structures, affecting only phytoplankton competing for inorganic nutrients. However, secondary effects on the food chain would be neglected. Management strategies three and four are the most appropriate if the exploitation of natural kelp beds is seriously considered. Unfortunately, both of these strategies require a broad knowledge of the structural and functional dynamics of complicated ecosystems--the boundaries of which are not well defined.

STRUCTURE AND FUNCTION OF ECOLOGICAL SYSTEMS

Populations

From a numerical perspective, population size can be considered the outcome of four ecological processes: natality, mortality, immigration, and emigration (Fig. 1). These processes are considered rate variables. The population size is the associated state variable, something that can be monitored or measured at any instant. The structural diagram of Fig. 1 can be translated into a mathematical form that represents a general model of population growth. For example, assuming overlapping generations and continuous growth, the following formula can be constructed:

$$\frac{dN}{dt} = bN - dN + I - E.$$

N is the population size, b is the specific natality rate (the rate per individual), d is the specific mortality rate, and I and E are the population immigration and emigration rates, respectively. This simple model implies that growth is change in population size and can be either negative or positive. Many special cases of population growth can be derived from this general model by imposing additional assumptions relative to the behavior of b and d. If b and d are

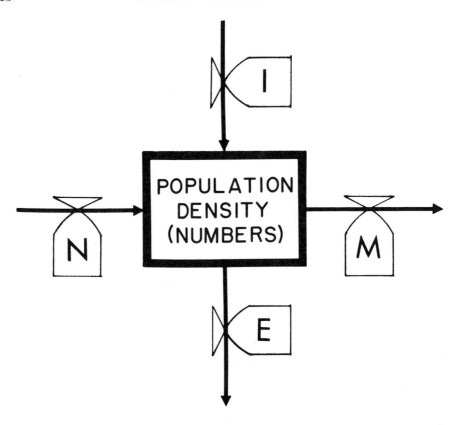

Figure 1. *A simple population model expressing density as numbers per unit of space. The associated rate variables are natality (N), mortality (M), immigration (I), and emigration (E).*

constants and I and E are equal to zero, growth is exponential; or if (b - d) is a linear function of population size with a negative slope, growth is logistic when E and I are unimportant. These simple models usually assume a stable age distribution, a crucially important structural aspect of population dynamics.

   A numerical approach to population dynamics is of great interest in certain instances, but a bioenergetic view is usually more pertinent when considering stocks of species exploited for commercial purposes. The population dynamics of *Iridaea cordata* are an illustration. Its life history involves the alternation of morphologically similar haploid and diploid generations. The gametophytic generation produces gametes, the spermatia and carpogonia, and after gametic union the mitotic division of the zygote nucleus leads to the formation of a carposporial conceptacle. Carpospores are released from the cystocarp which eventually develop into a free-living, asexual, diploid plant, the tetrasporophyte. The tetrasporophyte produces tetrasporangia, and, after meiosis, tetraspores are released which germinate into gametophytes. The quality of carrageenan varies with the stage in the life history (5). Tetrasporangial plants produce

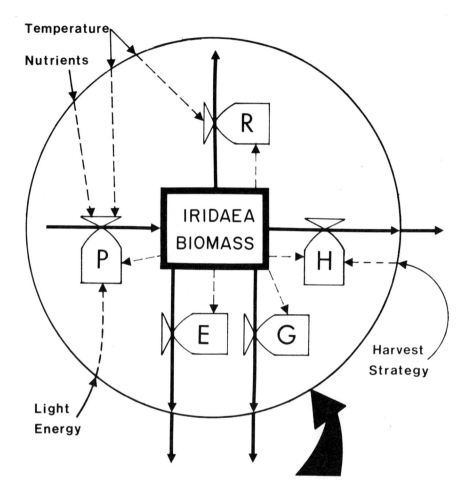

Iridaea Processes

Figure 2. *A simple model of a population of Iridaea cordata based on energy flow. The associated rate variables are photosynthesis (P), respiration (R), export (E), grazing (G), and human exploitation (H).*

only lambda-carrageenan, while the gametophytic generation yields kappa-carrageenan, the type most highly valued by industry. Moreover, a recent investigation of the structure of four populations of *I. cordata* indicates that natural populations tend to be dominated by the tetrasporangial stage (3).

Fig. 2 represents a simple structural model of an *Iridaea* population based on energy flow[1]. The large circle represents all the

---

[1]This model is based on studies of *Iridaea* by Dr. Judith E. Hansen, University of California, Santa Cruz.

relevant processes associated with *Iridaea*. The components within the circle represent the more important details of the system itself. The components outside the circle represent system inputs (i.e., anything that influences the system but is not in turn influenced by it). Within the system boundaries the rectangle is a state variable, the biomass of the population, and the "butterfly" symbols represent rate variables.

If the structural diagram of Fig. 2 is to be translated into mathematical form, it becomes necessary to derive interaction equations and estimate parameters that provide a satisfactory representation of the system relative to a set of well-defined objectives. The model structure itself evolves from such objectives long before a mathematical version is programmed on a computer. If the modeling effort is central to the research program, it provides an invaluable guide for field work by establishing research priorities. Various statistical procedures (e.g., the methods of multiple regression or nonlinear curve fitting) are employed to generate mathematical functions and parameters from field data. Therefore, parameter estimation becomes a problem of paramount importance, and model form is often dictated by constraints imposed by the estimation problems associated with sampling methods available in the field.

Of the rate variables in the model, postmortum decomposition and plant respiration can be combined into one called *respiration*. In the field it is difficult to physically separate microbial respiration from the respiration of healthy plants. Couplings between *Iridaea* and associated microbial decomposers are tight. The simple version of the *Iridaea* system also pools mechanical loss and organic leakage into another rate variable called *export*. If field data are available for parameter estimation, these processes could be treated individually just as well as two rate variables. The rate of human exploitation is a variable that can be manipulated either in terms of intensity or time. Temperature, light energy, and nutrients (nitrogen and carbon dioxide) are input variables that have already been studied in the field (Hansen, personal communication). A harvest strategy can be superimposed to manipulate the rate of exploitation.

The simple model of the *Iridaea* population might be appropriate for management strategy two if grazing is unimportant or excluded in the natural system under consideration. Hansen's population work (3) suggests that the principal grazers on *Iridaea cordata* include the genus *Lacuna* and other small gastropods that affect the plant population to a minor degree. However, the protection of population from primary consumer organisms in the ecological system often requires a large expenditure of energy and money. A simple population model conceivably could be adequate for algal populations or assemblages harvested from artificial substrates at intervals frequent enough to prevent the establishment of significant numbers of herbivores.

A working mathematical model similar to the illustration in Fig. 2 was described by McIntire (6). The state variable was the biomass of an entire assemblage of microalgae and associated heterotrophic microorganisms (i.e., a periphyton assemblage), treated collectively as a quasi-organism. This model illustrates the treatment of a collection of tightly coupled populations analytically as a single functional group. This approach works reasonably well with periphyton assemblages, because field measurements of primary production and community respiration usually correspond to the entire assemblage rather than to individual constituent populations. For example, McIntire and Wulff (7) have studied the rates of primary production and biomass accumulation associated with assemblages of microalgae in a marine laboratory ecosystem. Biomasses of about 170 g m^{-2} ash-free dry weight grew on acrylic plastic plates exposed for 29 days to an irradiance of 11,000 lux, 12 hr day^{-1}. Furthermore, the relationship between irradiance and the rate of gross primary production was established for assemblages physiologically adapted to different light levels and with different densities of biomass. Data of this kind are necessary to derive the mathematical functions and estimate the parameters in simple models involving plant populations.

If the effect of grazing on the plant population is appreciable, it may be necessary to move the processes associated with grazing inside the system boundary. In this case, a second state variable is identified for the grazer biomass, and the number of rate variables in the system increases from 5 to 10 (Fig. 3). The inclusion of grazer biomass within the system boundary is desirable when the plant and consumer processes are tightly coupled and strongly regulated by negative feedback loops between the various rate and state variables. With the addition of more variables, the problem of understanding system behavior becomes considerably more complicated, and more information is required for parameter estimation. In particular, food consumption rates in plant-herbivore systems are sometimes difficult to measure, especially if the consumers are small. Additional complexity is introduced if another item of interest turns out to be a secondary consumer feeding on the herbivore biomass. Management strategy three could possibly involve such a complication.

As the number of population interactions increase, the number of variables that must be accounted for rapidly approaches a limit beyond which the human mind can scarcely cope. Hence, ecological systems often are said to be multi-loop, nonlinear-feedback systems that behave in a counter-intuitive manner. To make matters more complicated, population interactions need not be of the plant-herbivore or predator-prey variety. Symbiotic relationships, amensalism or allelopathy (introduction of plant growth inhibitors), and interspecific competition are other important interactions that occur in complex ecological systems. Of these, competition and perhaps allelopathy may be of particular significance if the species composition or taxonomic structure of the plant assemblage has economic significance.

In the *Iridaea* system, additional state variables could represent different stages of the organism's life history. A partitioning of the population biomass into the gametophytic and tetrasporophytic

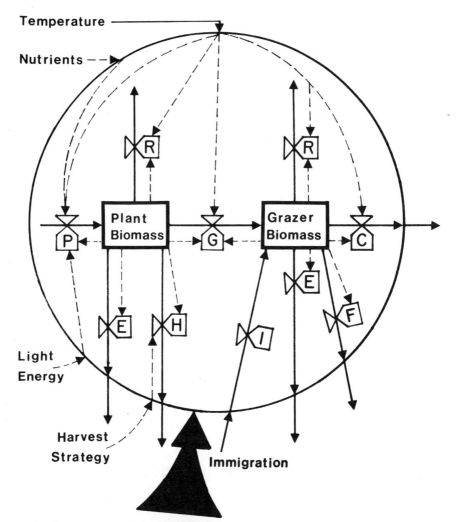

Plant – Herbivore Processes

Figure 3. *A model of a plant-herbivore system based on energy flow. The associated rate variables are photosynthesis (P), respiration (R), export (E), grazing (G), human exploitation (H), imports (I), fecal losses (F), and consumption by predators (C).*

generations might be desirable, particularly since the quality of the biochemical product is closely related to population structure. Such a partitioning creates a need for more field and experimental data and generates new estimation problems. Also, a partitioning into life history stages is usually only feasible when these stages are easily discernible in the field. With *Iridaea* this becomes a problem when the population is dominated by juvenile plants of less than 3 cm in length. It is nevertheless probable that modeling could help examine alternative management strategies relative to the intensity and time of harvest and the quality of the commercial product.

Communities and Ecosystems

The concepts of community and ecosystem are convenient abstractions that allow ecologists to hypothesize and theorize at complex levels of biological organization. Here, a community is considered to be any collection of co-occurring populations occupying an area of interest. To most ecologists, the term also implies a degree of structural and functional unity resulting from a history of coevolution within the system. Boundaries of communities are often more difficult to establish than the boundaries of populations, and there is a tendency for communities to change gradually along environmental gradients rather than exhibit sharp discontinuities compatible with discrete conceptualizations. Ecosystems are usually defined as relatively large communities functioning together with the abiotic components of the environment. Therefore, an ecosystem is the entire contents of a biotope, where the latter is a block of space in the biosphere that is interesting enough to stimulate investigation.

If exploitation is contemplated in a natural, relatively undisturbed ecological system (e.g., the natural kelp beds and intertidal algal communities of the Pacific Northwest), understanding at the community and ecosystem level of organization is absolutely essential for the successful implementation of either management strategies three or four. This contention can be argued on economic as well as aesthetic and biological grounds. Scientific theory associated with communities and ecosystems is poorly developed. However, ecologists have found it convenient to generalize in terms of both structural and functional attributes. For example, communities have taxonomic structure, trophic structure, and, at times, a predictable spatial arrangement of their constituent populations. Taxonomic structure is expressed by the number of taxa in the system (species richness), the kinds of taxa, and the relative abundances of the taxa. The latter property of taxonomic structure is often referred to as evenness, equitability, dominance, or redundancy. If the product of interest is confined to one or a relatively few species, the management strategy is usually to maintain a low diversity (i.e., low species richness and high dominance) in the system, usually at the cost of a high energy input. Trophic structure has to do with the functional groups of organisms that relate to patterns of energy flow. The latter may, or may not, be closely related to the taxonomic positions of the community constituents.

Bioenergetics at the community level of biological organization is an important topic relative to the exploitation of plant biomass and will be reviewed briefly. Using the following codification, formula may be constructed:

GPP = gross primary production
NPP = net primary production
NCP = net community production
 PR = respiration of autotrophic organisms

    HR = respiration of heterotrophic organisms
    CR = community respiration
Therefore, if there are no imports or exports,
    CR = PR + HR
    NPP = GPP - PR
    NCP = GPP - CR
    NCP = NPP - HR

Organic material is accumulating in the system when NCP > O, that is, when GPP/CR > 1. Communities with this characteristic are called autotrophic communities and are typical of ecological systems in which plant populations are the economic product of interest. Community succession in natural ecosystems usually involves a gradual change from an autotrophic community to a steady-state community characterized by NCP = O and GPP/CR = 1 over an ecological time scale. In other words, communities will not remain autotrophic indefinitely without artificial manipulation.

    The best strategy for the harvest of plant biomass, assuming species composition is unimportant, is to manage to maximize NPP and minimize HR. The maximization of NPP traditionally has involved the application of fertilizers, while the introduction of toxic chemicals is sometimes used to decrease HR. In either case, there is usually a great expenditure of energy and money, and the ecological system becomes closely coupled to an economic and social system.

MODELING AT THE ECOSYSTEM LEVEL OF ORGANIZATION

    The development of conceptual and more formal mathematical models of ecological systems at the ecosystem level of organization presents some unique problems, many of which are yet to be solved in a way acceptable to both field ecologists and modelers. Systems analysis in ecology from a rigorous perspective involves the translation of physical and biological concepts into mathematical form and the manipulation of the mathematical system thus derived. The principal tool of the systems analyst is the model, which in a broad sense is simply a statement of relationships. In its most formal, mathematical form, a model consists of variables, functional relationships, inputs, and parameters. Interaction equations express the relationships between variables in the system, and the inputs are introduced by expressions called forcing functions. Parameters are constants in the equations and forcing functions that must be estimated from experimental or observational data sets. Goals of model building usually include both description and prediction.

    The steps in constructing a mathematical model of an ecosystem include:

    1. The establishment of a system boundary compatible with a clear set of well-defined objectives.

    2. The identification of discrete subsystems within the boundary.

    3. The development of a tentative structural model--a systems diagram (e.g., Figs. 1, 2, and 3)--showing state and rate variables and other intermediate concepts.

    4. The accumulation and examination of experimental and observational data relevant to the structural model.

    5. The establishment of functional relationships between variables (equation writing).

    6. Estimation of parameters.

    7. Programming on a suitable computer system.

    8. The analysis of model properties.

Some of the model properties of particular interest are feedback and control, stability, and sensitivity, all of which are investigated by introducing perturbations to parameters and system inputs. Furthermore, an investigation of model properties often leads to changes in model form, and the procedure tends to be iterative, presumably increasing the understanding of the corresponding real-world system. Some of the more interesting insights into the real-world system have been obtained when the model fails to give the anticipated behavior or when the structure breaks down and no longer provides an adequate representation. In some respects, nothing succeeds like failure! Bizarre model behavior is an explicit monument to ignorance that is often hard to rationalize without further scientific inquiry.

Approaches to ecosystem modeling have varied from the development of very large complex models with many state variables and parameters to relatively simple models that attempt to represent coarse resolution dynamics. A good example of the variety of approaches to ecosystem modeling is available in a recent volume which describes the various modeling projects associated with the U.S. International Biological Program (12). W. S. Overton (11) has developed an especially good approach for ecosystem modeling in the Coniferous Forest Biome (I.B.P.). This approach conceptualizes ecosystems as hierarchically modular systems and involves the construction of systems and subsystems, each of which can be studied in isolation as long as the coupling structure is identified and its integrity maintained. These concepts have been incorporated by Overton (10) and White and Overton (14) into a general ecosystem paradigm called FLEX, based on the general systems theory of Klir (4). The FLEX paradigm is implemented at Oregon State University by the program FLEX2, a general model processor that accommodates both the holistic (FLEX mode) and the mechanistic (REFLEX mode) representations. The principal advantages of Overton's approach is that it allows the investigation of ecological systems at different levels of organization and provides a way to examine the behavior of subsystems in isolation within the structural framework of larger, more complex systems of which they are an integral part.

The ecosystem model is always an imperfect simplification and abstraction of the real-world system of interest. The model never includes everything that is present in the real system. Herein lies the difficulty of reconciling the differences between the modelers and the data collectors. Although modern-day electronic computing machines provide the scientist with an awesome capacity for number crunching, the field ecologist is often surprised to discover how quickly this capacity is approached when the complexities of entire ecosystems are represented in mathematical form. In fact, preoccupation with too much detail is the bane of the ecosystem modeler's struggle for a meaningful identification and representation of the system variables. A mechanistic model consisting of variables partitioned by taxonomic position is often unsatisfactory, particularly if dynamics revolve around energy flow in the system. A more reasonable approach in some instances is to picture an ecosystem as a nested, hierarchical system of biotic and abiotic processes hooked together by the relevant coupling variables. For example, the set of *Iridaea* processes within the circle in Fig. 2 could be coupled to other sets of plant processes--for example, through competitive interactions with the phytoplankton--and together these processes could represent the subsystems of a higher level of organization, the autotrophic processes. Autotrophic processes could be coupled to another subsystem at the same level called heterotrophic processes which in turn could be decomposed into its own subsystems, e.g., grazing, predation, and detritivory. This approach has been used to model lotic (flowing-water) ecosystems by McIntire *et al.* (8) and McIntire and Colby (unpublished manuscript).

While process modeling has certain advantages relating to the necessity for simplification, the approach often generates serious estimation problems. This is because most field and experimental data are obtained by scientists oriented toward the population level of biological organization. For example, how does one estimate the food consumption rate for the entire process of grazing in an ecosystem when such rates have been determined for only a few of the taxonomic entities involved? Progress in ecosystem theory and research depends to a large extent on an ability to rescale perspective from the population level of organization to a process or functional group perspective. New approaches to data collecting designed to measure rates of processing materials in large ecological systems must be found to support parameter estimates at the ecosystem and subsystems level. One new approach used by stream ecologists is to introduce leaf packs into a stream and measure the rate at which the material is converted into fine particle detritus by insects. This is called shredding and provides data from which the process rate can be estimated.

At the ecosystem level, there is some question as to just what the system state variables should represent. McIntire *et al.* (8) selected state variables on the basis of the various functional groups recognized by the current theory of energy transfer in lotic ecosystems. This approach--here referred to as the quasi-organism viewpoint--is based on the functional attributes of groups of organisms, while taxonomic position is essentially ignored. The state variable is the biomass at any instant involved in a particular process (e.g., primary production, grazing, and predation). If the population of a particular species or an individual organism is involved in more than one process, it is partitioned conceptually into functional groups, depending on the instantaneous proportions of the biomass involved in the corresponding processes. A refinement of the quasi-organism viewpoint is the designation of the state variables as capacities for processing. This approach recognizes that the quality of the biomass can affect the capacity to perform a system process. For example, if the species of grazing organisms change seasonally, food consumption rates per gram of biomass could exhibit corresponding changes. Under these circumstances, the state variable ought to be considered as capacity for grazing which is a function of biomass that changes with community composition. For resource exploitation, the state variable in the *Iridaea* subsystem could be the capacity to generate carrageenan. Carrageenan is obviously a function of plant biomass, but its quality and perhaps quantity per unit biomass could vary as the population age and life history stage changes with natural and seasonal changes or with exploitation. While functions relating process capacity to bio-mass may be difficult to derive, the concept of capacity as state variables in large ecosystem models provides a mechanism for incorpora-ting quality as well as quantity into the system model. If succes-sion or evolutionary change must be represented in the model, the establishment of functional relationships between process capacity and biomass is mandatory.

Output from simulation models is often expressed in the form of plots of state variables against time. While output in this form can provide adequate prediction of system behavior, it often falls short of explaining important mechanisms regulating ecosystem processes. In other words, values for state variables can go up and down, but it is not always obvious why such variation occurs. A new approach to the investigation of system behavior currently being developed by McIntire and Colby can help alleviate this problem.

Take, for instance, mechanisms regulating the process of grazing on *Iridaea* and assume that there is adequate representation of both the *Iridaea* and grazing processes in a mathematical simulation model. The growth rate per gram of biomass (or capacity) at time k associa-ted with grazing when *Iridaea* is present in unlimited supply is given by:

$$g_o(k) = \frac{1}{B(k)} \left[a_1 D(k) - R(k) \right] ,$$

assuming no losses from predation or export. In the equation, B is the biomass or capacity associated with grazing, D is the demand for food (i.e., the rate of consumption when food is in unlimited supply), a_1 is the fraction of the demand that is assimilated, and R is the rate of respiration. The specific growth rate g_o is analogous to the intrinsic rate of natural increase, a theoretical concept in popula-tion biology (1). In biological terms, g_o is a growth rate per unit weight or capacity in an environment with unlimited resources and free from negative effects from other processes. It is a function of density-independent factors only (e.g., temperature). If the *Iridaea* biomass is not in unlimited supply, the specific growth rate in the absence of predation and export is:

$$g_1(k) = \frac{1}{B(k)} \left[a_1 C(k) - R(k) \right] ,$$

where C is the actual food consumption rate. If the negative effects of export and the process of predation on grazing, in that order, are added, the equations become:

$$g_2(k) = \frac{1}{B(k)} \left[a_1 C(k) - R(k) - E(k) \right] \text{ and }$$

$$g_3(k) = \frac{1}{B(k)} \left[a_1 C(k) - R(k) - E(k) - P(k) \right],$$

where E is the rate of export and P is the flow to the process of predation. Therefore, at time k the limiting effects of food supply, export, and predation are $g_0(k) - g_1(k)$, $g_1(k) - g_2(k)$, and $g_2(k) - g_3(k)$, respectively. To analyze mechanisms accounting for the dynamics of the process of grazing, simply plot g_0, g_1, g_2, and g_3 against time and examine the areas between the curves relative to the plot of the corresponding state variable against time.

Modeling, with or without computer simulation, should be an essential part of the ecosystem research that must be conducted before engaging in responsible, large-scale exploitation of marine plant biomass in the Pacific Northwest. A similar view from a broader perspective is expressed by E. N. Hall, Chapter 20 in this volume. Moreover, it is important that modeling is not relegated to a peripheral or subordinate status within the research program, as its principal benefits depend on a central orientation from which research priorities and hypotheses can emerge. Unfortunately, leadership and budget control for large ecosystem research is often dependent on individuals more sympathetic with data collecting than with data analysis. Insufficient cognizance of the role of the system modeling and analysis can prove to be costly.

Ecologists appear to be notorious data collectors. Indeed, there are laboratories engaged in ecological research that could easily spend 5 years or more extracting information from data they have already collected! Sometimes, it is not always clear why the data were obtained in the first place; or, if the objectives were clear in the past, they are no longer relevant to the present.

A reasonable approach to ecosystem research related to the exploitation of marine plant biomass includes: (1) the identification of a well-defined research goal and the establishment of specific objectives that relate to that goal, (2) an exhaustive review of existing information that relates to the various specific objectives, (3) the development of tentative structural and conceptual models of ecosystem dynamics that identify system boundaries and variables, (4) the establishment of research priorities based on the tentative model and the knowledge of existing information, (5) data collection and analysis, and (6) the integration and the analysis of system behavior at various levels of organization. Steps (2) and (3) must proceed concurrently and be performed by biologists with a systems orientation and modelers with a biological orientation. Step (6) is the conclusion and should generate understanding of the mechanisms that regulate and control the relevant ecosystem processes. Whether a formal mathematical model is required for (6) depends on the objectives of the research program relative to the complexity of the ecological system under investigation. Plant systems generated on artificial substrates suspended over the continental shelf might require a less formal analysis than a large natural ecosystem, unless the artificial, man-created ecological systems are coupled to complex social and economic systems as is true of E. N. Hall's approach (Chapter 20).

ECOLOGICAL SYSTEMS AND SOCIETY

Until recent years, professional biologists were usually highly trained, specialized individuals with a relatively narrow view of the world. Even at the simple levels of biological organization, one could spend an entire lifetime trying to unravel the complexities of such systems as cells, organelles, or biochemical pathways. Specialists worked reasonably well during a period when resources and energy were in unlimited supply. Now, positive feedback models as manifested by the exponential growth of the human population of the world (doubling time ca 36 years in 1974) and the exponential decline of natural resources are no longer compatible with the realities imposed by the finite nature of the biosphere. Almost overnight affluent countries like the United States have begun to disintegrate. A relatively comfortable existence is threatened by the increasing generations of the future. The suddenness of the crisis is no surprise to those familiar with the mathematical properties of exponential growth. The new challenge to responsible biologists is not simply to understand cells, individuals, populations, communities, or even entire ecosystems in isolation, but rather to understand how such ecological systems relate to the social, economic, and industrial systems of the world. Unfortunately, most scientists are not professionally trained to think this way. Help is needed from generalists--the systems scientists.

The publication of The Limits to Growth by Meadows *et al.* in 1972 (9) marked the first attempt to expose the educated public to systems science from a global perspective. The world model which was very effectively presented consisted of social, economic, industrial, and ecological subsystems coupled in such a way as to provide predictions of population size, resources, food availability per capita, pollution, and industrial output per capita during the next 100 years. The Limits to Growth had an impact similar to Rachel Carson's Silent Spring (2). Its predictions were frightening and a threat to traditional economic models that equated growth to prosperity. In spite of an enormous effort by various factions to destroy the credability of The Limits to Growth, the world model served its purpose well by helping to focus the public's attention on serious world-wide problems.

Much as Meadows *et al.* (9) addressed the population problem, E. N. Hall's Chapter 20 in this volume approaches the problem of exploiting the marine plant biomass from a systems perspective. Although the feasibility of Mr. Hall's proposal needs evaluation, his conceptual model is composed, as it should be, of ecological, economic, social, industrial, and political subsystems, the dynamics of which were projected over the next 125 years. Mr. Hall's proposal

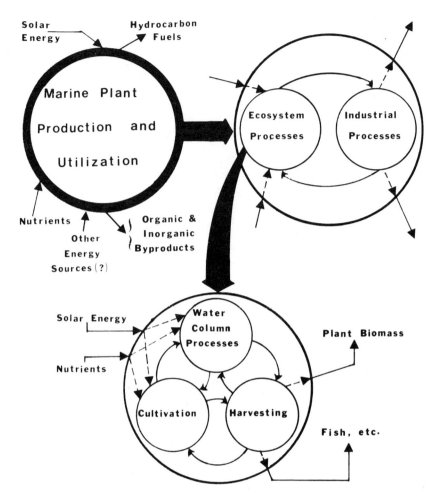

Figure 4. *A schematic representation of a Marine Plant Production and Utilization system showing the hierarchical composition of the Ecosystem Processes subsystem.*

can also be interpreted by two systems diagrams (Figs. 4 and 5), which are illustrative of the concepts developed in the previous section of this chapter and emphasize the examination of systems within a hierarchical structure at different levels of resolution. Figs. 4 and 5 are based primarily on Hall's Fig. 7 and are intended to put his model into a slightly different structural form.

In Fig. 4 the entire system under consideration is arbitrarily called Marine Plant Production and Utilization. Holistically, the system receives inputs of solar energy and nutrients from the sea and theoretically generates hydrocarbon fuels and various organic and inorganic byproducts. In the early stages of development the system also receives energy subsidies from sources other than solar radiation. A dynamic understanding of the behavior of the total

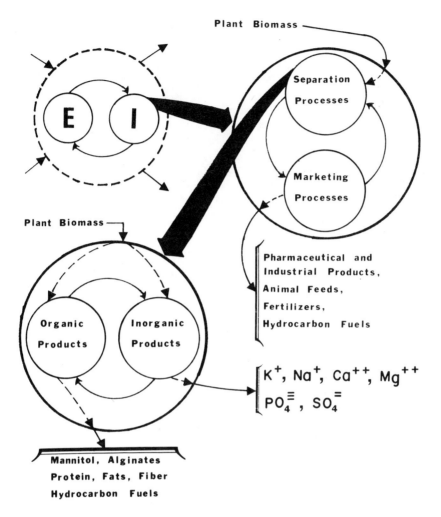

Figure 5. *A schematic representation of the mechanistic structure of the Industrial Processes and Separation Processes subsystems in the Marine Plant Production and Utilization system.*

marine plant system is required to deal with questions related to the global steady-state equilibrium which is of course a larger and more complex system.

The marine plant system can be examined mechanistically in terms of two coupled subsystems: Ecosystem Processes and Industrial Processes. It is important to identify relevant coupling variables between these two subsystems. For example, such variables might include a flow of plant biomass from Ecosystem Processes to Industrial Processes and flows of hydrocarbon fuels and nutrients (fertilizers) from Industrial Processes to Ecosystem Processes. Presumably fuels would be needed to power the machinery necessary for cultivation and harvesting. After the coupling variables are carefully identified,

Ecosystem Processes and Industrial Processes can be uncoupled and examined in terms of their holistic behaviors or mechanistically in terms of their coupled subsystems. Understanding at the level of the total ecosystem and the total industrial system is necessary for the identification of mechanisms relating to project feasibility and for decisions relating to possible couplings with the social and economic systems which are not included in this diagram. Such generalizations may only have significance after coupling variables are identified. Such variables represent the inputs and outputs associated with the holistic behavior of the Ecosystem Processes and The Industrial Processes.

Ecosystem Processes can be partitioned into coupled subsystems referred to here as Cultivation, Water Column Processes, and Harvesting. The latter is included as a subsystem of Ecosystem Processes because its couplings are stronger with the biological system than with the industrial system. Cultivation includes all relevant processes associated with artificial substrates. Water Column Processes include physical, chemical, and biological processes that directly or indirectly support Cultivation. Harvesting couples to both of these systems, because other products of interest to society also may be generated from the fabricated habitat. The inclusion of Water Column Processes in the model allows for the investigation of possible conflicts resulting from exchanges of materials between an artificial substrate and the water mass.

The subsystems of Ecosystem Processes also could be dissected further and examined mechanistically. For example, Cultivation could be uncoupled and modeled as a system of state, rate, and control variables such as those illustrated in Figs. 2 and 3. Coupling variables with Harvesting and Water Column Processes must be identified before Cultivation can be examined in isolation. They represent inputs and outputs indirectly affecting the Cultivation subsystem.

Industrial Processes (I) can be conveniently partitioned into two subsystems: Separation Processes and Marketing Processes (Fig. 5). Separation Processes include the physical procedures and operations involved in the processing of the 35 tons $acre^{-1}$ $year^{-1}$ of marine algae projected by Mr. Hall. Marketing Processes include the commercial packaging, distribution, and advertising of the final products that are eventually transferred to consumers (e.g., pharmaceutical products, animal feeds, fertilizers, and hydrocarbon fuels). The raw materials that represent outputs from Separation Processes can be partitioned into Organic Products and Inorganic Products as subsystems.

The boundaries arbitrarily selected for the system illustrated in Figs. 4 and 5 are obviously too restrictive to reflect the potential impact of a new technology on society. The social, economic, legal, and political ramifications of a project of this magnitude almost defy comprehension, largely because the current conceptual models of how the world system operates have stressed suboptimization over relatively short periods of time--of a decade or so. Now the life style and values of individuals in the social systems of the world must be optimized within the next 50 to 100 years. Inefficient use of resources and lost time in planning strategies for the future are no longer tolerable. Systems science and modeling can provide organization and direction that will increase the probability that the critical deadlines will be met before rapidly expanding social demands cause a catastrophic global collapse.

REFERENCES

1. Birch, L. C. 1948. The intrinsic rate of natural increase of an insect population. J. An. Ecol. 17:15-26.
2. Carson, R. 1962. Silent Spring. Houghton-Mifflin, Co., Boston, Massachusetts. 368 pp.
3. Hansen, J. E., and W. T. Doyle. 1976. Ecology and natural history *Iridaea cordata* Rhodophyta; (Gigartinaceae): population structure. J. Phycol. 12:273-278.
4. Klir, G. J. 1969. An Approach to General Systems Theory. Van Nostrand-Reinhold, Princeton, New Jersey. 323 pp.
5. McCanless, E. L., J. S. Craigie, and J. E. Hansen. 1975. Carrageenans of gametangial and tetrasporangial stages of *Iridaea cordata* (Gigartinaceae). Can. J. Bot. 53:2315-2318.
6. McIntire, C. D. 1973. Periphyton dynamics in laboratory streams: a simulation model and its implications. Ecol. Monogr. 43:399-420.
7. McIntire, C. D., and B. L. Wulff. 1969. A laboratory method for the study of marine benthic diatoms. Limnol. Oceanogr. 14:667-678.
8. McIntire, C. D., J. A. Colby, and J. D. Hall. 1975. The dynamics of small lotic ecosystems: a modeling approach. Verh. Internat. Verein. Limnol. 19:1599-1609.
9. Meadows, D. H., D. L. Meadows, J. Randers, and W. W. Behrens III. 1972. The Limits to Growth. Universe Books, New York. 205 pp.
10. Overton, W. S. 1972. Toward a general model structure for a forest ecosystem. *In:* Franklin, J. F., L. J. Demster, and R. H. Waring (eds.) Proceedings, Symposium, Research on Coniferous Forest Ecosystems. Pacific Northwest Forest and Range Experiment Station, Portland, Oregon, pp. 34-47.
11. Overton, W. S. 1975. The ecosystem modeling approach in the coniferous forest biome. *In:* Patten, B. C. (ed.) Systems Analysis and Simulation in Ecology, Vol. III. Academic Press, New York, pp. 117-138.
12. Patten, B. C. (ed.) 1975. Systems Analysis and Simulation in Ecology, Vol. III. Academic Press, New York. 601 pp.
13. Ricker, W. E. 1958. Handbook of Computations for Biological Statistics of Fish Populations. Fisheries Research Board Canada Bull. 119. 300 pp.
14. White, C., and W. S. Overton. 1974. User's Manual for FLEX2 and FLEX3 General Model Processors. Forest Research Laboratory, Oregon State University, Corvallis, Bull. 15. 181 pp.

- 19 -
Engineering of Structures in the Ocean
JOHN H. NATH AND ROBERT A. GRACE

The farming of benthic algae in the ocean will require
structures for plant support and farm activities. Although some
preliminary work has been done on design, construction, and testing
of a potential support platform, many engineering unknowns remain.
This paper presents an overview of the various types of offshore
structures that are now operational--and might be considered for
ocean farms. Fabrication, launching, deploying, and mooring problems
for such structures will be reviewed and the influence of environmental factors such as wind, waves, currents, and earthquakes will
be outlined. A discussion of the general design process involved in
the engineering of ocean structures will focus briefly on types of
platforms for marine farms.

Naturally occurring stands of kelp are harvested off the coast
of California and elsewhere. After processing, components are used
in human foods, industrial products, pharmaceutical products, and
fertilizers. Seaweeds are also processed to yield animal food supple
ments and can produce methane gas to serve as an energy source. The
potential value of macroscopic, marine, benthic algae which might be
cultivated in the ocean under a bountiful supply of solar energy has
prompted serious consideration of ocean farming to supplement food
supplies and conventional energy sources, such as petroleum and natural gas.

In order to make a significant contribution to energy and food
demands, extensive areas of ocean bottom must be modified to some
degree (34). There is clearly insufficient ocean area within the
10 m and 25 m depth contours, ideal for benthic algae, to support
the production anticipated. The bottom at these depths is seldom
ideal, in terms of firmness and stability, for seaweed attachment.
Thus a solid substrate should be maintained at 10 m to 25 m below
the still water level regardless of the overall water depth.

This paper will provide background on state-of-the-art offshore

335

structures that might be considered in ocean farming. They may be
fixed to the bottom and extend through the sea-air interface, may
be floating, may be a blend of these two extremes in the case of
"hybrid" structures, or they may be built on the sea bed but submerged. Engineering considerations for the design of such structures for performance and safety in the marine environment are also
involved. There are problems due to corrosion, fouling, fatigue,
winds, waves, currents, ships, and fishing gear. Interaction of
such influences and the various components of the farm, both animate
and inanimate, must be considered. There are problems of materials
selection, sizes of members, types and weights of anchors, as well
as of locating the farm in suitable waters and over suitable bottoms.

The effectiveness of the ocean farming concept will be judged
chiefly upon whether or not items derived from seaweeds can be produced at a profit. Liquidating the very high capital cost of the
supporting structure will be a primary consideration. This paper
addresses economic questions within the ocean farming concept, but
a detailed review of costs is not possible at this time.

OCEAN STRUCTURES

A brief review of the types of ocean structures now in existence will indicate what can currently be constructed. The mining
of energy-rich products such as petroleum can return the very high
investment offshore structures require and still produce a profit.
Such structures may be economically questionable for seaplant farming, but they do illustrate an important construction capability.

Bottom-supported, Surface-piercing Structures

The most common installation of this type is the platform used
for exploration and production in offshore oil fields. Such structures are usually made of steel and supported by piles driven into
the sea floor. Living and working areas on various decks cap the
platform. Many have been installed in shallow waters around the
world.

The deepest installation to date is in 260 m of water in the
Santa Barbara Channel in California. The Exxon Corporation spent
$67 million for the deep platform installed in 1976. The platform
jacket is 264 m high and the total platform height when topped with
a three-deck structure is 288 m. Total jacket weight is 11,000
metric tons; the base is 71.5 x 52 m with the top 38 x 14 m; the
eight legs are 1.37 m in diameter; and twenty steel piles are driven
as deep as 107 m into the bottom material to pin the structure to
the sea floor.

Burmah Oil Co. operates a steel platform in the North Sea in
161 m of water. The base is 85.5 x 82 m and the deck 61 x 82 m.
The jacket for the platform is 185 m high and the total platform,
to the top of a flare stack, is 295 m high. The jacket alone weighs
22,000 metric tons.

Shell Oil Co. is constructing a structure 372 m high that will
be installed in the Gulf of Mexico. With towers, this platform will

be as high as the Empire State Building. The platform base dimensions are 116 x 122 m, and the weight will be 41,000 metric tons.

In addition, large structures are built from reinforced concrete.
The enormous size of such structures is illustrated by the dimensions
of the following platform that has been installed in 104 m of water
in the North Sea. The base, set on the sea floor, is a gigantic
cellular concrete caisson 39 m high. In horizontal extent the base
is square and 70 m to a side. The steel platform deck is supported
by two concrete columns each 80 m high.

Single-leg structures have also been installed, e.g., off
Alaska. This approach is being planned for offshore terminal use
for discharge of oil from tankers (23). Such an apparatus can be
articulated at the ocean floor rather than fixed. Stability is
achieved from buoyancy tanks near the surface. Pipes from a re-
finery on shore run to the terminal, and hoses are used to connect
to ships.

A further type of offshore structure is the guyed tower, which
is more lightly built than the customary offshore platforms and
steadied by guy lines attached to heavy anchors. One such system is
currently under test by Exxon Corp. in the Gulf of Mexico. It is
113 m high, set in 91 m of water, and used as a model for a proto-
type to be ultimately installed in waters 450 m deep or greater in
the North Sea (10).

Enormous steel oil storage tanks also have been installed in
various parts of the world. A set of three 500,000-barrel tanks
is located in 49 m of water in the Persian Gulf. These tanks have
an inverted conical base and a cylindrical upper portion that passes
through the water surface. The diameter of the base is 82 m and the
tanks are 64 m high (11). A still larger tank, with capacity of one
million barrels, has been installed in the North Sea. This cylindrical tank, having a perforated wall, is 90 m high and located in 70 m
of water (12).

A final class of bottom-supported structures that extends above
the water surface involves breakwaters and jetties. Although such
installations may be made from steel sheet piling, the usual case
involves the dumping of rock. Depending upon the severity of anticipated wave conditions, such structures may be protected by large
concrete units such as dolosse or tribars (32).

Floating Structures

The offshore oil industry is most heavily involved in floating
structures. Work barges are equipped with giant cranes to assist
in installing fixed platforms and other offshore facilities. Pipe
barges are involved in various ways in laying pipe along the ocean
floor.

The semi-submersible is a favorite offshore drilling platform
because of its stability in waves. An upper deck mounts living
quarters but it serves chiefly as a working platform. Primary items
are a drilling derrick, storage for drilling rods, and machinery.
Three or more vertical columns support the deck above the water

surface on the flotation units, which are below the surface. In
some cases, the floats may be interconnected. Water can be pumped
into or out of the floats to adjust the draft of the platform.
Shallow draft is desired when the platform is towed to location
whereas deeper draft assures better motion stability--as long as
the underside of the deck clears the highest waves. Station keeping
is achieved from mooring lines or dynamic positioning.

Tankers may discharge or take on crude oil at floating mooring
buoys. The design of such installations either involves a circular
unit floating in the sea-air interface or an elongated spar buoy
that floats in a nearby vertical attitude with the bulk of its
length submerged. The spar buoy design even extends to special
research ships that proceed to a desired ocean location in horizontal attitude, then are upended when on station. An example of
such vessels is the 116 m long FLIP, operated by the Scripps Institution of Oceanography.

Floating breakwaters are another class of structure that may
be of particular use to ocean farming. The past decade has seen a
considerable amount of research on this approach to reducing the
level of wave action in ports and harbors (1).

Hybrid Structures

Combinations of bottom-mounted and floating structures exist.
A jack-up platform has legs that can be raised and lowered with
respect to the deck structure, which floats. The platform is
towed in floating mode to a desired location, the legs are then
dropped to rest on the sea bed, and the deck portion is then jacked
clear of the water.

A second type of hybrid structure involves tension-leg platforms. A floating rig, usually a semi-submersible, is positioned
at a desired location. Vertical tension cables, connected to
anchors on or in the sea bed, are used to pull it deeper into the
water. Such structures are not practical in shallow depths; their
use is generally confined to water 120 m or deeper.

Submerged, Bottom-mounted Structures

There are thousands of miles of pipelines lying on the sea
bed or buried in it. Outfall pipes, carrying sewage or wastewaters
from manufacturing and power companies, comprise an important group
of submarine pipelines. Steel pipelines are the most common; concrete is frequently used for large outfalls.

Oil production platforms that rise above the water surface are
subject to wind loads and direct wave effects. There has been a
movement over the past decade to design completely submerged production units that rest on the sea floor in deep water. Controls
for such systems can either be internal or external, observable
and adjustable by divers or by small, manned submarines. In the
former case a diving bell is lowered to the production unit, mates
with it, and technicians enter to work.

The diver habitat is another bottom-based structure. There have been numerous experiments, e.g., Sealab I and II, Hydro-Lab (17), and Tektite, wherein divers have occupied a structure set on the sea bed for a period of several days or more. Although such an operation is entirely feasible, the offshore oil and gas industry prefers to transport its divers under pressure each day from a work barge on the surface to the work site on the bottom in a diving bell.

The final type of submerged installation involves vertical cables stretched tightly between a clump anchor on the sea floor and a submerged float (usually spherical) between the bottom and the surface. Such systems support oceanographic measuring instruments.

OPERATIONS

A few factors of engineering importance in the operation of ocean structures may be useful in this review. In some cases there is little latitude in locating ocean engineering structures. For instance, an oil field production platform must be positioned over or near a previously drilled exploratory well, and submarine pipe-lines between platforms essentially have predetermined routes.

If there is a choice in the positioning of an ocean structure, it should be outside of shipping lanes and remote from fishing grounds where the potential for collision is high. Of particular importance is information on bottom conditions and on waves and swell. If possible, an ocean farm structure should be located away from severe waves. The wave climate involves both the heights and periods of waves and extends to the duration over which such conditions last.

The nature of the sea bed is critical. This may be less true for floating structures than for those attached to the sea bottom, but anchors must have sufficient holding capacity for a platform when it is subjected to high loadings. A rock bottom, or one with rock outcroppings, might be completely unsuitable unless receptacles for anchors were drilled or trenches cut in the bottom by blasting. The following characteristics are often important: the strength of the substrate must be adequate to support weight; the material cannot be subject to extensive erosion under heavy wave action in the water depths involved; the particle size of soft bottoms must be such that the phenomenon of liquefaction (complete loss of strength due to pressure buildup) does not occur; and bottom slopes and material sizes should not be such that slides are possible.

The characteristics of materials well below the bottom may be as important as those of the surface material. Piles driven deeply into the sea bed, to provide vertical support for a bottom-based structure and partial stabilization against horizontal displacements, can be properly designed only with adequate engineering data on the bottom material to the depths involved. Seismic profiling, done from a survey ship, generally must be supplemented by actual drill-ing and coring at the site.

It is usually difficult to install even small experimental structures in the ocean. Enormous logistical problems for complex deployment operations exist for massive offshore structures. The submerged research structure described by Grace and Nicinski (14) consisted of a pipe 5.3 m long and 0.41 m in diameter set on a steel base composed of structural steel that had horizontal dimensions of 2.4 x 4.9 m. The weight of the entire structure was only 5.5 metric tons. However, six hours were required by an experienced crew to transport the structure on a lift vehicle 150 m from the fabrication shop to a loading dock, to secure the structure against the hull of a 17 m work boat, and then for that vessel to proceed 1000 m to the test site in 11 m of water and lower the structure to a predetermined location on the bottom. Waves prolonged the time taken for the work boat to get the structure into position.

The very large Exxon Santa Barbara Channel platform was fabricated on shore in two parts which were towed separately on barges to a joining location. They were linked and welded while floating horizontally. The platform was then towed to the final location, tipped upright under tight control, and positioned on the sea floor. Its piles were later driven into the sea bed. The deployment operation required weeks.

Occasionally structures can be built in the dry adjacent to the waters they are to occupy. The 500,000-barrel oil storage tanks referred to earlier were built in the dry, but within a walled-off area below the level of the water. When each tank was completed, water was allowed to fill the area. The tank was towed to the final location and sunk into the desired position. Such operations are time consuming and expensive.

The traditional means of keeping a ship, buoy, or floating platform on station is to use mooring lines (or cables or chain) extending to anchors set well away from the floating plant. The scope (the ratio of the length of the mooring line to the depth) is usually from 5 to 12 for most anchors and most types of bottom material. Conventional anchors are designed to dig into the sea floor when pulled horizontally. Thus an anchor is set a considerable distance from the floating structure it is holding. Clump anchors are also often utilized; these rely completely on dead weight. In addition, anchors can be explosively imbedded into bottom materials.

Steel cable (wire rope) is customarily used to connect sizable floating platforms to their anchors. Such cable is usually spooled on a winch and can be taken up or let out as the platform drifts from its desired location. Drilling platforms may have eight or more winched anchor systems; pipe barges might have six; floating dredges, four.

Floating platforms or ships can also be maintained on station with dynamic positioning. The desired station is established using some form of accurate high seas positioning such as satellite navigation. Three acoustic transmitters are placed on the sea bed at that location in a triangular array. Acoustic receivers on the hull of the vessel pick up the sound pulses generated. Differences in arrival times can be translated by appropriate electronic control

circuitry into the vector error of the vessel's position. Special
propulsers mounted on the hull are then brought into play to return
the vessel to its position. On the Sedco 709 semi-submersible, for
example, there are eight 3000-HP controllable-pitch, azimuthing
thruster units (29). Complete computer control of such systems is
possible.

THE MARINE ENVIRONMENT IN AN ENGINEERING CONTEXT

The oceanographer is a scientist who gathers data on the marine
environment. The ocean engineer uses such data combined with engineering principles to design ocean installations. Wind, waves, currents, storm surge, gravity (pressure, buoyancy), earthquakes, and
corrosion must be considered in designs. Although some data have been
obtained on winds and waves during great storms, design waves are normally obtained theoretically from long records of winds. Past data of
wind and waves can be extrapolated in a probabilistic manner to obtain
estimates of rare but possible occurrences (2). Extreme wind and wave
information for different areas along the U.S. coast has been published
(26,31). Severe design storms can be employed (3,5).

Design currents can be estimated from storm winds (4). Steady
currents flowing past a pile for an offshore structure cause vortices
to be shed alternately from one side and then the other. This
oscillatory transverse loading is similar, although of higher frequency, to the oscillatory, in-line loading due to waves. This
cycling of loading in most structures creates fatigue and must be
predictable for safe design. A member cycled many thousands of
times at appreciable stress levels does not have the strength of
the uncycled member.

In the marine environment corrosion fatigue is also a possibility. Corrosion can penetrate into minute cracks opened up during
the member's cycling. A very important class is galvanic corrosion.
When two different metals are electrically connected and immersed
together in an electrolyte such as sea water, there is a flow of
electricity. The anode becomes corroded. Positively charged metallic ions leaving the anode cause the corrosion. Electrons freed
by the release of the metallic ions pass from one metal to the other
along the electrical connection. A galvanic series lists metals by
their tendencies to be anodic with respect to other metals. Cathodic
protection can be provided for a given metal (otherwise anodic)
either by attaching to it a more anodic (less noble) metal (such as
zinc) or by impressing a voltage difference against the natural
galvanic cell potential (18).

The term fouling refers to the growth of marine organisms on
structures or, possibly, within them. There are literally thousands
of such organisms that will attach themselves to structures and
grow. For example, when ship hulls are fouled with barnacles, the
resistance to motion through the water is increased. Cleaning of
the hull becomes a necessity. This can be done either by moving the
ship into a dry dock or by having divers with special tools clean
the hull in the water at dockside. The latter approach, which is

relatively new, is much less costly and time-consuming than the former. Growths on fixed structures effectively increase the size of members. This increases the hydrodynamic loads that must be resisted by the structure. Attempts to prevent fouling involve treating the material and/or its surface with a substance that is toxic to potential fouling organisms. Paints that slowly give up metallic copper ions in the solution are the most effective. The usual material for offshore structures is steel, and the overall protection of such surfaces against corrosion and fouling involves substances such as epoxy-base paints containing metallic additives.

Predicted earthquakes can provide the most important information for design load for structures built in deep water. A primary effect from earthquakes is the shaking of the foundation, and then the structure, in the relatively still water. The hydrodynamic forces on the structure can be very large, but if it is loosely moored to the bottom the forces are minor. A secondary effect from an earthquake may be the creation of a tsunami wave when the ocean bottom deforms drastically, such as a large subsidence. As such long waves grow in height as they shoal, they can create very large forces on nearshore structures.

ENGINEERING DESIGN

A major role of the engineer is to predict forces and associated structural response. In addition, the engineer is usually expected to locate the structure, and this may involve the interplay of several disciplines. An engineer is concerned with safety and complying with the few codes that exist for ocean work, with maintaining reasonable stresses throughout the structure, and with compliance with local laws and regulations. There is also concern to some degree with aesthetics. Particular concern is given to the economics of the construction, location, and operation of the structure. In engineering design an attempt is made to minimize total annual cost while still maintaining a satisfactory situation in the structure and its environment.

The analysis of structural response can be divided into two categories: 1) static analysis wherein the forces are not assumed to fluctuate with respect to time; 2) dynamic analysis wherein the forces not only fluctuate with respect to time but the structural response is a function of the frequency of fluctuation of the forces. If the natural periods of vibrations in the structure are well below the range of periods of the force oscillations, the structure may be designed with a static analysis. Otherwise, the dynamic amplification of the structural response must be considered, and this significantly complicates the analysis.

Much reliance is placed on mathematics in developing the basic equations which describe the processes involved. Only a few solutions can be obtained in closed form. Most ocean structural analysis is highly nonlinear and the engineer frequently resorts to numerical solution of the basic equations. He is also frequently called upon to make quick approximations for the purpose of estimates.

A typical offshore structure, whether fixed to the sea floor or floating, is a matrix of circular cylindrical members in various arrangements. The analysis involves the determination, for particular fixed sizes, lengths, and orientations, of the stresses occurring in each member. The loads are made up of the weights of each component; water pressure forces from gravity, waves, currents, and earthquakes; forces from guy wires or anchor cables; and wind forces (30). Comprehensive numerical computer programs have been developed for such analyses.

Hundreds of good recent papers describing recent advances in the analysis of ocean structures exist. The reader is urged to delve into the representative bibliographic references which are provided at the end of this chapter.

For determining wave forces alone, structures are divided into those that are small with respect to the wave lengths and those that are large. Large structures introduce appreciable changes to the wave field whereas small structures do not. Analysis of large structures can be carried out by using diffraction theories (13) and computer solution and/or by testing a small model of the structure in a wave tank.

Water motion under waves involves both velocities and accelerations (7, 8, 9). The Morison equation used to determine wave-induced forces on small structures considers both aspects (19, 21, 24, 28, 33). It includes the wake formation effects (velocity-dependent) like the familiar concept in treating steady flows of real fluids using the drag force (16). It also includes the acceleration-dependent effects through adaptation of the inertial force concept for accelerating ideal fluids. Forces and water motion are connected through force coefficients. The drag force involves effects of the turbulent wake behind an object. Analytical approaches break down in such cases. Thus experiments must be used to establish force coefficients. However, because of the effects of acceleration, physical modelling is considerably more complicated than the aerodynamicist's use of small-scale wing models in wind tunnels for determining drag and lift forces. There are many other complications in determining wave force coefficients. One which must be taken into account is the proximity of nearby boundaries (25, 35, 36).

MARINE PLANT FARM STRUCTURES

A few efforts have been made in the United States to construct support structures on which kelp is grown. These experiments have been small. Larger farms exist in Japan, but in protected waters. In order to design and construct reliable and relatively low-cost farm systems for the continental shelf in deeper waters, a considerable amount of design and experimentation will be required. Coordinated efforts by engineers and botanists will be necessary. For example, current flowing through the kelp will create forces on the structure which will need to be considered in the structural design. The same current will provide nutrients which will be scrubbed from the water by the plants--to the detriment of those at the downstream end. The engineer will need to configure the structure to minimize forces and the botanist will want to configure the structure to maximize absorption of nutrients. A systems analysis including these topics and all others will be required.

Experimental structures to support marine plants have been designed, fabricated, and tested. This research was funded jointly by the U.S. Energy Research and Development Administration and the American Gas Association. The California Institute of Technology and the Naval Undersea Centers in Hawaii and San Diego performed the work. Although the experiments were aimed primarily at determining whether kelp could be grown on artificial substrates, with or without artificial upwelling of nutrient-rich bottom water, they also clearly demonstrated that the substrate structure must be well engineered to withstand the rigors of the marine environment.

A 3-ha experimental farm 1000 m off the northwest tip of San Clemente Island, California, was the first attempt. The depth at this site was about 90 m and the farm was placed in January 1974 (25). The farm involved 100 to 150 *Macrocystis pyrifera* plants attached to submerged ropes strung across an anchored structure. In January 1975, the anchor at one corner of the structure came loose, the farm floated to the surface, and the whole system was completely destroyed --presumably by a passing ship.

A much smaller second farm was installed off Corona del Mar, California, but it simply disappeared after several months. Another effort off Catalina Island, California, was destroyed by waves in less than a month. An upwelling experiment was damaged by a Navy barge which struck the upper structure. Leaks appeared in the upwelling pipe, and the single kelp plant chafed against the structure so that its growth could not be evaluated (25).

These problems caused a pause in further work. One of the two sponsors, the American Gas Association, selected a new contractor, General Electric Company, which has subcontracted with Global Marine Development, Inc., to continue work with a fabricated structure designed but never tested by the Naval Underseas Center. Testing is to be conducted 6.5 km offshore from Corona del Mar, California, in water about 300 m deep. The quarter-acre module (QAM) which has been proposed looks like an inverted clothes line floating in the water as shown in Fig. 1. The six arms are tapered poles, 16 m long, made of prestressed concrete, attached to an extension of the spar buoy at the center through large gimbaled rings. Connecting the arms are concentric rings of polypropylene line except for the outermost ring which is wire rope. The QAM is capable of handling approximately 100 plants attached to the lines at 3 m centers.

The QAM is an example of a rigid member structure. Another preliminary design on a tension grid structure also has been made (20). It is much like a fish net, which in fact was used for a physical model test. Kelp bearing lines were stretched across an orthogonal grid of cables. Positive buoyancy for these cables was provided by taut-line buoys, and the whole structure was maintained on station by anchor lines. Such structures should have a built-in system of redundancy and regular diver inspections of the entire installed

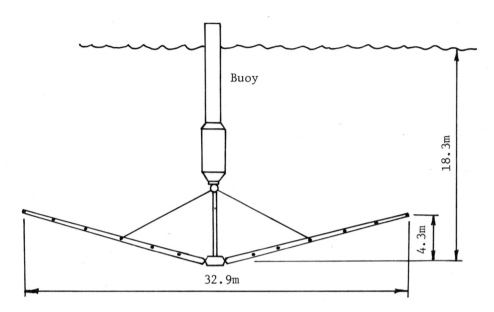

Figure 1. *Schematic of Quarter Acre Kelp Farm Module (QAM) (20).*

system must be considered indispensible. The mooring system may
prove a particularly critical problem for such offshore floating
platforms.
    The most serious difficulty for any offshore farm structure
will probably involve hydrodynamic loads coupled with strength decreases due to corrosion. Hydrodynamic loads come from both currents and waves with the latter the more important in exposed waters
where offshore marine plant farming would likely be done.
    Although farms for marine plants could extend over many square
kilometers of ocean surface, they would be composed of a myriad of
components that would be small with respect to the lengths of design
waves. Drag and inertial force coefficients for such structures
can come only from experimentation. Although wave force coefficients
have been established for very regular, stationary structures such
as the sphere (15) and the cylinder (14,22,28,35,36), irregular
structures and assemblages of elements considered for marine farms
would have to be tested. Although some information on wave-kelp
interaction has been gathered (6), more experimentation is required.
There are two means of proper testing, either using waves in the
sea or using large wave flumes. A small wave flume cannot be used
to test such structures because of the impossibility of satisfying
modelling criteria for both drag and inertia effects at the same
time (27).
    Ocean testing usually involves full prototype sizes (14,33).
However, sometimes a scale model can be constructed and positioned
in a desired location which accurately simulates the design condition.
Distinct problems in these procedures do exist (14,15,33). The logis-
tics of carrying out wave force experiments in the sea is difficult.

and the wave conditions cannot be selected, terminated, or rerun. Furthermore, many operational problems arise, such as watertightness, corrosion, passing ships, inquisitive divers, and marine creatures. Great care and extensive experience are required in carrying out such work successfully.

One way to partially nullify such problems is to do experimental work at large ocean engineering institutions on land. The Wave Research Facility at Oregon State University contains a wave flume 104 m long, 4.6 m deep, and 3.7 m wide. Wave periods from about 1 second to 5 seconds are utilized and wave heights up to 1.7 m can be generated for a 2.4-second wave (35).

A most promising approach will be to combine the large flume and ocean testing technologies. Final proving of design in the sea can be attempted after thorough preliminary laboratory testing of the many alternatives.

SUMMARY

There has been considerable experience in the offshore oil industry and in the U.S. Navy with large, complicated structures placed and successfully operated in the ocean. The construction of offshore marine plant farms composed of many component structures of moderate size and collectively covering a large ocean surface area is probably feasible from the engineering data available. However, such structures must be economical so that societal profit can be realized from the venture. Therefore, careful engineering design must be proposed and tested for economy, reliability, and safety. This calls for the exercise of analytical and experimental tools. It will not be a simple task, and adequate resources must be made available if such a venture is to be a success.

REFERENCES

1. Adee, B. H. 1976. A review of developments and problems in using floating breakwaters. *In:* Proceedings, Eighth Annual Offshore Technology Conference, Houston, Texas, May. Offshore Technology Conference, 6200 N. Central Expressway, Dallas, Texas. 2:225-236.
2. Benjamin, J. R., and C. A. Cornell. 1970. Probability, Statistics and Decision for Civil Engineers. McGraw-Hill Book Co., New York, pp. 271-286.
3. Bretschneider, C. L. 1972. A non-dimensional stationary hurricane wave model. *In:* Preprints, Fourth Annual Offshore Technology Conference, Houston, Texas, May. Offshore Technology Conference, 6200 N. Central Expressway, Dallas, Texas. 1:51-68.
4. Bretschneider, C. L. 1967. Estimating wind driven currents over the continental shelf. Ocean Industry 2(6):45-48.

5. Bretschneider, C. L. 1973. Design hurricane waves for the Island of Oahu, Hawaii, with special application to Sand Island ocean outfall system. Look Lab/ Hawaii 3(2):35-53.
6. Charters, A. C., M. Neushul, and C. Barilotti. 1969. The functional morphology of *Eisenia arborea*. *In:* Margalef, R. (ed.) Proceedings, Sixth International Seaweed Symposium, Madrid, Spain. Dirección General de Pesca Marítama, Ruiz de Alarcón, 1 Madrid, Spain, pp. 89-105.
7. Dalrymple, R. A. 1974. A finite amplitude wave on a linear shear current. J. Geophys. Res. 79:4498-4504.
8. Dean, R. G. 1965. Stream function representation of nonlinear ocean waves. J. Geophys. Res. 70:4561-4572.
9. Dean, R. G. 1974. Evaluation and Development of Water Wave Theories for Engineering Application. U. S. Army Corps of Engineers, Coastal Engineering Research Center. Spec. Rep. No. 1. 2 vols.
10. Deep water, cost-saving platform. 1976. Ocean Industry 11(5):49-51.
11. Dubai's offshore facility: a preview of future systems. 1971. Ocean Industry 6(3):42-43.
12. Ekofisk One becomes an island in the North Sea. 1973. Ocean Industry 8(8):21-24.
13. Garrison, C. J., and V. S. Rao. 1971. Interaction of waves with submerged objects. ASCE J. Water., Harbors, Coastal Engr. Div. 97(WW2):259-277.
14. Grace, R. A., and S. A. Nicinski. 1976. Wave force coefficients from pipeline research in the ocean. *In:* Proceedings, Eighth Annual Offshore Technology Conference, Houston, Offshore Technology Conference, 6200 N. Central Expressway, Dallas, Texas. 3:681-694.
15. Grace, R. A., and G. T. Y. Zee. 1977. Further tests on ocean wave forces on spheres. ASCE J. Water., Port, Coastal, Ocean Div. (in press)
16. Hoerner, S. F. 1965. Fluid Dynamic Drag. Hoerner Fluid Dynamics, Brick Town, New Jersey.
17. Hydro-Lab Journal. 1973-current. The Perry Foundation, Inc., Riviera Beach, Florida.
18. International Nickel Company. 1966. Guidelines for Selection of Marine Materials. International Nickel Company, Seattle, Washington.
19. Ippen, A. T. (ed.) 1966. Estuary and Coastline Hydrodynamics. McGraw-Hill Book Co., Inc., New York, pp. 341-403.
20. James, A. L., and D. W. Murphy. 1976. Kelp Support Substrate Structures for Use in the OFEF Project. Final Report prepared at the U. S. Naval Undersea Center, Kaneohe, Hawaii, for the Ocean Food and Energy Farm Project. 158 pp.
21. Morison, J. R., M. P. O'Brien, J. W. Johnson, and S. A. Schaaf. 1950. The force exerted by surface waves on piles. Am. Inst. Mining, Metallurgical, and Pet. Eng., Pet. Trans. 189(2846):149-154.
22. Nath, J. H., T. Yamamoto, and J. C. Wright. 1976. Wave forces

on pipes near the ocean bottom. *In:* Proceedings, Eighth Annual Offshore Technology Conference, Houston, Texas. Offshore Technology Conference, 2600 N. Central Expressway, Dallas, Texas. 1:741–747.

23. Occidental's North Sea oil terminal gets fixed single-point moorings. 1976. Ocean Industry 11(10):80–81.

24. Ocean Test Structure Technical Plan. 1974. Exxon Production Research Co., Houston, Texas.

25. Pollack, A. 1976. Sea farm experiment under fire. San Diego Union, 28 August: Bl, B5.

26. Quayle, R. G., and Fulbright, D. C. 1975. Extreme wind and wave return periods for the U. S. Coast. Mariners Weather Log 19(2):67–70.

27. Rance, P. J. 1969. The influence of Reynolds number on wave forces. *In:* Proceedings, Symposium on Wave Action. Delft Hydraulics Laboratory, Delft, The Netherlands. 4:13.

28. Sarpkaya, T. 1976. Vortex Shedding and Resistance in Harmonic Flow about Smooth and Rough Circular Cylinders at High Reynolds Numbers. Naval Postgraduate School, Monterey, California, Rep. No. NPS 59SL76021.

29. Sims, C. N. 1976. Sedco 709--First dynamically stationed semi-submersible. Ocean Industry 11(12):30–32.

30. Stockard, D. M. 1976. Effects of pile-soil-water interaction on the dynamic response of a seismically excited offshore structure. *In:* Proceedings, Eighth Annual Offshore Technology Conference, Houston, Texas, May. Offshore Technology Conference, 2600 N. Central Expressway, Dallas, Texas. 3:617–631.

31. Thom, H. C. S. 1971. Asymptotic extreme-value distributions of wave heights in the open ocean. J. Mar. Res. 29(1): 19–27.

32. U. S. Army, Corps of Engineers, Coastal Engineering Research Center. 1973. Shore Protection Manual. Fort Belvoir, Virginia, 3 volumes.

33. Wiegel, R. L., K. E. Beebe, and J. Moon. 1957. Ocean wave forces on circular cylindrical piles. ASCE J. Hydraulics Div. 83(HY2):1199.

34. Wilcox, H. A. 1976. The ocean food and energy farm project. Suppl. Calypso Log 3(2):1–6.

35. Yamamoto, T., and J. H. Nath. 1977. High Reynolds number oscillating flow by cylinders. *In:* Proceedings, Fifteenth International Conference on Coastal Engineering, Honolulu, Hawaii, 1976. American Society of Civil Engineers, 345 E. 47th, New York 10017. (in press)

36. Yamamoto, T., J. H. Nath, and L. S. Slotta. 1974. Wave forces on cylinders near plane boundary. ASCE J. Water., Harbors, Coastal Engr. Div. 100(WW4):345–359.

SELECTED BIBLIOGRAPHY ON MARINE ENGINEERING

Mariculture Experiments and Structures

Bardach, J. E., J. H. Ryther, and W. O. McLarney. 1972. Aquaculture; The Farming and Husbandry of Freshwater and Marine Organisms. John Wiley & Sons, Inc., New York. 868 pp.

Jensen, A. 1971. The nutritive value of seaweed meal for domestic animals. *In:* Nisizawa, K. (ed.) Proceedings, Seventh International Seaweed Symposium, Sapparo, Japan. John Wiley & Sons, New York, pp. 7-14.

Ribakoff, S. B., G. N. Rothwell, and J. A. Hanson. 1974. Platforms and housing for open sea mariculture. *In:* Hanson, J. A. (ed.) Open Sea Mariculture. Dowden, Hutchinson & Ross, Inc., Stroudsburg, Pennsylvania, pp. 295-333.

Bottom-supported, Surface-piercing Structures

Exxon's offshore oil platform nearly doubles water depth record. 1977. Civil. Engr. 47(3):51-54.

Havers, J. A., and F. W. Stubbs, Jr. 1971. Handbook of Heavy Construction. 2nd Ed. McGraw-Hill Book Co., Inc., New York.

Lamb, B. 1971. Long-legged platform walks into ocean job. Construc. Methods Equip. 53(4):81-84.

McClelland, B., J. A. Focht, Jr., and W. J. Emrich. 1967. Problems in design and installation of heavily loaded pipe piles. *In:* Proceedings, Conference on Civil Engineering in the Oceans. American Society of Civil Engineers, 345 E. 47th, New York 10017, pp. 601-634.

Schempf, F. J. 1976. Giant platforms scheduled for deep water. Ocean Industry 11(5):70, 73, 74.

Floating Structures

Largest floating storage barge. 1971. Ocean Industry 6(11):25.

Meals, W. D., *et al.* 1969. Rigging, tackle, and techniques. *In:* Myers, J. J., C. H. Holm, and R. F. McAllister (eds.) Handbook of Ocean and Underwater Engineering. McGraw-Hill Book Co., Inc., New York, pp. 4-32 to 4-90.

Pipeliners will be busy in '76. 1976. Ocean Industry 11(2):107-109.

Roorda, A., and J. J. Vertregt. 1963. Floating Dredges. The Technical Publishing Co., H. Stam N. V., Haarlem, The Netherlands.

Thomson, W. T. 1965. Vibration Theory and Applications. Prentice-Hall, Inc., Englewood Cliffs, New York. 384 pp.

Station Keeping

Bemben, S. M., and E. H. Kalajian. 1969. Vertical holding capacity of marine anchors in sand. *In:* Proceedings, Civil Engineering in the Oceans II, Miami Beach, Florida, December.

American Society of Civil Engineers, 345 E. 47th, New York 10017, pp. 117-136.

Berteaux, H.O. 1976. Buoy Engineering. John Wiley & Sons, Inc., New York. 314 pp.

Chapman, C. F. 1976. Piloting, Seamanship and Small Boat Handling. Motor Boating, New York. 639 pp.

Koster, J. 1974. Digging In of Anchors into the Bottom of the North Sea. Delft Hydraulics Laboratory, Delft, The Netherlands, Pub. No. 129.

Nath, J. H. 1971. Dynamic response of taut lines for buoys. Mar. Tech. Soc. J. 5(4):44-46.

Smith, C. E., T. Yamamoto, and J. H. Nath. 1974. Longitudinal vibration in taut-line moorings. Mar. Tech. Soc. J. 8(5):29-35.

Snyder, A. E. 1969. Winches and deck machinery. *In:* Myers, J. J., C. H. Holm, and R. F. McAllister (eds.) Handbook of Ocean and Underwater Engineering. McGraw-Hill Book Co., Inc., New York. 4-90 to 4-123.

Snyder, R. M. 1969. Buoys and buoy systems. *In:* Myers, J. J., C. H. Holm, and R. F. McAllister (eds.) Handbook of Ocean and Underwater Engineering. McGraw-Hill Book Co., Inc., New York. 9-81 to 9-115.

Submerged, Bottom-mounted Structures

Cayford, J. E. 1966. Underwater Work. Cornell Maritime Press, Inc., Cambridge, Massachusetts. 258 pp.

Danforth, L. J. 1966. The feasibility of an offshore underwater oil-drilling platform. *In:* Proceedings, Offshore Exploration Conference, Long Beach, California. M. J. Richardson, Inc., 2516 Via Tejon, Palos Verdes Estates, California, pp. 159-233.

Goss, W. M. 1970. Big 14,400-ft subsea line pulled. Oil Gas J. 68(47):87-91.

Grace, R. A. 1978. Marine Outfall Systems: Planning, Design and Construction. Prentice-Hall, Inc., Englewood Cliffs, New Jersey.

NEMO--A new undersea observatory. 1966. Undersea Tech. June:39.

OSI unveils underwater production system. 1968. Offshore 28(11):83-92.

Santa Barbara outfall underway. 1976. Calif. Builder Engr. 82(3):18-20.

Submerged production stations. 1968. World Oil July:104.

Subsea completions no longer experiments. 1976. Offshore 36(13):198-201.

Subsea production systems. 1976. Ocean Industry 11(7):39-47, 50-56.

Unique system set for 1000-foot waters. 1968. Oil Gas J. 66(32):94.

Materials, Corrosion, and Fouling

Control of Corrosion on Steel, Fixed Offshore Platforms Associated with Petroleum Production. 1976. National Association of Corrosion Engineers, Houston, Texas, Pub. Standard RP-01-76, April.

Davis, J. G. 1972. Cathodic protection of offshore facilities. World Dredging Mar. Construc. 8(5):14-17, 37.

Davis, J. G., G. L. Doremus, and F. W. Graham. 1972. The influence of environment on corrosion and cathodic protection. J. Pet. Tech. 24(3):323-328.

La Que, F. L. 1968. Materials selection for ocean engineering. *In:* Brahtz, J. F. (ed.) Ocean Engineering. John Wiley & Sons, Inc., New York, pp. 588-632.

La Que, F. L. 1975. Marine Corrosion: Causes and Prevention. John Wiley & Sons, Inc., New York. 332 pp.

Proceedings, Third International Congress on Marine Corrosion and Fouling. 1972. U.S. National Bureau of Standards, Gaithersburg, Maryland, October.

Saroyan, J. R. 1969. Protective coatings. *In:* Myers, J. J., C. H. Holm, and R. F. McAllister (eds.) Handbook of Ocean and Underwater Engineering. McGraw-Hill Book Co., Inc., New York. 7-37 to 7-76.

Tuthill, A. H., and C. M. Schillmoller. 1967. Selection of marine materials. J. Ocean Tech. 2(1):6-36.

Woods Hole Oceanographic Institution. 1952. Marine Fouling and Its Prevention. Prepared for the Bureau of Ships, U.S. Navy Dept. U.S. Naval Institute, Annapolis, Maryland.

Gathering Pertinent Marine Data

Baird, W. F., *et al.* 1971. Canada's wave climate and field measurement program. *In:* Preprints, Third Annual Offshore Technology Conference, Houston, Texas, April. Offshore Technology Conference, 6200 N. Central Expressway, Dallas, Texas. 2:163-170.

Brackett, R. L., and Parisi, A. M. 1975. Hand-held Hydraulic Rock Drill and Seafloor Fasteners for Use by Divers. U.S. Navy, Civil Engineering Laboratory, Port Hueneme, California. TR-824, August.

Dill, R. F., and D. G. Moore. 1965. A diver held vane-shear apparatus. Mar. Geol. 3:323-327.

Emrich, W. J. 1971. Performance study of soil sampler for deep penetration marine borings. *In:* Sampling of Soil and Rock. American Society for Testing and Materials, Philadelphia, Pennsylvania, Spec. Tech. Pub. 483, pp. 30-50.

Flemming, B. W. 1976. Side-scan sonar: a practical guide. Int. Hydrographic Rev. 53(1):65-92.

Fletcher, G. F. A. 1969. Marine-site investigations. *In:* Myers, J. J., C. H. Holm, and R. F. McAllister (eds.) Handbook of Ocean and underwater Engineering. McGraw-Hill Book Co., Inc., New York. 8-18 to 8-31.

Grace, R. A. 1970. How to measure waves. Ocean Industry 5(2):65-69.

Harrison, W., and A. M. Richardson, Jr. 1967. Plate-load tests on sandy marine sediments, Lower Chesapeake Bay. *In:* Richards, A. F. (ed.) Marine Geotechnique. University of Illinois Press, Urbana, pp. 274-290.

Harris, D. L. 1972. Wave estimates for coastal regions. *In:* Swift, D. J. P., D. B. Duane, and O. H. Pilkey (eds.) Shelf Sediment Transport. Dowden, Hutchinson, and Ross, Inc., Stroudsburg, Pennsylvania, pp. 99-125.

Hersey, J. B. 1963. Continuous reflection profiling. Chapter 4. *In:* Hill, M. N. (ed.) The Sea, Vol. 3, The Earth Beneath the Sea. Wiley-Interscience, New York, pp. 47-72.

Hogben, N., and F. E. Lumb. 1967. Ocean Wave Statistics. Her Majesty's Stationery Office, London, England. 263 pp.

Horrer, P. L. 1967. Methods and devices for measuring currents. *In:* Lauff, G. H. (ed.) Estuaries. American Association for the Advancement of Science, Washington, D.C., Pub. No. 83. 757 pp.

McLeod, W. R. 1976. Operations experience with a wave and wind measurement program in the Gulf of Alaska. *In:* Proceedings, Eighth Annual Offshore Technology Conference, Houston, Texas, May. Offshore Technology Conference, 6200 N. Central Expressway, Dallas, Texas. 2:719-733.

Nath, J. H., and F. L. Ramsey. 1976. Probability distributions of breaking wave heights emphasizing the utilization of the JONSWAP spectrum. J. Phys. Oceanogr. 6:316-323.

Noorany, I. 1972. Underwater soil sampling and testing--a state-of-the-art review. *In:* Underwater Soil Sampling, Testing, and Construction Control. American Society for Testing and Materials, Philadelphia, Pennsylvania, Spec. Tech. Pub. 501, pp. 3-41.

Russell, T. L. 1963. A step-type recording wave gage. *In:* Ocean Wave Spectra. Prentice-Hall, Inc., Englewood Cliffs, New Jersey, pp. 251-257.

Scott, R. F. 1970. In-place ocean soil strength by accelerometer. ASCE J. Soil Mech. Found. Div. 96(SM1):199-211.

Smith, D. T., and W. N. Li. 1966. Echo-sounding and seafloor sediments. Mar. Geol. 4:353-364.

Tidal Current Tables 1976, Pacific Coast of North America and Asia. 1975. U.S. Dept. of Commerce, National Oceanic and Atmospheric Administration, Washington, D.C. National Ocean Survey, Rockville, Maryland 20852. 255 pp.

Tide Tables 1976, High and Low Water Predictions, West Coast of North and South America, Including the Hawaiian Islands. 1975. U.S. Dept. of Commerce, National Oceanic and Atmospheric Administration, Washington, D.C. National Ocean Survey, Rockville, Maryland 20852.

van Haagen, R. H. 1969. Oceanographic instrumentation. *In:* Myers, J. J., C. H. Holm, and R. F. McAllister (eds.) Handbook of Ocean and Underwater Engineering. McGraw-Hill Book Co., Inc., New York. 3-57 to 3-105.

Wind and Current Loads on Offshore Structures and Components

Aiston, S. T., and J. H. Nath. 1969. Wind drag coefficients for bluff offshore ocean platforms. *In:* Proceedings, First Annual Offshore Technology Conference, April. Offshore Technology Conference, 6200 N. Central Expressway, Dallas,

Texas. 1:713-720.

Gerrard, J. H. 1961. An experimental investigation of the oscillating lift and drag of a circular cylinder shedding turbulent vortices. J. Fluid Mech. 11:244-255.

Jones, W. T. 1976. On-bottom pipeline stability in steady water currents. *In:* Proceedings, Eighth Annual Offshore Technology Conference, Houston, Texas, May. Offshore Technology Conference, 6200 N. Central Expressway, Dallas, Texas. 2:763-778.

Sainsbury, R. N., and D. King. 1971. The flow induced oscillation of marine structures. *In:* Proceedings. The Institution of Civil Engineers, Great George St., Westminister, London, SW1, England. 49:269-302.

Wave-induced Loads on Offshore Structures and Components

Chowdhury, P. C. 1972. Fluid finite elements for added-mass calculations. Int. Shipbuilding Progr. 19(217):302-309.

Dalrymple, R. A. 1975. Waves and wave forces in the presence of currents. *In:* Proceedings, Civil Engineering in the Oceans III, University of Delaware, Newark. American Society of Civil Engineers, 345 E. 47th, New York 10017, pp. 999-1018.

Dalton, C., and R. A. Helfinstine. 1971. Potential flow past a group of circular cylinders. ASME J. Basic Engr. 93D:636-642.

Dean, R. G., and P. M. Aagaard. 1970. Wave forces: data analysis and engineering calculation method. J. Pet. Tech. 22(3):368-375.

Garrison, C. J., and P. Y. Chow. 1972. Wave forces on submerged bodies. ASCE J. Water., Harbors, Coastal Engr. Div. 98 (WW3):375-392.

Garrison, C. J., F. H. Gehrman, and B. T. Perkinson. 1975. Wave forces on bottom-mounted large-diameter cylinder. ASCE J. Water., Harbors, Coastal Engr. Div. 101(WW4):343-356.

Grace, R. A. 1977. Near bottom water motion under ocean waves. *In:* Proceedings, Fifteenth International Conference on Coastal Engineering, Honolulu, Hawaii, July, 1976. American Society of Civil Engineers, 345 E. 47th, New York 10017. (in press)

Hogben, N., J. Osborne, and R. G. Standing. 1974. Wave loading on offshore structures--theory and experiment. *In:* Proceedings, Symposium on Ocean Engineering at the National Physical Laboratory, London, England. Royal Institution of Naval Architects, London, England, Pap. No. 3.

Hogben, N., and R. G. Standing. 1974. Wave loads on large bodies. *In:* Proceedings, International Symposium on Dynamics of Marine Vehicles and Structures in Waves at University College, London, England. Institution of Mechanical Engineers, London, England, Pap. 26.

Hogben, N., and R. G. Standing. 1975. Experience in computing wave loads on large bodies. *In:* Proceedings, Seventh Annual Offshore Technology Conference, Houston, Texas, May. Offshore Technology Conference, 6200 N. Central Expressway, Dallas, Texas. 1:413-431.

Hudspeth, R. T., R. A. Dalrymple, and R. G. Dean. 1974. Comparison of wave forces computed by linear and stream function methods. *In:* Proceedings, Sixth Annual Offshore Technology Conference, 6200 N. Central Expressway, Dallas, Texas. 2:17-32.

Nath, J. H., and D. R. F. Harleman. 1969. Dynamics of fixed towers in deep water random waves. ASCE J. Water., Harbors Div. 95(WW4):539-556.

Plate, E. J., and J. H. Nath. 1969. Modeling of structures subjected to wind waves. ASCE J. Water., Harbors Div. 95(WW4):491-511.

Silvester, R. 1974. Coastal Engineering. 2 vols. Elsevier Scientific Publishing Co., Amsterdam, The Netherlands.

Wiegel, R. L. 1964. Oceanographic Engineering. Prentice-Hall Book Co., Inc., Englewood Cliffs, New Jersey. 532 pp.

Yamamoto, T., and J. H. Nath. 1976. Hydrodynamic forces on groups of cylinders. *In:* Proceedings, Eighth Annual Offshore Technology Conference, Houston, Texas, May. Offshore Technology Conference, 6200 N. Central Expressway, Dallas, Texas. 1:759-768.

Operations

Aagaard, P. M., and C. P. Besse. 1973. A review of the offshore environment--25 years of progress. J. Pet. Tech. 25(12): 1355-1360.

Bea, R. G. 1971. How sea floor slides affect offshore structures. Oil Gas J. 69(48):88-92.

Bea, R. G., H. A. Bernard, P. Arnold, and E. H. Doyle. 1975. Soil movements and forces developed by wave-induced slides in the Mississippi Delta. J. Pet. Tech. 27(4):500-514.

Blumberg, R. 1966. Introduction--Offshore risks. *In:* Hurricane Symposium. American Society for Oceanography, 1900 "L" St. N.W., Washington, D.C., Pub. No. 1, pp. 294-303.

Christian, J. T., *et al.* 1974. Large diameter underwater pipeline for nuclear power plant designed against soil liquefaction. *In:* Preprints, Sixth Annual Offshore Technology Conference, Houston, Texas, May. Offshore Technology Conference, 6200 N. Central Expressway, Dallas, Texas. 2:597-606.

Cross, R. H. 1974. Hydrographic surveys offshore--error sources. ASCE J. Surveying Mapping Div. 100(SU2):83-93.

Demars, K. R., and D. G. Anderson. 1971. Environmental Factors Affecting the Emplacement of Seafloor Installations. Naval Civil Engineering Laboratory, Port Hueneme, California, Tech. Rep. R744.

Henkel, D. J. 1970. The role of waves in causing submarine landslides. Geotechnique 20(1):75-80.

Ingham, A. E. (ed.) 1975. Sea Surveying. 2 vols. John Wiley & Sons, Inc., New York.

Komar, P. D., and M. C. Miller. 1973. The threshold of sedimentary movement under oscillatory water waves. J. Sediment. Petrol.

43(4):1101-1110.
United States Coast Pilot 7, Pacific Coast, California, Oregon,
     Washington and Hawaii. 1975. 11th Ed. U. S. Dept. of
     Commerce, National Oceanic and Atmospheric Administration,
     Washington, D. C. National Ocean Survey, Rockville, Mary-
     land 20852. 395 pp. + 25 tables.

Engineering Design

Bijker, E. W. 1974. Coastal engineering and offshore loading facili-
     ties. *In:* Proceedings, Fourteenth Coastal Engineering Con-
     ference, Copenhagen, Denmark. American Society of Civil
     Engineers, 345 E. 47th, New York 10017. 1:45-65.
Brown, R. J. 1973. Pipeline design to reduce anchor and fishing
     board damage. ASCE J. Trans. Engr. 99(TE2):199-210.
Clough, R. W., and J. Penzien. 1975. Dynamics of Structures.
     McGraw-Hill Book Co., Inc., New York. 634 pp.
Garrison, C. J., and R. B. Berklite. 1972. Hydrodynamic loads
     induced by earthquakes. *In:* Preprints, Fourth Annual
     Offshore Technology Conference, Houston, Texas, May.
     Offshore Technology Conference, 6200 N. Central Expressway,
     Dallas, Texas, 1:429-442.
Madayag, A. F. 1969. Metal Fatigue: Theory and Design. John Wiley
     & Sons, Inc., New York. 425 pp.
Nath, J. H., and M. P. Felix. 1970. Dynamics of single point moor-
     ing in deep water. ASCE J. Harbors Coastal Engr. Div.
     96(WW4):815-833.
Nath, J. H., and S. Neshyba. 1971. Two point mooring systems for
     a spar buoy. ASCE J. Water., Harbors, Coastal Engr. Div.
     97(WW1):125-135.
Nath, J. H., and T. Yamamoto. 1974. Forces from fluid flow around
     objects. *In:* Proceedings, Fourteenth Coastal Engineering
     Conference, Copenhagen, Denmark, June. American Society of
     Civil Engineers, 345 E. 47th, New York 10017. 3:1808-1827.
Posey, C. J. 1971. Protection of offshore structures against under-
     scour. ASCE J. Hydraulics Div. 97(HY7):1011-1016.
Smith, C. E., T. Yamamoto, and J. H. Nath. 1974. Longitudinal vi-
     brations and tautline moorings. Mar. Tech. Soc. J. 8(5):
     29-35.
Whitaker, T. 1970. The Design of Piled Foundations. Pergamon Press,
     London, England. 188 pp.
Yamamoto, T., J. H. Nath, and C. E. Smith. 1974. Longitudinal
     motions of taut moorings. ASCE J. Water., Harbors, Coastal
     Engr. Div. 100(WW1):35-50.

- 20 -

The Ocean Resource Challenge

EDWARD N. HALL

Traditionally, human history has been divided into Eras. Frequently these have been related to religious events, political transformations, and decisive military actions. In this chapter, to highlight certain critical aspects of today's society, human history will be treated as embracing three Eras: *Pre-Agricultural*, *Careless*, and *Steady State*. During the half million years or so of Pre-Agricultural society, humans were food-gathering creatures organized into wandering tribal groups which interacted only to a very limited extent. Within the tribal groups, primitive techniques employed for food gathering assured extremely low populations which in turn permitted environmental stability through the interplay of natural control phenomena. In such a world the large land area required for the survival of each person held the total human population to less than 10 million.

About ten thousand years ago the first great technical achievement of mankind, agriculture, transformed human societies, allowing stable human communities, occupational specialization, great population increases, a sharp rise in intellectual pursuits, and a profound enhancement of engineering activities. As a consequence, a worldwide human population much in excess of 100 million could be supported.

During all but the last hundred years, man has lacked the skill to ruin the earth or any substantial part of it. The scale of his activities was so miniscule relative to those of natural phenomena that the equilibrium of the biosphere was not threatened. Of the 100 billion tons per year of carbon dioxide produced by oxidation of organic matter on land, man's contribution was insignificant. Moreover, the greatest cities built during these thousands of years rarely exceeded 50,000, or about 5,000 persons per square mile. The lack of engineering skills essential to the erection of high buildings, the distribution of water, the collection

and disposition of waste, telecommunications, and rapid transportation prevented ill-considered and excessive urban growth. A most
significant indicator of mankind's gentle impact on the established
equilibria of the earth during most of the first two eras of human
history is the magnitude of energy transformation which prevailed.
To the middle of the 18th century, mean energy conversion rates at
a tenth of a kilowatt per person were rarely exceeded by any society.
At a population approximating 100 million, this could amount to no
more than a total of 10^7 kilowatts worldwide, an insignificant fraction of the 2×10^{14} kilowatts of solar flux impinging on this planet.
Limited by technical ignorance, the social subsystems of mankind
during the Pre-Agricultural Era and the prevailing Careless Era
have remained intensely parochial. Nations and cities contended
with supply and disposal problems and ignored the impacts of such
activities on their global neighbors. The long and successful tradition of careless conduct, plus the social, economic, and ethical
values engendered by it, constitutes one of the greatest threats to
the perpetuation of the human species on this earth today.

The Careless Era a long transient bridging the Pre-Agricultural
and the coming Steady State Era, is drawing to a close. Steady State
is the best humanity can look forward to. By the end of the 21st
century, barring an ecological catastrophe, this may be approached
with a world population of about 20 billion and energy conversion
needs and aspirations approximating 6×10^{18} BTU annually--about 16
times today's figure. Food requirements can be expected to rise by
a factor of 5 to 10 and demand for non-fuel minerals by something less.
Satisfaction of energy needs appears most pressing, because success
in generating more energy can facilitate satisfaction of burgeoning
food and mineral aspirations. Failure may well preclude such an
achievement completely.

Few alternative energy conversion technologies capable of assuming a substantial percentage of the burden currently borne by fossil
fuels can be perceived. Each has attractive and unattractive aspects.
All valid candidates must be pursued to maximize the probability of
success in meeting the unprecedentedly massive demands for new energy.
The time remaining for this accomplishment is short--less than 50
years if major social disruption is to be averted. This chapter is
devoted to an exploration of one potential new technology--the use of
energy-rich raw materials in benthic marine algae on an unprecedented
scale. The enormity of the biomass crop needed for the new technology and the magnitude of the area needed to support it direct attention to the ocean above the continental shelves. Moreover, success
in exploiting the continental shelves would be one means of redressing
the consequences of man's unbalanced activities on land and sea.

Proposals for alternatives to the energy conversion system of the
world, which is supported predominantly by fossil fuels today, are
bedeviled by unrealistic accounting procedures. Improper allowances
for depletion associated with fossil fuels render alternative systems,
for which true depletion or replacement costs are included, economically non-competitive at this time. To contend with this apparent
handicap, huge subsidies of extended duration will be required, or

the products of the renewable systems will have to be distributed through more commercially valuable channels than fuel alone for several decades to come.

THE PLANETARY RESOURCE INVENTORY

The development of agriculture started mankind on a long path toward population growth and affluence. The slope of the growth transient was increased sharply during the 18th and 19th centuries with more efficient chemical-to-mechanical energy transformations, shifts to electrical energy, and accelerated exploitation of massive quantities of fossil fuels. Fossil-fueled engines gave man the means to disturb naturally controlled equilibria that had prevailed since the earth was young. Today, the ten-thousand-year transient which comprises what I would name the Careless Era has nearly bridged the period of exponential growth and inevitably will soon shift into an Era best called the Steady State.

Earth is a bounded region in the universe. Within it, nothing can increase exponentially indefinitely. Human populations must ultimately be limited by the critical materials found on earth of which they are composed. Energy transformation rates similarly must be limited by the heat transport properties of the atmosphere, the effective solar flux reaching the surface of the planet, and the ability to maintain tolerable surface temperatures. Such open-ended activities as mineral and fuel extraction and waste disposal must become closed cycle activities regulated to achieve an equilibrium. For example, Alvin Weinberg of Oak Ridge estimated that the phosphorus in the crust of the earth is sufficient to sustain no more than 20 billion people at one time! Phosphorus may not turn out to be the limiting element for man, and Weinberg's estimate could be wrong by as much as several hundred percent still without altering the pattern of mankind's future significantly. The choice is a Steady State society or disaster as the limits of resources are reached.

To explore the potential of a Steady State world, Weinberg's figure has been reduced to 15 billion as a concession to the other forms of life which might be allowed to exist with humanity. If this number is correct within broad limits, the next century should witness the greatest deceleration of population growth, plus the most rapid acceleration of energy conversion rates and mineral extraction in history. Such a scenario is more likely than the false perspectives derived from open-ended exponential figures so commonly cited today regarding growth of population, energy transformation, and waste disposal, as well as food and mineral consumption. Fig. 1 shows an idealized pattern of exploitation and population growth.

Recognition that critical activities are not accelerating at rates indicated by open exponentials but are changing into a steady state pattern requires that plans be made for the future rather than to simply regard it with dismay. The general character of a stable human society varies little in models even with wide variations in key parameters. A crude model based largely on the postulated chemical composition of the earth's crust, on trends of energy consumption by the

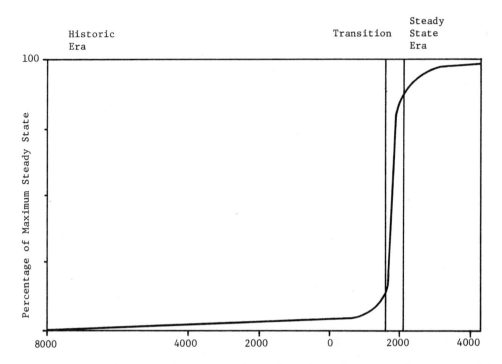

Figure 1. *An idealized perspective plotting the concomittant world-wide increases in population, mineral extraction, fuel extraction, and solid waste disposal from the beginnings of recorded time to the present and extrapolated to the year 4000. The open-ended exponential growth of population and exploitation of resources cannot continue beyond the brief, current transitional or Careless Era.*

wealthy, and on the observed rates of saturation of American society by energy-transforming gadgetry, is displayed in Table I.

STRATEGY FOR A STEADY STATE INVENTORY

If satisfaction of food, mineral, and energy requirements of the human population moving into the Steady State Era can be regarded as the most important objective confronting mankind today, attention should be directed toward supplying energy requirements first. The model anticipates that population will increase by a factor of 4. Global per capital energy conversion requirements also seem to be rising by a factor of 4. Thus, worldwide energy demands can be expected to rise by a factor of 16. While the direct effect of population increase can be expected to be the same on food and minerals, it is probable that per capita increases will be less drastic.

Increased food requirements should be partially met by diversified and improved food quality and distribution. Improvements in plant food taste, texture, and appearance can be expected to reduce

Table I. *Possible equilibrium parameters for a future Steady State Earth and the United States of America in 2100 A.D.*

EARTH

Population -- 15 Billion

Power Consumption @ 13 kw per capita = 2×10^{11} kw

Total Daily Solar Flux = 2×10^{14} kw

Total Annual Energy Conversion Requirement = 6×10^{18} BTU

UNITED STATES OF AMERICA

Population -- 500 Million

Power Consumption @ 13 kw per capita = 6.7×10^9 kw

Total Annual Energy Conversion Requirement = 2×10^{17} BTU

---

dependence on animal foods. As food chemistry creates vegetable products which cannot be distinguished from meat, meat production may diminish, decreasing plant crop requirements. The conversion of cellulosic materials into food without the relatively inefficient conversion by metabolism of domesticated animals should alleviate further the problem of increasing the world's food crops. The greater diversity of diet which can be expected if poverty is reduced should lead, moreover, to a more effective use of protein foods by humans as amino acid profiles more closely approach metabolic requirements. Food production will still have to increase 5 to 10 times to ensure a nutritionally satisfied worldwide society during the Steady State Era. This will require substantial increases in rates of worldwide energy transformation to support the necessary intensive agricultural activities. A substantial energy increase will be essential to meet future food needs.

Predicting the world's essential mineral needs and aspirations is more difficult. U.S. Geologic Survey estimates indicate exhaustion of reserves for many of our essential minerals including ores for iron, copper, zinc, and lead within 30 to 50 years. However, the significance of the term "reserve" must be recognized. Reserves are generally considered to be masses of minerals which can be recovered economically using today's techniques. In almost all cases, the amount of such material available through alternate techniques, *frequently requiring higher energy expenditure*, can be greatly increased. Substitution of plastics, glasses, and special fibers can lead to a marked reduction in projected demand for conventional critical materials.

Despite the favorable factors, failure to act promptly to alleviate massive shortages in food, minerals, and energy could result in utter chaos. Fig. 1 indicates the unprecedented magnitude of the undertakings required and the limited time available.

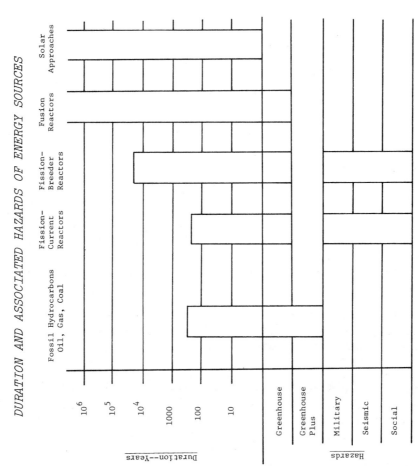

Figure 2. *An idealized conception of the duration of energy sources available for earth in a future Steady State Equilibrium. The hazards of each source are summarized in non-quantitative bars. Tides, geothermal, wind, and ocean thermal sources require further assessment. The population is extrapolated to be 15 x 10⁹; the power consumption to be 13 kw per capita; and the total power demand to be 2 x 10¹¹ kw.*

Failure to meet mounting energy requirements will lead inevitably to a deficiency in food and mineral requirements as well. The inexorable and rapid deterioration of the current world's major energy resource, fossil fuels, will make energy recovery from that source increasingly expensive. The task of changing the world's rapidly expanding energy conversion system from reliance upon fossil fuels to some other energy source is so enormous that vigorous efforts to satisfy the massive energy requirements of the future must be undertaken *immediately*.

MODES OF ENERGY GENERATION

Fig. 2 is an effort to portray, in simple terms, the pros and cons of the major modes by which man may satisfy his energy aspirations in the future. The most optimistic picture one can draw from geology is that the world can expect little more than another two centuries of primary reliance on fossil fuels (Fig. 2). Even this assumes that polar deposits of gas and oil will prove much larger than currently known and will be economically recoverable, that more coal will be recovered from future deposits than in the past, that tertiary oil recovery will be economically feasible on a massive scale, and that the enormous deposits of oil shales and sands known to exist will become economically exploitable. A less optimistic prediction is that known fossil fuel deposits will last 50 to 75 years (2,7). Furthermore, continued heavy reliance on combustion of fossil fuels will expose human society to the hazards of *"greenhouse"* and *"greenhouse-plus."* Greenhouse, here, refers to an increased surface temperature of the earth resulting from fossil fuel combustion. The relative opacity of the carbon dioxide in the atmosphere to transmission of infra-red radiation in the frequency range of the earth's albedo is largely responsible. Greenhouse-plus results from combustion of the enormous amounts of fossil fuels at an accelerated rate. This will increase the concentration of carbon dioxide in the earth's atmosphere and thus further exacerbate the initial greenhouse effect. Ignorance of long-term solar changes and the complexity of the earth's atmospheric system preclude any estimate of how rapidly the earth's surface temperature would rise, but rise it almost certainly will.

The United States and 41 other nations are now heavily committed to light-water-fission-nuclear power plants (5). In this country, 66 nuclear plants are licensed to operate with a capacity of approximately 42 gigawatts, somewhat more than 8% of the nation's total electrical capacity. The United States is apparently committed to building 228 of these stations with a combined capacity of 226 gigawatts. Commitment by the other nations is of the same total magnitude.

The massive commitment to light-water-fission-nuclear systems should not be construed to indicate that a single adequate replacement for fossil fuel combustion has been identified. As indicated (Fig. 2), uranium fuels for these systems promise to last about as long as hydrocarbons. (These numbers are disputed, with some groups claiming that economic uranium supplies can be expected to last beyond a thousand years.) The greenhouse effect due to increased energy conversion rates would equal that due to prospective hydrocarbon operations. However, the adverse effects of greenhouse-plus would be absent, since atmospheric carbon dioxide would not be increased. Additional hazards associated with such systems can be perceived as indicated. The relatively simple separation of plutonium from spent fuel of such power plants could lead directly to widespread nuclear arms proliferation. The full effects of seismic and other poorly defined stresses on the integrity of containment vessels and emergency shutdown systems are not known. The security of nuclear materials at plant sites and during transit is endangered by the risk of would-be highjackers and traffic accidents.

The fission breeder (Fig. 2) refers to approaches similar to the fast-neutron-liquid-metal system heavily stressed in United States energy planning during the last 10 years. Because of the much improved neutron economy of this system as contrasted with light-water-fission devices, electrical energy output per pound of uranium or thorium should be improved many fold. Consequently, reliance on such systems could provide man with adequate energy resources for perhaps ten thousand years. Unfortunately, the deficiencies associated with the simple fission systems pertain as well to the breeder reactors under consideration. Moreover, the amount of radioactive material which will have to be transported annually will increase materially, further aggravating the social hazard.

Fusion is a most attractive alternative (4). If one assumes that the cycle can be operated by a deuterium/deuterium reaction with the raw materials derived from sea water, the energy demand postulated in the model could be supported into the indefinite future. The only obvious hazard would be the greenhouse effect which could cause a rise in the earth's temperature and lead to the melting of the polar ice caps. Despite clear definition of postulated data and mechanisms, no controlled useful energy has been developed by a fusion system. The mechanisms proposed imply that the smallest units capable of fully demonstrating feasibility would be 3 gigawatts or larger, about three times that of the largest of today's central power stations. The necessity of performing experimental work at this level constitutes a significant handicap. Table II shows some problems currently associated with nuclear systems.

This discussion of nuclear energy conversion systems is not intended to inhibit efforts to develop such systems. The stark picture portrayed in Fig. 1 and Table I suggests that neglect of any reasonably promising approach to satisfying the world's energy demands would be ill-advised. However, limitation of the nation's efforts to solve its prospective energy problems to nuclear options alone would be equally ill-advised.

SOLAR ENERGY--A PROBLEM OF ENTRAPMENT

Solar energy represents a promising partial alternative to a future nuclear power-driven world. Conversion of solar flux to biomass to fuel is one approach requiring no quantum improvement in scientific or engineering skill to allow economic implementation. Major

Table II. *Consideration of the impact of the attainment of the Lawson criterion on fission and fusion reactors*[1,2].

Proposed System	Thermal Efficiency Cooling Water Requirements	Neutron Economy Insuff. Reserves	High Capital Costs	Radioactive Hazards--Siting & Construction Delay	Disposition of Radioact. Waste	Difficult Development
Light Water Reactor	2	3	2	3(f)	2	0
High Temperature Gas Reactor	0	3	1(a)	3(f)	2	0
Liquid-Metal Fast-Breeder Reactor	1	0(c)	2(a)	3(f)	2	2(b)
High-Temperature Gas-Cooled Breeder-Reactor	0	0(c)	2	3(f)	2	1
Fusion Reactor	0	0(d)	?	0(g)	0	3(e)

[1]Figures indicate magnitude of engineering problems as follows: 0-None, 1-Significant, 2-Severe, 3-Critical.
[2]Letters indicate remarks as follows: (a) For High Temperature Gas Reactor

	Capital Cost	Kwhr Cost
Steam Cycle	$330/kw	8.5 mills
Closed Brayton	275/kw	7.6 mills

(b) sea wolf problems remain;   (c) rate of doubling is critically low from 7-15 years;   (d) rate of doubling is excellent--about 1 month;   (e) could improve greatly within 8 years;   (f) reactor inventory of 2-10 tons of hazardous radioactive material;   (g) ¼ gram of radioactive material in reactor.

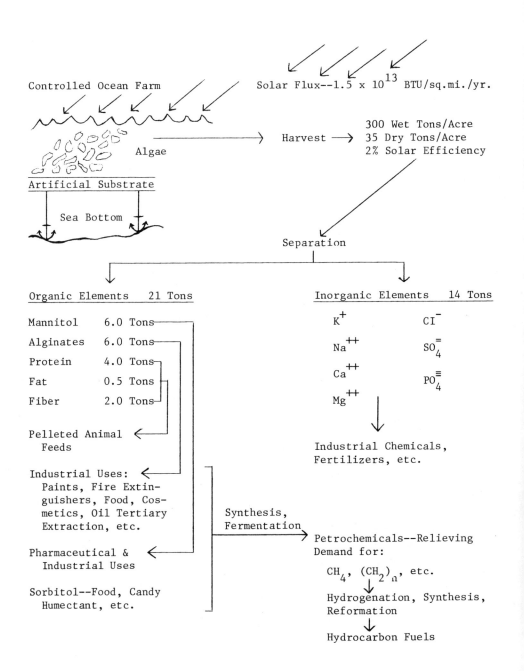

Figure 3. *A conceptual flow chart with some approximate parameters relating to the cultivation and processing of a marine plant biomass created by controlled mariculture.*

elements of one such system are illustrated in Fig. 3. The fuels so produced promise continuing exploitation of the enormous investment in infrastructure associated with energy conversion, transmission, and use.

The scale of the undertaking required to exploit the solar flux is reflected in some simple figures. If we assume a world population of approximately 20 billion and a daily per capita energy conversion rate of 13 kilowatts (the present USA rate is 10 kilowatts), world requirements will amount to 2.7×10^{11} kilowatts or 8×10^{18} BTU annually. While this is an immense figure, it constitutes less than .2% of the solar flux (2×10^{14} kilowatts) impinging on this planet. One must assume that man's ingenuity is sufficient to cope with such favorable terms, even though massive areas of the earth may be required for entrapment because of the diffuse nature of this radiation. If the population of the United States is projected to be 500 million, at the per capita consumption equal to that assumed for the rest of the world, our national energy requirement will be 2×10^{17} BTU per year, or about three times the present value.

Estimating the mean insolation on the United States to be 1.5×10^{13} BTU per square mile per year, at 100% conversion and transmission efficiency, the area to satisfy the projected gross requirement would be 15,000 square miles. Assuming that a net efficiency of 20% can be realized by photovoltaic devices, the area needed would be 75,000 square miles or roughly that of a medium-sized state such as Nebraska. Capital, maintenance, and operating costs of such a system would be substantial. Alternatively, the use of green plants to convert solar into chemical energy with a comprehensive efficiency (sun to plant to fuel) of 1% would require 1.5 million square miles to meet the nation's needs. Improvements in cultivation, processing, and genetics could perhaps raise this to 5%, with a reduction in the area needed to .3 million square miles. It should be noted, moreover, that the energy developed by this approach would be chemical and take the form of hydrocarbon fuels. Capital and operating costs of converting such energy to forms desired by mankind have traditionally been a small faction of those associated with electrical systems. For this approach, capital and operating costs associated with the primary conversion element, the farm, which is critically massive in terms of area, could be expected to be relatively low. Nevertheless, in the USA such an area would constitute 20% to 100% of all the land now devoted to agriculture.

The massiveness of such an undertaking can be appreciated by comparing the tonnages of hydrocarbons prospectively required by the United States to the tonnage of the world's major crops today. Currently, the sum of the world's crops of wheat, rice, corn, potatoes, and barley is 1.5 billion tons. This constitutes about 60% of the world's total food for human consumption. To support the projected energy requirements of the United States alone, more than 5 billion tons of hydrocarbon would be needed. Obviously, greatly improved agricultural techniques, dramatically enlarged areas of cultivation, or both, would be required.

The enormity of the annual biomass crop required to assume the energy burden currently borne by fossil fuels demands that all

promising approaches be investigated. Five sources of fuel are
apparent: 1) conversion of agricultural wastes, 2) conversion of
urban wastes, 3) fresh water plants, 4) terrestrial plants, and
5) marine plants. The attractiveness of the first three, enhanced
by their providing solutions to pressing pollution problems, is dimmed
by the fact that they can satisfy but a small proportion of the world's
future energy needs. The fourth candidate is promising and significant, but limited by competition with agriculture for arable land.
The fifth approach, cultivation and conversion of marine plants to
convenient fuels, though limited at present by lack of specific knowledge can promise not merely to help satisfy burgeoning energy requirements, but to alleviate difficult problems stemming from food and
mineral shortages, as well as pollution.

Great increases in human population, as well as growing per capita rates of energy exchange, during the 20th century have distorted
long-term worldwide distribution of key materials. Typically, fertilizers, essential minerals, and organic residues, extracted or cultivated, are leached out of land areas and deposited in the sea.
Consequently, great areas of ocean are becoming contaminated and
terrestrial resources of essential raw materials are being depleted.
Cultivation, harvesting, and processing of selected marine plants
appears the most readily achievable means of redressing this balance,
since it has been demonstrated that marine organisms concentrate
within their tissues many essential materials found in huge and growing quantities, but at totally uneconomic concentrations, in the
world's oceans.

The massive diversion of arable land to energy generation through
biomass could be reduced if a substantial part of the new crop were
raised in the sea. Moreover, sea farming as contrasted with land
farming guarantees constant availability of the major plant nutrient--
water. On many areas of the extensive continental shelves of the
United States, because of natural upwelling, phosphate is relatively
abundant, potash is always in excess, and much of the nitrate requirement is present a good part of the year. For this reason as well as
the ease of anchoring and the minimization of transport and harvesting
costs, farm sites on continental shelves should be sought as a first
priority. Deficiencies of nutrients in the areas under cultivation
could be compensated when necessary by fertilizing with nitrates and
phosphates recovered from processing operations. Past experience
indicates that over extended periods, potash and phosphates may exceed
needs and could be marketed for terrestrial crops. In areas of nitrate
deficiency, incremental needs could be met by a Haber process conversion of a small fraction of the harvested crop.

Sea farming could alleviate mankind's perpetual need for essential
fertilizers--nitrates, phosphates, and potash--without which terrestrial agriculture would decline to the limited yields per acre common
in the 19th century.

Continued heavy reliance on fossil fuels combustion will invite
inadvertent, unfavorable climatic changes caused by the greenhouse
effect. Heavy dependence upon fission or fusion will invite the same
problems by increasing the heat transport requirement through the

atmosphere. Dependence upon contemporaneous photosynthesis to meet man's need for energy, however, should have little or no gross effect on the earth's surface temperature since heat released from combustion of plant products will balance the solar energy absorbed by the plants. Carbon dioxide released by combustion of photosynthetic products or their derivatives will not increase the carbon dioxide in the atmosphere, since roughly equal amounts will be continually synthesized into new fuels.

CONVERTING TO A RELIANCE ON BIOMASS

Conversion of this nation's energy system from primary reliance on fossil fuels to one capable of satisfying human aspirations indefinitely will be a massive undertaking, which, to avoid persistent and crippling dislocations, should be accomplished in the next half century. Unless new promising approaches are investigated, evaluated objectively, and vigorously pursued, the advantages of our early appreciation of urgency will be lost and disaster will result. In this time of technical uncertainty, excessive support for any single approach to the detriment of viable alternatives must be avoided.

The problem of conversion is further complicated by the questionable bookkeeping procedures of the fossil fuel age. Maintenance of a stable society over extended periods requires that obsolescence, attrition, and depletion be accounted for appropriately. Despite the routine (and occasionally notorious) appearance of depletion allowances on the balance sheets of hydrocarbon extraction companies, the term as used is not consistent with good accounting practice. Appropriate sinking funds permit construction of new buildings when deterioration has rendered old structures uneconomic. The depletion allowance for the fossil fuel extractor, however, permits the write-off of the value of the real estate, the tools with which extraction has been accomplished, and the labor involved, but does not cover renewal of the resource. If the mean time required by nature to reproduce the resource of coal, oil, or gas at the site of extraction were assumed to be only 20,000 years, which is much lower than actual, an investment in that site at the start of natural production, of $1.00 at 1% interest compounded annually, would be worth 2.68×10^{86} dollars at the end of the regeneration period. Even if the process of oil, gas, and coal formation were continuing today at the same rate and at the same sites as in past eons, depletion allowances would have to be astronomical to ensure continuing availability of these fuels. In short, the price we pay for naturally occurring hydrocarbons does not include a true and full allowance for depletion and renewal but is maintained at a level far below its worth as a continually available good.

This anomaly exercises a profound influence on the course the nation must follow during the next half century in establishing truly renewable energy resources. The price of energy derived from such a system will reflect the actual costs of depletion. It cannot be expected to be competitive with naturally occurring hydrocarbons. Unless massive efforts are initiated now to develop such renewable resources, however, there will be an interim period of vast disruption during which adequate supplies of reasonably priced energy will be available neither from new undeveloped systems nor from the old non-renewable natural sources. This dilemma can be resolved in two general ways. Vast subsidies can be provided over several decades permitting the development and growth of new systems until they are competitive with naturally occurring hydrocarbons. Or, new energy systems can be developed, which for several decades will be oriented primarily toward production of more valuable products than energy. Then, as capital and operating costs drop during the transition period with increasingly larger-scale efforts and experience and as the costs of natural hydrocarbons rise, a relatively painless transition may be effected. Ocean farming offers the potential for non-subsidized development, coupled with a steadily increasing competence to pro-duce massive energy resources when required. It also offers 1) a reasonable probability of economic self-sufficiency, 2) the potential for exploiting huge new areas without infringing on regions currently devoted to conventional agriculture, and 3) more freedom from the constraints imposed by latitude and altitude when engaging in the cultivation of terrestrial crops.

Because the carbohydrates involved in the metabolic and struc-tural systems of red and brown algae are easily extractable as mater-ials of substantial economic value, sale of these and their deriva-tives (Fig. 3) could contribute significantly to support of ocean farm development during the protracted interval before fuels derived from them would be competitive. The alginic acid and related chemi-cal structures of brown and red algae are similar to cellulose, the major constituent of terrestrial vegetation. The linear polymetric structure of cellulose, its close association with chemically in-tractable materials, and its high molecular weight render it awkward and expensive as a raw material for many chemical syntheses. More-over, alcohol sugars, the common stored form of carbohydrate in many algae, are about as easy to extract and purify as the starch typically stored by most terrestrial plants. For those terrestrial plants such as sugar cane and sugar beet, in which the stored carbohydrate takes the form of mono and di rather than polysaccharides, extraction costs are lower, but market values of the alcohol sugars and alginates sub-stantially exceed those of the conventional sugars and starches.

The initial development of sea farming for products of higher value than energy involves a lapse of time before economic competi-tiveness with other forms of energy can be assured. This deferred energy approach may appear strange and unattractive. However, during the extended interval of energy non-competitiveness, the increasing substitution of biomass for natural gas and petroleum as a petro-chemical feed stock should progressively release conventional fuel to help meet the nation's energy needs. Moreover, any resolution of the energy dilemma without recourse to such an approach is difficult to conceive. It seems probable that ocean-farm energy-conversion procedures will become competitive sooner than fusion systems. Few alternative systems, however, are likely to defray many or perhaps all interim development expenses during the period before energy competitiveness has been achieved.

THE NEW MARICULTURE

Although men have for millennia cultivated and harvested certain algal crops, largely for fertilizers and feeds, the biology of benthic intertidal and subtidal plants, coupled with the geometry of the earth's continental shelves, assures that crops will be inadequate to meet human needs, unless radically new cultivation techniques are developed. If substantial contributions are to be made to human society, marine crops will have to be raised on carefully designed substrates in areas of the sea where depths are too great to permit reproduction and growth of selected plants (8). Cultivation of crops on such substrates constitutes a profound departure from all previous mariculture experience and will require considerable development and analytical effort.

Conversion of marine crops into useful products will involve extensive chemical transformation. Techniques frequently proposed on the basis of ancient tradition and practice characteristic of underdeveloped societies, such as those of India and China, generally involve aerobic or anaerobic fermentation (6). While such procedures are appropriate when applied to heterogeneous or uncontrolled materials, such as urban waste, direct chemical synthesis as stressed by the petrochemical industry would seem preferable where feasible. Reaction dwell times, and consequently land areas and capital costs associated with processing, are greatly reduced by such an approach (1,3). This is particularly appropriate when large concentrations of pertinent organic products and raw materials can be extracted from plants in a relatively pure state at reasonable cost.

A development strategy leading to commercial exploitation of marine crops may now be described. Promising candidates for cultivation should be selected on the basis of nutritional requirements, potential symbiosis, solar energy conversion efficiency, harvestability, toughness, and chemical composition. Sites can then be identified along the continental shelves at which established upwellings provide a flow of nutrients approximating the needs of the selected plants. Design, development, and tests of substrates capable of withstanding severe storm conditions at selected sites on the continental shelves must be pursued vigorously. Achieving successful designs combining mechanical adequacy with extremely low long-term costs is as necessary as identifying ideal plants and culture sites. A full knowledge of the mechanical properties of the plants selected, to permit analysis of interactions between water, substrate, and plants, also will be essential. Inevitable transfers of water-borne spores, as well as hazards associated with disease, predation, and toxin susceptibility, make unlikely the dependence upon only one species in the plantations. Experimentation with several species on several experimental substrates should indicate appropriate mixes of species. An effective program aimed at commercial exploitation would therefore include research tasks covering analysis and design of substrates, cultivation tests, harvesting, processing, product definition, marketing, and studies of legal and societal considerations.

Task 1 -- Cultivation

Cultivation research will require emphasis on biology, species selection, hydrodynamics, and plantation siting. Promising subtidal and intertidal species capable of remaining totally submerged indefinitely should be thoroughly examined. Programs of hybridization, selection, and genetic engineering should be established. The following criteria should be included in species selection:

1. *Growth Rates*. The projected mass of plant that can be harvested annually per acre should be known. This would be expressed as a function of light intensity, temperature, nutritional conditions, plantation configuration, age, and other pertinent parameters.

2. *Nutritional Requirements*. Survival, growth, and quality in terms of concentration of specific constituents should be emphasized. Effectiveness of appropriate compounds of nitrogen, phosphorus and minor and trace materials must be examined under varying environmental conditions. Plant nutrient uptake as related to age, harvest cycle pattern, nutrient concentration, temperature, flow and turbulence, light intensity, composition and concentration of toxins, and plant condition should be understood. Trace elements must be included in this analysis.

3. *Pathology*. Procedures to understand relevant diseases and their frequency and symptoms as affected by significant parameters include concentrations of disease organisms, responses of the plant, temperature effects, nutritional and age effects, as well as disease response to light intensity.

4. *Toxin Susceptibility*. Vulnerability to toxins, such as heavy metal ions, and pollutants must be examined. Means of contending with these by chelation, genetic selection, and hybridization should be examined.

5. *Implantability*. The establishment of colonies of plants on artificial substrates must be studied. The time and effort required for establishment and maturation of plantations should be estimated and assessed.

6. *Harvest Analysis*. Based on nutrition, illumination, temperature, age, season, and other significant variables, a harvest strategy must be established. Investigation of procedures to assure minimum loss of valuable products during and immediately following harvesting should be pursued.

7. *Predation*. Identification of predatory organisms and potential crop damage should be related to environmental conditions, harvesting cycle, plant age, and health. Measures to contend with predators should be explored.

8. *Long-Term Survivability of Plantation*. Such qualities as reproducibility, productive life expectancy, deterioration of productivity with age, duration of juvenile phases, and symbiosis must be considered.

9. *Mechanical Survivability*. The limits of mechanical stress which the plant can bear must be explored. This research should include efforts to determine tensile strength, modulus, and moments of inertia on plant parts in various excitation modes. Plastic

plant simulators should be fabricated which will then be subjected
to a spectrum of storm conditions while associated with experimental
substrates in a facility like the Environmental Fluid Dynamics
Research Facility at Oregon State University. Responses to a broad
spectrum of storm conditions which might require years of exploration to determine under natural conditions could be explored in
months. The effect of variations in substrate design, depth, density of planting, patterns and intensity of turbulence and wave
action, and rates and gradients of currents on survivability should
be estimated.

10. *Coastal Compatibility*. Compatibility with shipping, fishing, and recreation should be known. The design and depth of substrates, the tendencies of structures to foul beaches, and aesthetic
methods of exercising effective surveillance must be examined from
the point of view of both the biologist and the social ecologist.

11. *Plantation Location*. Locations of plantations should be
determined from studies of coastal shelf extent and bottom characteristics, nutrient availability, storm frequency and intensity,
currents, temperatures, anchorability, climatic effects, and the
effect on fishing, recreation, and navigation. Climatic effects
refers to inadvertent climate changes which might be induced by the
presence of extensive marine plantations at the selected sites.

12. *Hydrodynamics*. Studies will focus on the ability of artificial substrates to survive under wide ranges of anticipated sea
states and to accommodate plant requirements over long periods.
Diverse construction materials and designs must be thoroughly explored. Strength, durability, densities, anchoring problems, and
costs should be studied. Structural stresses should be analyzed
on various substrate designs under varying current and storm conditions. Conclusions should be confirmed by testing with the appropriate technique of fluid dynamics. Plant losses associated with
stresses such as chafing or tearing related to substrate designs,
planting patterns, and storm conditions should be estimated. Tentative conclusions should be tested by plant simulators attached to
experimental model substrates in a fluid dynamics laboratory. The
degree of turbulence attenuation in specified plantation areas
should be estimated and tested. Flow patterns through various proposed plantation configurations should be explored. Effectiveness
of various plantation designs as functions of external currents,
macro and micro turbulence, and varying nutrient distributions
should be tested. Nutrient concentration profiles should be monitored in three dimensions. Similar investigations should determine
the flow and absorption of nutrients as affected by the manner of
introduction, turbulence, temperature, density differentials, and
the composition and concentration of nutrient supplements. Anchoring and depth stabilizing techniques must be explored in the laboratory as well as under actual operating conditions. Methods of
chemical and biological monitoring of plantation waters should be
surveyed and tested. Nutrient introduction techniques should be
explored.

Task 2 -- Harvesting

Harvesting techniques should be developed and tested to establish costs and equipment requirements, and to obtain design criteria for specialized equipment to be used at sea and along the coast.

1. *Yields.* Maximum harvest yields should be established as a function of planting density, plantation configuration, and cutting time, frequency, and depth.

2. *Handling.* Investigations should be conducted in crop handling and treatment during harvesting to minimize transport of sea water and loss of economically valuable products. Maximum carbohydrate and protein yield should be sought. Early development and test of specialized harvesting equipment on natural beds of selected plants should be undertaken.

3. *Equipment.* Harvesting equipment and techniques should make maximum use of expensive machinery. Site selection and harvester designs should permit year-round use of such equipment by concentrating on different geographic areas as seasonal conditions permit.

4. *Weather.* Conditions of weather must be correlated with growth to schedule operations ensuring both high yield and maximum safety.

Task 3 -- Processing and Marketing

Processing and marketing are based on the composition of plants harvested, the spectra of potentially marketable products, and the current value per acre of selected mixes of major components and their derivatives at prevailing prices. The elasticity and saturability of markets for these products should be determined. Marketable derivatives and chemical intermediates associated with these products are illustrated in Fig. 3. This is but one example of an almost infinite number which can be postulated.

1. *Product Mixes.* From such data, a series of product mixes composed of feeds, fuels, fertilizers, and specialty chemicals should be developed to increase income progressively in the future.

2. *Market and Product Shift.* Current study strongly indicates that early farms will be operated most profitably by maximum production of specialty pharmaceuticals and chemicals such as mannitol, alginates, and their derivatives. As the markets for such materials become more nearly saturated, a shift to straight chemical synthesis where possible in order to yield higher percentages of feeds, fuels, and industrial chemicals will become more attractive. Early profitable operation of ocean farms should thus be possible. As the cost of natural hydrocarbons increases and ocean farming and processing techniques become more refined, direct contributions to the fuel supply in the form of higher alcohols or hydrocarbons should become substantial.

3. *Processing Techniques.* Processing techniques and equipment must be selected, location of equipment determined, and the capital and operating costs associated with designated products must be explored. Where fermentation procedures seem appropriate, as for instance in the treatment of residual cellulosics after removal of

the less refractory carbohydrates, aerobic versus anaerobic fermentation or procedures involving both must be considered. Employment of fixed enzymes as an alternative to microbial digestion should be explored.

4. *Petrochemistry*. Every effort should be made to apply the established procedures of petrochemistry such as hydrocracking and reformation to minimize cost and land needed for factories. Experimental processing equipment will have to be designed, fabricated, and tested. Processing equipment specifications will have to be prepared. Tentative product specifications and tests based upon perceived requirements, coupled with processing and raw material limitations, will have to be developed.

5. *Wastes*. The nature and control of sludges with respect to nutritional and toxological qualities must be studied, and, if feasible, composition and processing specifications established. The use of these materials as pelletized feeds and fertilizers should be explored.

6. *Equipment*. Process equipment should be designed with versatility to allow adaptation of farm and processing plant output to varying market demands. Processing and sale of intermediate products should be studied.

7. *Microbial Digestors*. Improved plant and microorganic strains involving selection, mutation, and genetic engineering should be developed.

Task 4 -- Product Definition

Engine fuels, animal feeds, and fertilizers will have to be tested. The potentials of competitive products must be carefully assessed and compared. Legal and custom-imposed restraints on transportation and use of prospective products imposed by the governmental agencies such as the ICC and FDA, etc., will have to be studied and accommodated.

Task 5 -- Legal and Societal Considerations

Site jurisdictions and navigational constraints will need to be established in close conjunction with the governments of the coastal states concerned. Federal government, navy, and coast guard relations must be considered. Through public relations and educational activities concerned populations must be made aware of ongoing and planned activities.

SUMMARY AND CONCLUSIONS

Needless to say, the many tasks and subtasks exemplified here will require the detailed analysis of many minds and establishment of a very large and integrated research effort of significant cost. The major perceptual problem facing the leadership of the nation's current energy program is one of *scale*. The needs are so enormous and the solutions must be so innovative and radical that many shrink from the steps that must be taken to ensure that the nation and the

world with it does not founder. Time is not with us. Unless major
research and development programs for developing alternative sources
of energy are begun promptly, society may later lose the economic
strength to mount them at an appropriate scale at all. We indeed
face "the moral equivalent of war." Wars are not won without courage
and the will to exploit a salient when it is discovered. One has
been discovered that science clearly has the opportunity to exploit
if provided the resources and the leadership.

REFERENCES

1. Appell, J. R., *et al.* 1971. Converting organic wastes to oil,
    replenishable energy resource. U. S. Department of the
    Interior, Mines Bureau, Pittsburgh, Pennsylvania 15213.
    20 pp.
2. Energy Through the Year 2000. 1972. U. S. Department of the
    Interior, Washington, D.C. 20240, pp. 16–33.
3. Feldmann, H. F. 1973. Hydrogasification of municipal solid
    waste and cattle manure to pipeline gas. *In:* Abstracts of
    Papers, 165th National Meeting of the American Chemical
    Society, Dallas, Texas, April. American Chemical Society,
    1155 Sixteenth Street, N.W., Washington, D.C. 20036.
    FUEL 002.
4. Kintner, E. 1977. Status and prospects for continued progress
    in magnetic fusion energy research and development. *In:*
    Energy Technology IV. Government Institutes, Inc., 4733
    Bethesda Ave., Bethesda, Maryland 20014, pp. 167–191.
5. Levinson, M. 1977. The status of fission power. *In:* Energy
    Technology IV. Government Institutes, Inc., 4733 Bethesda
    Ave., Bethesda, Maryland 20014, pp. 192–198.
6. McCarty, P. L. 1964. Anaerobic waste treatment fundamentals.
    Part four--process design. Public Works 95(12):95–99.
7. Sutherland, L. H. 1977. Natural gas technology. *In:* Energy
    Technology IV. Government Institutes, Inc., 4733 Bethesda
    Ave., Bethesda, Maryland 20014, pp. 58–77.
8. Wilcox, H. A. 1976. Ocean farming. *In:* Proceedings of a
    Conference on Capturing the Sun through Bioconversion.
    Washington Center for Metropolitan Studies, 1717 Massachu-
    setts Ave., N.W., Washington, D.C. 20036, pp. 255–276.

- 21 -
Legal Considerations in the Maritime Setting
RICHARD P. BENNER

An array of international, federal, state, and local laws
and legal doctrines faces the culturist who wishes to use the marine
biomass of the Pacific Northwest for production of food or energy.
This chapter is intended as an introduction to the quagmire of laws
and regulations that must be understood and obeyed if a new industry
is to be established on the ocean bottom.

The oceans and tidal waters until recently have not been subject to man's dominion. Man always had been able to use the oceans,
but only temporarily, for navigation, fishing, and recreation. The
oceans and tidal waters became a "commons" for the use of all the
dominion because, in the words of Lord Hale in the 18th century,
those waters were not "maniorable" and were uniquely suited for common use.

This situation changed rather suddenly in the twentieth century
following the industrial revolution and the development of the nation-
state. Today there are many more users of the oceans, many more uses,
and consequently many more conflicts. The complexity of law results
in part from the large number of users and the failure of law to keep
pace with technology. The idea of the oceans and tidal waters as a
"commons" has endured. Arvid Pardo of Malta stirred the international
community in the 1960s with his exhortation that the oceans and seabed
were the "common heritage of mankind" and ought not to be exploited
for the benefit of any one nation. There is a corresponding tradition
in American law to respect tidal waters as common property. Common
law left over from colonial times holds that coastal states hold title
to submerged and submersible lands and the waters above them in trust
for use by all citizens of the state. The difficulty that the mariculturist faces is that he generally needs exclusive use of what has
been traditionally viewed as a common resource. The use he wishes to
make of the common resource is not among those for which the resource
has been traditionally protected. In short, the culturist is on the

377

frontiers of the law, bucking tradition and the inertia of legal systems. Fewer than half a dozen states in the U.S. have mariculture laws. Oregon is not one of them. The United Nations Law of the Sea Conferences have not been able to resolve the international conflicts standing in the way of new law governing use of mineral and living resources in the oceans. To find a way through the maze of overlapping jurisdictions and laws, one must determine first where the mariculture is going to take place and second what the activity is going to be.

WHERE MARICULTURE IS TO TAKE PLACE

There are two general rules that apply to the location of the activity.
1. There is no place where no law applies, so there is no use in trying to find it.
2. Federal law applies everywhere to U.S. citizens. For example, a U.S. citizen cannot take a marine mammal from any waters without a federal permit regardless of national jurisdiction.

Scientists abhor boundaries as nature abhors a vacuum, but lawyers thrive on them and lawyers write the laws. Boundaries largely determine which government has jurisdiction in an area. The number of permits required for any activity in coastal waters is an inverse function of the number of miles from the U.S. baseline or "Inland Waters." Figs. 1 and 2 illustrate the boundaries encountered in determining jurisdiction.

High Seas

Everything from the territorial limits of the U.S., 3 miles offshore, to the territorial limits of any other nation is considered High Seas (Fig. 1). International law applies and states that the High Seas are common property over which no nation can exercise sovereignty. Freedom of the High Seas includes navigation, fishing, laying of cables and pipelines, and overflights. These freedoms are not absolute. Fishermen make temporary exclusive use of certain areas and obstruct use by others. France and the United States have restricted areas for nuclear testing. But there have been no claims of ownership. Deep Sea Ventures, Inc., a subsidiary of Tenneco, may provide the first test of private ownership by claiming a seabed area in the Pacific for mining manganese and nickel.

Continental Shelf

Although the High Seas are common property, the seabed underlying the High Seas may not be. This creates a paradox. According to the 1958 Convention on the Continental Shelf, a coastal nation can exercise complete jurisdiction over exploration and exploitation of the natural resources of its continental shelf. The coastal nation cannot exercise complete sovereignty, but the control it can exercise is very close to that. The U.S. exerts its control through the 1953 Outer Continental Shelf Lands Act. Some coastal states also

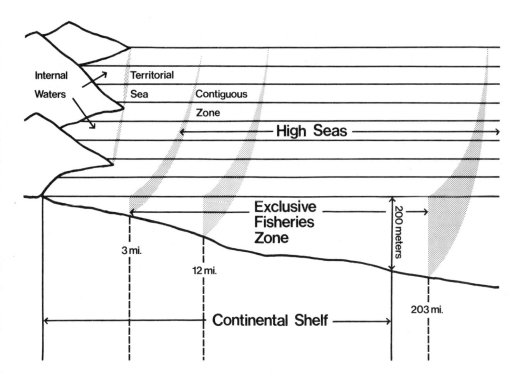

Figure 1. *Illustrated are the boundaries of the High Seas, the Exclusive Fishery Zones, and the Continental Shelf which are encountered in determining governmental jurisdiction.*

have made claims to the outer continental shelf beyond the 3-mile territorial sea, but the Supreme Court has repeatedly rebuffed them, most recently in *U.S. v. Maine* in 1975.

A coastal nation can exercise control seaward from its coast to a depth of 200 meters, or beyond to any depth permitting exploitation (Fig. 1). In other words, the limits to control by a coastal nation are undetermined. If Deep Sea Ventures demonstrates that it can raise manganese nodules from the ocean floor, the Convention of the Continental Shelf will be obsolete. By means of such new technology, the U.S. can exercise control over the entire floor of the Pacific to the exclusion of other nations. The mariculturist should be aware, therefore, that his activity on the High Seas may be subject to U.S. law if it includes any exploitation of the seabed beneath.

The Exclusive Fishery Zones

Exclusive fishery zones have not yet been completely countenanced by international law. The 1958 Fisheries Convention authorized declaration of conservation zones without specifying distances. The U.S. declared a 12-mile zone in 1966 over which the U.S. would exercise

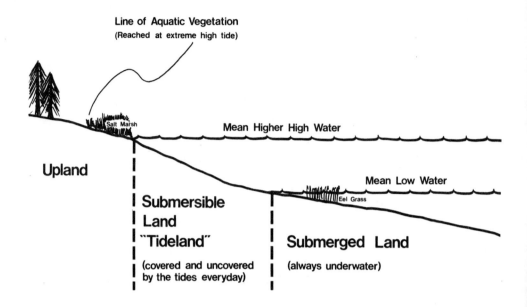

Figure 2. *Illustrated are the boundaries of the Upland, Submersible Lands--"Tideland," and Submerged Lands which are encountered in determining governmental jurisdiction.*

complete control of all types of fishing. Foreign fleets must have permission to fish in the zone. The zone was, in fact, only 9 miles wide when measured from the previously recognized territorial limits of the U.S. The U.S. now exercises complete control over all fishing within a 200-mile zone measured seaward from U.S. territorial limits (Fig. 1). The 1976 Fishery Management and Conservation Act, which established the wider zones, also declares U.S. control over anadromous fish originating from U.S. water throughout their migratory range *except* when the fish migrate into the territorial sea or conservation zone of another nation.

So, within the fishery zone, the mariculturist faces both international and federal laws. He may also be subject to state laws if he takes certain species. Decisions on the *Skiriotes* case dealing with sponges off Florida's coast and on the recent case dealing with crabs off the Alaska coast recognize the authority of coastal states to manage fisheries of unique importance if there is no conflict with federal law.

The Contiguous Zone

The U.S. contiguous zone extends 9 miles beyond territorial

limits (Fig. 1). The 1958 Convention on the Territorial Sea and Contiguous Zone gives coastal states limited jurisdiction over customs, immigration, sanitary, and fiscal matters. The contiguous zone is still High Seas subject to international law, but the coastal nation controls the seabed and much activity on the surface.

The Territorial Sea

The territorial sea of the U.S. is a 3-mile-wide strip measured seaward from the coast line (Fig. 1). The U.S. exercises complete sovereignty over the seabed, the surface, and all mineral and living resources. Foreign nations have very limited rights within the U.S. territorial sea. Foreign ships do have the right of innocent passage and are entitled to adequate warning of navigational hazards. The federal government shares sovereignty with coastal states. The federal government had authority under the Constitution to exercise full control over the territorial sea and owned the seabed until 1953. Well-organized representatives of the coastal states forced an unwilling Congress to relinquish the U.S. title and most of its control to the states in the Submerged Lands Act of 1953, after oil was discovered off the California coast. The Submerged Lands Act gave the coastal states title to the seabed under the Territorial Sea and the right to manage all mineral and living resources, but the federal government reserved its full commerce, defense, and foreign affairs powers under the Constitution. In reality, coastal states have only the illusion of sovereignty. Very little can happen in or on the territorial sea without the approval of at least one federal agency.

There is a new element, however, which allows the coastal states to make the most of their "limited sovereignty." The 1972 Coastal Zone Management Act said federal actions in a state's coastal zone must be consistent with the state's coastal management program to the maximum extent practicable. For example, the Army Corps of Engineers cannot grant Deep Sea Ventures, Inc., a permit to build a mineral extraction facility 2 miles off the southern Oregon coast if such an activity does not conform to Oregon's coastal program.

The Inland Waters

Inland waters are waters landward of the baseline used to measure all of the other zones. Bays, estuaries, rivers, streams, and lakes generally are classified as inland waters and are free from international law but are still subject to the full panoply of state and federal laws (Fig. 1). Generally, the state owns the land under navigable waters. In most states, including Oregon, tidelands, or submersible lands as they are called in Oregon law, can be sold to private persons. However, tidelands sold by the state are encumbered with a trust or easement in the public to continue navigation, fishing, and recreation over them. In other words, the private owner can use his tidelands only in ways that are consistent with those uses or that don't unreasonably interfere with them.

Both state and private tidelands are also subject to the

"navigational servitude" which is really the exercise by the federal government of its navigation, commerce, and flood control powers under the Commerce Clause of the Constitution. The Army Corps of Engineers can dredge through tidelands to deepen a navigation channel without permission, though the Corps may have to compensate. Since tidelands in inland waters can be privately owned, mariculturists working in this area must deal with common law property doctrines.

THE EXCLUSIVE USE OF SURFACE ON THE HIGH SEAS WITHIN THE U.S. FISHERIES AND CONTIGUOUS ZONES

With a general background understanding of the various zone boundaries and jurisdictions, it should be useful to examine scenarios relating to specific activities and locations in the zones especially relating to Oregon waters. For example, a mariculture operation requiring exclusive use of an ocean surface for a floating structure anchored to the seabed 10 miles off the Oregon coast would be on the High Seas outside U.S. territorial waters, but within the U.S. exclusive fishery zone and contiguous zone and would be anchored to the U.S. continental shelf (Fig. 1).

International Laws

Because the High Seas are controlled by international law, a violator risks being taken to the World Court. No sovereign claim is possible, and the operation cannot interfere with navigation, fishing, overflight, or the right to lay pipes or cables.

As observed earlier, neither of these principles is absolute. Many nations have temporarily subjected areas of the High Seas to their dominion. Mariculture is arguably among the uses protected by freedom of the High Seas as is fishing. As long as the area taken for use is not too large and interference with other High Seas uses is not unreasonable, it appears that mariculture has a place in the existing international scheme. However, mariculture is new. It is not one of the traditional uses of the High Seas. Commentators have expressed the hope that if a conflict between mariculture and another High Seas use reaches the World Court, the decision should be based on which use provides the greatest benefit to mankind rather than some notion of historical priority.

Deep Sea Ventures, Inc., has claimed a seabed area 1/2 the size of Ireland in the Pacific. Deep Sea Ventures, Inc., will not be without competitors for long. Hopefully the impasse at the Law of the Sea Conference will break before control of the seabed becomes more complicated by such ventures.

Federal Laws

In the U.S. exclusive fishery zone, fishing, if not done by a U.S. citizen, requires a permit from the Department of Commerce. Taking any "unique species" subject to state jurisdiction outside state boundaries may also require a state permit. Sponges and crabs are, therefore, best avoided. Endangered species and marine mammals

are also protected by federal legislation requiring permits from
the Interior and the Marine Mammal Commission. The U.S. contiguous
zone is subject to U.S. customs, immigration, fiscal, and sanitary
laws. A permit from the Environmental Protection Agency (EPA)
under the 1961 Oil Dumping Act is necessary to discharge oil. To
discharge any other "pollutant" such as fish waste or anything but
sea water, another permit from EPA under the 1972 Federal Water Pollution Control Act is required. If a facility will obstruct navigation, and even a single piling will, a permit from the Army Corps
of Engineers under the Rivers and Harbors Act is needed.

The jurisdiction of the Corps outside U.S. territory was un-
successfully challenged several years ago in the unusual case of
U.S. v. Ray. Lewis Ray and Acme Contractors decided to turn two
submerged coral reef formations 4 1/2 miles off the coast of Florida
into a sovereign nation known as Atlantis, Isle of Gold. Ray and
Acme began dredging sand from the seabed and covering the reefs to
bring them above sea level. The Corps sought to enjoin their activity because they had failed to obtain a permit. The Corps disliked
the prospect of a sovereign nation in the middle of its jurisdiction.
The government also charged Ray with trespass. Ray answered that he
couldn't be charged with trespass by the U.S. because the U.S. did
not own the reefs. They were outside U.S. territory. As to the
Corps' jurisdiction, Ray argued there wasn't any outside the U.S.
territorial sea.

The Court of Appeals agreed with the government on both counts.
The Court said the 1958 Convention on the Continental Shelf gave the
U.S. the exclusive right to exploit the resources of its shelf. Anyone besides the U.S. would have to get a permit from the U.S. to
exploit shelf resources. The Court pointed to the 1953 Outer Continental Shelf Lands Act. The Act expressly extended the Corps of
Engineers' jurisdiction to prevent obstructions to navigation by
artificial islands or fixed structures on the continental shelf.
So a permit from the Corps is required, even though the U.S. does
not own the seabed under its contiguous zone. And even though the
U.S. does not own the seabed, a lease from the Department of the
Interior under the Outer Continental Shelf Lands Act is necessary
to anchor any facility to the seabed.

Under the 1975 Deepwater Port Act the Secretary of Transportation can proclaim a safety zone around a deepwater port installation
to prohibit activities which might interfere with operations. The
Secretary of Commerce can establish a marine sanctuary under the
1972 Marine Sanctuaries Act for preserving ecological or aesthetic
values and prohibit inconsistent activities within the sanctuary.
Finally, if proposed mariculture involves a major facility or activity significantly affecting the human environment, an environmental
impact statement (EIS) will be required. The federal courts have
required impact statements for all federal leases of the seabed for
extraction of oil.

State Laws

If a mariculture operation is planned outside Oregon's territorial limits, it is largely free of state regulation unless the fishing of anadromous species is involved. For the same boundary reason, there is no local law to apply.

THE EXCLUSIVE USE OF SURFACE WITHIN THE U.S. TERRITORIAL SEA

International Laws

If an anchored mariculture device is planned within the Territorial Sea, it is completely controlled by the coastal nation. International law applies only to the right of innocent passage and the measures necessary to assure adequate publicity to obstructions in navigation.

Federal and State Laws

The primary difference between activity in the Territorial Sea and activity within the Fisheries and Contiguous Zones is that within the Territorial Sea the state and federal governments share sovereignty. The state of Oregon owns the seabed and natural resources of the Territorial Sea and holds them in trust for the people of Oregon. Exclusive use of any portion requires a lease from the Division of State Lands under the Oregon leasing law, ORS Chapter 274. No lease is required from the federal government.

Actually, it is not entirely clear that a floating pen anchored to the bottom does require a state lease. Oregon has no law which expressly requires a lease for exclusive rights to a surface area of Oregon waters. Chapter 274 governs only sale and lease of submerged and submersible lands. But the Division of State Lands has taken the position that exclusive use of a surface area is a *de facto* exclusive use of the seabed under the surface area and requires a lease. Even if the Division is challenged on that position, any contact with the seabed by permanent anchors, for example, is likely to trigger the lease requirement.

Oregon law authorizes lease of the seabed under Oregon's territorial sea for harvesting kelp (ORS 274.885 *et seq.*). Oregon law also authorizes the Fish and Wildlife Commission to issue permits for operation of salmon hatcheries (ORS 507.700 *et seq.*). But there is no Oregon mariculture law dealing with other plant or animal species or with rights to areas of water surface. Those involved in mariculture might well assist in drafting a mariculture law for Oregon, perhaps following the Florida model.

With the exception of a possible federal fishing permit, activity in the Territorial Sea also remains subject to all of the federal laws mentioned in the first example. But in addition, a permit from the Oregon Division of State Lands is required if the facility involves dredging or the deposit of 50 cubic yards or more of material pursuant to the Oregon Fill and Removal Law (ORS 541.605 *et seq.*). Any discharge other than fresh water or seawater requires a permit from the Oregon Department of Environmental Quality (DEQ). However, the U.S. Environmental Protection Agency has delegated its permit issuing authority to Oregon, so a permit from the state and federal government for the same activity won't be necessary unless the discharge is oil.

If the mariculture installation requires some structure on or the removal of material from the "ocean shore," a permit from the Oregon Department of Transportation under the Beach Bill will be needed (ORS 390.605 *et seq.*). The "ocean shore" is the area of wet and dry sand along the coast between extreme low tide and the line of vegetation. The Department will issue a permit only if it finds the activity will not interfere with public safety, aesthetics, and recreation.

THE EXCLUSIVE USE OF SUBMERGED AND SUBMERSIBLE LAND ON AN ESTUARY WITHIN INLAND WATERS

If the mariculture is to be further inland than the first two hypothesized fixed pens and were to be installed on the submerged and submersible lands of an estuary and if canals and fills must be established, the facility will be within inland waters and outside of international jurisdiction. Ownership of estuarine tidelands can reside in the state, in the federal government, or in a private person. The mariculturist will have to procure a lease from someone. It is important to remember that the lease, or title should the tidelands be bought, is subject to the public trust and cannot be used in a way which unreasonably interferes with public uses of waterways. The lease or title is also subject to the federal navigational servitude under the Commerce Clause of the U.S. Constitution and can be taken at any time for navigation, commerce, or flood control purposes. Holding and improving tidelands is risky.

International Laws

There is no international right-of-innocent-passage law for inland waters. Only in a rare case would mariculture in an estuary invoke any doctrine of international law. However, persons may be held liable for damage to foreign nationals caused by release of mutant species into the ocean.

Federal Laws

Federal laws cannot be avoided. Any structure on or over tidelands will require a permit from the Army Corps of Engineers as an obstruction to navigation under the Rivers and Harbors Act. Until recently the Corps viewed its authority under that Act narrowly and required a permit only if the structure actually interfered with navigation. But the Corps' role has been expanded by a statute and court decision to cover fills all the way to mean higher-high-water on the Pacific Coast and even to affect fills of diked marshland in San Francisco Bay (Fig. 2). Permits are necessary from the Corps for depositing dredge or fill materials into navigable waters under the Federal Water Pollution Control Act (FWPCA). Congress defined

"navigable waters" much more broadly in this Act than courts have
defined it under the Rivers and Harbors Act. Congress reasoned that
the way to stop pollution of the nation's waters was to get to its
source. It defined "navigable waters" to include all waters. The
courts have responded by holding that dredge and fill activities in
mosquito canals, mangroves, and saltmarsh all the way to the line of
salt water vegetation require a Corps permit under the FWPCA. Discharge of any pollutant, including hot water, into navigable waters
as defined above requires a permit from the Environmental Protection
Agency. This can be the same permit obtained from the Department
of Environmental Quality.

Some locations may be within an estuarine sanctuary established
under the Federal Coastal Zone Management Act. Oregon has the only
estuarine sanctuary thus far, South Slough of Coos Bay. A site may
be within a wildlife refuge under one of the dozens of federal acts
establishing refuges, or it may be within a "waterfront safety zone"
declared by the Secretary of Commerce under the 1972 Ports and Waterways Safety Act. Work within any of these areas must be certified
by the appropriate federal agency.

State Laws

Any dredge activity or fill greater than 50 cubic yards requires
a permit from the Oregon Division of State Lands. The standards for
obtaining a fill permit are high. The state, even if it does not own
the tidelands, holds an easement over the tidelands in favor of the
public. The Division will issue a fill permit only if it finds that
the fill does not unreasonably interfere with state policy to preserve
its waterways for navigation, fishing, and recreation; that the fill
accords with sound principles of conservation; that the fill is consistent with existing waterway uses; and that the fill conforms to
the local comprehensive plan. The Division interprets those standards
rigorously, and, in the North Bend airport case, required return of
the spoils island, which had been used to extend a runway, to tidal
influence as mitigation for what it found to be unreasonable interference with the productivity of the Coos Bay estuary.

Of course, any mariculture which might produce energy anywhere
within Oregon's territorial limits requires a permit from the State
Department of Energy.

Local Laws

Activity in Inland Waters is also subject to the control of
cities, counties, and special districts. Every major estuary on the
Oregon Coast is within a port district. Ports have authority to
regulate activities on waters within their boundaries, including the
construction of docks, wharves, and piers. The purpose of the regulation is to keep the waters open for navigation. A fixed pen or
other structure over navigable waters in a port district will require
a permit from the port.

Every estuary is also within either a city or a county. In
many states and under Oregon law, cities and counties have primary

land use planning jurisdiction. Senate Bill 100, the Oregon Land
Use Planning Law, requires every city and county to adopt a comprehensive plan and zoning ordinances for implementing the plan. All
land and water areas must be included. Once a plan and zoning ordin-
ance are adopted, any person establishing a mariculture facility must
comply even if the site is leased from the state. Compliance generally means obtaining a permit from a city or a county.

The local comprehensive plan assumes greater importance in
Oregon law than may seem apparent. A key objective of Senate Bill
100 was to coordinate the many state and local plans and actions
affecting land use. To this end, Senate Bill 100 created the Land
Conservation & Development Commission (LCDC) and directed it to
establish statewide goals for land use planning in Oregon. The
goals are implemented at the local level in city and county comprehensive plans. Any state agency or special district whose plans.
or actions affect land use must also follow the goals. In other
words, the Division of State Lands must apply the goals when it reviews an application for a fill permit. Therefore, goals become
the statewide standards and local plans are the implementing
measures. Once LCDC reviews a local plan and finds that it complies
with the statewide goals, actions of state agencies will be measured
against the plan. No state agency can act or issue a permit which
does not comply with a local plan that has been approved by LCDC.
For example, the Division of State Lands cannot issue a permit to
fill an area of Coos Bay estuary if Coos County's comprehensive
plan does not authorize fill at that location.

Those interested in engaging in maricultural activities would
be best advised to start at the local level to ensure that cultural
plans comply with a city or county comprehensive plan. If the project does comply, that doesn't guarantee success in obtaining all
permits, but if the project does *not* comply, the culturist certainly
cannot obtain other state or federal permits.

SUMMARY AND CONCLUSIONS

There are a few bright spots amid the imposing matrix of state,
federal, and local laws outlined here. Congress and many state
legislatures, including that of Oregon, have recognized that a vast
regulatory structure discourages even the most persistent entrepreneur. Two of the more significant recent efforts, besides Senate
Bill 100, to bring some sanity into this maze should be mentioned.

Congress added what is known as the "federal consistency" clause
to the Coastal Zone Management Act of 1972 to give assurance to coast·
al states that actions by federal agencies would not undermine a
state's coastal program. The federal consistency clause states that,
to the maximum extent practicable, each federal agency conducting
or supporting activities directly affecting the coastal zone shall
do so in a manner consistent with state programs approved by the
Secretary of Commerce. Thus, the Army Corps of Engineers cannot
issue permits to dredge or fill which do not conform to Oregon's
coastal program. Also, since cities and counties have the primary

responsibility to implement Oregon's coastal program, federal activities must conform to the local comprehensive plan to the maximum extent practicable. Both state and federal laws accord a very high status to the local comprehensive plan. Again the entrepreneur should begin his plans in cooperation with local authorities. Another helpful development is that Oregon now has a "one stop" permit system. Under a new 1975 law, any person can submit one application to the Executive Department requesting issuance of all state permits. If the activity complies with applicable laws, the applicant receives the permit in a single procedure.

LCDC adopted coastal goals in 1975. The coastal goals which became effective January 1977 are the primary elements in Oregon's coastal zone management program which was approved by the Secretary of Commerce on April 6, 1977. Like the original statewide goals, the coastal goals will be implemented by coastal cities and counties. Plans and actions by local governments and state agencies which affect land use must comply with the coastal goals. There are four goals and each may affect a mariculturist's activity. The Shorelands Goal requires cities and counties to protect shorelands of special ecological significance and other shorelands valuable as sites for water dependent uses, such as port facilities or mariculture. The Estuarine Resources Goal requires LCDC to classify all estuaries on the Oregon Coast according to the most intensive use permitted. Within each estuary cities and counties must establish discrete units for management at different levels of activity, consistent with the overall LCDC classification. The goal limits fills in estuaries to those uses which must be adjacent to the water and requires mitigation where harm is done to estuarine productivity. The Beaches and Dunes Goal limits development of unstable sand areas and prohibits most construction on active foredunes. The Ocean Resources Goal requires analysis of long-term costs and benefits and information collected by agencies with jurisdiction in Oregon's Territorial Sea. Little input has been made to these developing goals by those interested in the maricultural activities reviewed in this volume.

If the restrictions appear overwhelming, trust that the law is there to help. If there is any doubt about the competence of the law to resolve conflicts among users of ocean and tidewater resources, the precedent shown in the following quote, found in a law review article, offers reassurance.

> Once, (says an author; where I need not say)
> Two Travellers found an oyster in their way.
> Both fierce, both hungry, the dispute grew strong,
> While, scale in hand, Dame Justice passed along.
> Before her each with clamor pleads the laws
> Explains the matter and would win the cause.
> Dame Justice, weighing long the doubtful right,
> Takes, opens, swallows it before their sight.
> The cause of strife removed so rarely well
> "There! take (says Justice) take ye each a shell.
> We thrive at Westminister on fools like you.
> It was a fat oyster--live in peace--Adieu."

Index

Index

Zinc, 222
Zonarol, 280
Zonation, West European, 252

Zonation, European rocky shore, 254
Zooplankton, 47
Zostera, 261